Endohedral Metallofullerenes

Endohedral Metallofullerenes

Fullerenes with Metal Inside

Hisanori Shinohara
Department of Chemistry and Institute for Advanced Research,
Nagoya University, Japan

Nikos Tagmatarchis
Theoretical and Physical Chemistry Institute,
The National Hellenic Research Foundation, Greece

Contents

Harold Kroto
Tallahassee, USA
January 2015

Foreword

At the end of the hectic couple of weeks in 1985 during which C_{60} Buckminsterfullerene was discovered, as I was leaving Houston I suggested to Sean O'Brien (one of the wonderful group of students involved in the discovery) that if the molecule really was a fullerene cage it should be possible to put an atom inside it. I suggested on the basis of my familiarity with ferrocene that an iron atom might be a good bet. Unfortunately that was not a good choice as Sean could not get a mass spectrometric signal for such a species. The next day Jim Heath tried lanthanum which went in successfully and led to the second fullerene paper on La@C_{60}. I am not sure that Sean ever forgave me for suggesting iron and indeed I think that to this day no iron-containing fullerene has been discovered, which is actually rather interesting. During the period after the discovery, I discovered the fact that a small fullerene tetrahedral C_{28} should be metastable and perhaps further stabilized by adding four hydrogen atoms, one to each of the four carbon atoms at the tetrahedral apexes. I suggested it might be a sort of tetravalent carbon cluster "superatom." Later on, this tetravalent hypothesis was supported by the Rice group's detection of a metastable endohedral cluster U@C_{28} in which the tetravalency was interestingly satisfied from within by the tetravalent uranium atom. Thus was the amazing novel field of endohedral fullerene chemistry born. I do not think that for one second we ever imagined that these experiments would result in such an exciting and wide-ranging field of unique chemistry which this compendium beautifully highlights.

Harold Kroto
Tallahassee, USA
January 2015

Preface

Since the first experimental discovery of C_{60} fullerene by Kroto, Smalley and co-workers in 1985, the term nanocarbons or carbon nanomaterials (including carbon nanotubes and graphene) has been emerged and prevailed quite rapidly in the area of nanoscience and nanotechnology as well as of materials science in general. One of the crucial factors of this prevalence is the discovery of macroscopic synthesis of fullerenes by Kräetschmer, Huffman and co-workers in 1990.

In fact, during the past decade, fullerenes have already been practically applied as important constituents of various composite materials such as solar cells, reinforcing materials for sporting goods, lubricants of car engine oils and even cosmetics. In particular, practical applications of certain functionalized fullerenes as a brand-new solar cell battery have been a competitive and heated R&D target in chemical and electric industrial companies worldwide. Unlike the conventional silicon-based solar cells, solution-based fullerene solar cells possess unique and novel characteristics of paintable and flexible performance.

One of the most intriguing outgrowths of fullerenes is the so-called "endohedral metallofullerenes", fullerenes with metal(s) encapsulated. One may easily think that the metallofullerenes might exhibit salient electronic and magnetic properties which are totally different from those of the conventional (empty) fullerenes: the presence of a atom even within a fullerene may drastically alter its electronic properties. The origin of the metallofullerene can be traced back to 1985, a week after C_{60} was experimentally discovered by the Sussex-Rice research team. Their idea is quite simple: if C_{60} has a spherical hollow shape as the first Nature paper advocates, then C_{60} can encage a metal atom inside. Indeed, in their second paper, they performed laser-vaporization cluster-beam time-of-flight mass spectrometry using lanthanum-doped graphite disks and observed an enhanced peak due to $La@C_{60}$ in a mass spectrum, suggesting that the La atom is safely entrapped inside C_{60} as expected.

Up until early 1990's, however, there had been a heated and controversial discussion as to whether or not a metal atom is really entrapped inside of the fullerene or rather externally bound from outside. The first experimental verification of the endohedral nature was brought by the Nagoya-Mie research team in 1995 when they performed synchrotron X-ray powder diffraction on a chromatographically purified metallofullerene sample.

Now that the existence of "endohedral metallofullerenes" had been confirmed, there was an outbreak of research in the area of this brand-new fullerene family. This book primarily deals with the research and development of the metallofullerene after 1995, where synthesis, purification, structures, electronic/magnetic properties and some of the important applications of different types of metallofullerenes are fully and chronologically described.

The authors acknowledge fruitful collaborations and discussion with a number of our present and former graduate students, postdocs and laboratory research staff. We must apologize for not mentioning them individually. We also thank our publishers for their encouragement and patience. Last but, of course, never least, we want to thank our families for their continuing support.

Hisanori Shinohara, Nagoya
Nikos Tagmatarchis, Athens
April 2015

Personal Reflection – Nori Shinohara

At the breakfast table at a small hotel on the shore of Lake Konstanz in Germany, Rick Smalley, who was a professor at Rice University at the time, showed me a slide for a presentation. It was Wednesday, September 12, 1990. He asked, "Nori (my nickname), do you know what the fine black powder at the lower right of this slide is?" It was C_{60} powder! I was momentarily stunned by what I saw on the slide that Smalley handed me and could not immediately understand what had happened [1].

Smalley and I were by chance staying at the same hotel while attending the 5th International Symposium on Small Particles and Inorganic Clusters (ISSPIC 5) (Figure 1) that was being held at the University of Konstanz in Germany, September 10–14, 1990. The organizers were the cluster physicists Eckehalt Recknagel (a University of Konstanz professor at the time) and Olof Echt (a University of New Hampshire professor).

The first announcement of the easy and astonishing macroscopic synthesis of C_{60} by sublimation of graphite was made at this symposium, which is well-known in the fields of clusters and ultra-fine particles. Furthermore, the announcement of the century concerning macroscopic synthesis of fullerenes by Wolfgang Krätschmer was an unannounced presentation that took less than 10 min if I remember correctly. I was also attending the symposium for a presentation of my own research on laser spectroscopy of benzene molecular clusters.

It was a coincidence that Smalley, who would later receive the Nobel Prize for the discovery of fullerenes in 1996 with Harry Kroto and Bob Curl, was also staying with his wife at the small and stylish hotel at which I was staying, the Villa Hotel Barleben am See in Konstanz, the location of the University of Konstanz where the symposium was held.

Figure 1 Abstract for ISSPIC 5 held at the University of Konstanz, Germany, September 10–14, 1990

It was the middle day of the conference, September 12. "Solid C_{60} Isolated" was written on Smalley's slide displaying the C_{60} powder.

Fullerenes are the third allotrope of carbon after graphite and diamond, and C_{60} is the most typical molecule among them. In 1985, Rick Smalley, Harry Kroto, and co-workers discovered C_{60} (by chance) using laser-vaporization cluster-beam mass spectroscopy [2], but no one had succeeded at macroscopic synthesis at the time.

Smalley had planned to talk about metal and semiconductor clusters, the research into which he was putting his energy at the time under the title of "ICR Probes of Cluster Surface Chemistry" at this international symposium. However, when Smalley went up to the podium, he abruptly announced that he would talk about C_{60} rather than cluster surface chemistry.

Smalley started this talk with "The first discovery of C_{60} five years ago in 1985 was through joint research between Rice University and the University of Sussex. However, Dr Krätschmer *et al.* of the Max Planck Institute, Heidelberg, have very recently discovered a very simple method for producing large amounts of C_{60}." Smalley abandoned the 50 min lecture he was invited to give and hurriedly had Krätschmer give an unannounced presentation on this important discovery.

According to the discovery by Krätschmer and co-workers [3], C_{60} could easily be produced in large quantities using arc discharge (at this point, more precisely, resistive heating) on graphite. All of the researchers at the symposium were surprised at this. Furthermore, this unannounced presentation was the beginning of all the subsequent research into nano-carbons (e.g., fullerenes, carbon nanotubes, nanopeapods, graphene, etc.), as well as full-scale research on nanotechnology.

After Krätschmer's presentation, most of those attending the symposium returned home to carry out additional testing. From that day forward, there was an explosion of research groups at hundreds of locations worldwide enthusiastically plunging into C_{60} research. This 10 min, unannounced presentation proclaimed the start of global research into fullerenes, carbon nanotubes and, after that, nanocarbons. This was the biggest discovery in the last 50 years, maybe even the discovery of the century, and brought with it a completely new trend in materials science and technology.

That summer I was 36, and my own research also underwent a big change.

As soon as research on fullerenes was ignited, there was the abnormal situation of 30 or more related papers appearing in a single day. Because of this, researchers were driven by a sense of fear, and there was no rest for the weary. There was a sense that the work being done today would be overshadowed the next morning. In fact, though it is only half true and I am half joking, we said that the researchers with the dark circles under their eyes at international conferences were the ones working on fullerenes at the time. It was an absurd period for me, doing things such as staying overnight with students at the laboratory.

I realized that in the midst of this madness, however, many experiments I conducted and reports I wrote would probably be completely forgotten in 4–5 years' time. Therefore, I narrowed down my target to fullerene research and began joint research with Yahachi Saito (currently a Nagoya University professor) on endohedral metallofullerenes, where a metal is inserted into a fullerene cage [1,4,5].

The reason I selected metallofullerenes was that my research on water clusters prior to my research on fullerenes provided important hints for this research. Since water molecules create box-like cage clusters similar to fullerenes [6,7], a molecule or atom can be inserted into the central interior hollow space. Likewise, I thought that it might be possible to encapsulate a metal atom inside a carbon cage. I wondered if I could create a fullerene with completely new electronic and magnetic properties not found up to that time if a metal atom could be placed in this space, since the space in fullerenes is a complete vacuum.

My first report on metallofullerenes [4] was beaten out by Smalley's group, but this research went very well, and it turned out that my joint research with Saito advanced to be top in this field worldwide. In 1995, with the cooperation of Makoto Sakata (a Nagoya University professor at the time) and Masaki Takata [currently a Tohoku University professor], I obtained the first experimental proof of the encapsulation of metal atoms through synchrotron X-ray diffraction measurements (cf. Section 3.2.1).

Interestingly, the instant the metal atom entered the fullerene, two or three of the outer electrons of the metal atom transferred to the fullerene cage. This phenomenon is called intra-fullerene electron transfer, and we found that we could create fullerenes with very interesting electronic and electron transport properties that could not be obtained with normal fullerenes.

In the late summer of 1990, I was stunned by the unannounced lecture on macroscopic synthesis of C_{60} by Krätschmer at an international symposium in Konstanz, Germany, and thereafter plunged myself into nanocarbon research. Subsequently, my odyssey of a quarter century in nanocarbon research has taken me into fullerenes, metallofullerenes, carbon nanotubes, peapods and, recently, graphene-related materials. Virtually all of the topics on endohedral metallofullerenes described in this book started after the Konstanz Symposium.

I wonder what novel nanocarbon materials will appear next.

REFERENCES

[1] Shinohara H 2012 In pursuit of nanocarbons *Chem. Record* **12** 296
[2] Kroto H, Heath J R, O'Brien S C *et al.* 1985 C_{60}: Buckminsterfullerene *Nature* **329** 529
[3] Krätschmer W, Fostiropoulos K, Lamb L D and Huffman D R 1990 Solid C_{60} – a new form of carbon *Nature* **347** 354
[4] Shinohara H, Sato H, Saito Y *et al.* 1992 Mass spectroscopic and ESR characterization of soluble yttrium-containing metallofullerenes YC_{82} and Y_2C_{82} *J. Phys. Chem.* **96** 3571
[5] Shinohara H, Sato H, Ohchochi M *et al.* 1992 Encapsulation of a scandium trimer in C_{82} *Nature* **357** 52
[6] Shinohara H, Nagashima U, Tanaka H and Nishi N 1985 Magic number for water-ammonia binary clusters: enhanced stability of ion clathrate structures *J. Chem. Phys.* **83** 4183
[7] Nagashima U, Shinohara H, Nishi N and Tanaka H 1986 Enhanced stability of ion-clathrate structures for magic number water clusters *J. Chem. Phys.* **84** 209

1

Introduction

1.1 THE FIRST EXPERIMENTAL EVIDENCE OF METALLOFULLERENES

Endohedral metallofullerenes, the fullerenes with metal atom(s) inside, are an interesting class of fullerene materials because electron transfer from the encaged metal atom to the carbon cage has been known to occur and this dramatically alters the electronic and magnetic properties of the fullerenes.

Just a week after the first experimental observation of the "magic number" soccerball-shaped C_{60} in a laser-vaporized cluster beam mass spectrum by Kroto *et al.* [1], the same research group of Kroto, Smalley, and co-workers also found a magic number feature due to LaC_{60} in a mass spectrum prepared by laser vaporization of a $LaCl_3$ impregnated graphite rod [2]. They observed a series of C_n^+ and LaC_n^+ ion species with LaC_{60}^+ as a magic number ion in the mass spectrum (Figure 1.1) and concluded that a La atom was encaged within the (then hypothetical) soccerball-shaped C_{60}. This was obviously the first proposal of the so-called "endohedral metallofullerene" concept based on experiments. They first tried Fe with no success and found that La is a "correct" atom for encapsulation within fullerenes. It is interesting to note that even today Fe has not been encapsulated by fullerenes.

Further circumstantial (not direct) evidence that metal atoms may be encaged in C_{60} was also reported by the Rice group, showing that LaC_{60}^+ ions did not react with H_2, O_2, NO, and NH_3 [3]. This suggests that reactive metal atoms are protected from the surrounding gases and are indeed trapped inside the C_{60} cage.

Endohedral Metallofullerenes: Fullerenes with Metal Inside, First Edition.
Hisanori Shinohara and Nikos Tagmatarchis.
© 2015 John Wiley & Sons, Ltd. Published 2015 by John Wiley & Sons, Ltd.

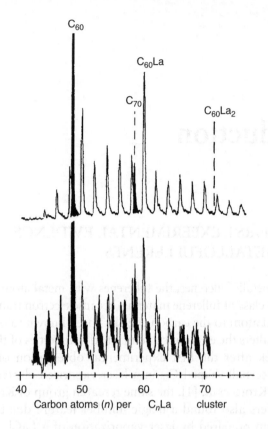

Figure 1.1 Laser-vaporization supersonic cluster-beam time-of-flight mass spectrum of various lathanum-carbon clusters. LaC_{60} is seen as an enhanced (magic number) peak. Reprinted with permission from Ref. [2]. Copyright 1985 American Chemical Society

In fact, the first direct evidence of the soccerball (truncated icosahedron) C_{60} was amply demonstrated in 1990 by a historical experiment done by Kräetschmer, Huffman, and co-workers. They succeeded in producing macroscopic quantities of soccerball-shaped C_{60} by using resistive heating of graphite rods under a He atmosphere [4,5]. The resistive heating method was then superseded by the so-called contact arc discharge method [6] since the arc discharge method can produce fullerenes order of magnitudes larger than that by resistive heating. Since then, the arc discharge method has become a standard method for fullerene synthesis.

The first production of macroscopic quantities of endohedral metallofullerenes was also reported by Smalley and co-workers [7]. They used high-temperature laser vaporization of La_2O_3/graphite composite rods together with the corresponding contact arc technique to produce various sizes of La-containing metallofullerenes. Contrary to the previous expectation, only

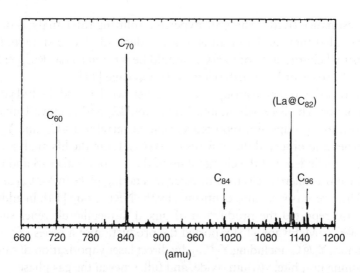

Figure 1.2 An FT-ICR mass spectrum of hot toluene extract of fullerene soot produced by high-temperature laser vaporization of a 10% La$_2$O$_3$/graphite composite rod. Reproduced with permission from Ref. [7]. Copyright 1991 American Chemical Society

the La@C$_{82}$ fullerene was extracted by toluene solvent even though La@ C$_{60}$ and La@C$_{70}$ were also seen in the mass spectra of the sublimed film from soot. In other words, the major La-metallofullerene with air and solvent stability is La@C$_{82}$, and La@C$_{60}$ and La@C$_{70}$ are somehow unstable in air and solvents.

Figure 1.2 shows a Fourier-transform ion cyclotron resonance (FT-ICR) mass spectrum of hot toluene extract of fullerene materials produced by laser vaporization of a 10% La$_2$O$_3$/graphite composite rod [7]. In addition to empty fullerenes, only the La@C$_{82}$ metallofullerene is seen and La@C$_{60}$ and La@C$_{70}$ are completely absent in the mass spectrum of the solvent extracts. The special behavior of the La@C$_{82}$ fullerene was soon confirmed by Whetten and co-workers [8]. However, they also observed that at relatively high loading ratios of La$_2$O$_3$ in composite rods a di-lanthanofullerene, La$_2$@ C$_{80}$, was also produced by the resistive-heating method and found to be another solvent-extractable major lanthanofullerene [8,9].

1.2   EARLY YEARS OF METALLOFULLERENE RESEARCH

The first important information on the electronic structure of La@C$_{82}$ was provided by the IBM Almaden research group. The charge state of the encaged La atom was studied by Johnson *et al.* [10] using electron spin resonance (ESR). The ESR hyperfine splitting (hfs) analysis of La@ C_{82} revealed that the La atom is in a trivalent +3 charge state and that the formal charge state of La@C_{82} should be written as La^{3+}@C_{82}^{3-}: three outer electrons of La transferring to the C_{82} cage [11].

Several other research groups extended their work to endohedral yttrium compounds. The Rice–Minnesota University [12] and Nagoya University [13] research groups also reported solvent-extractable Y@C_{82} and Y$_2$@C_{82} fullerenes and observed the ESR hfs of Y@C_{82}. From the hfs analyses both groups concluded that the charge state of the Y atom is also +3 and that a similar intra-fullerene electron transfer was taking place in Y@C_{82} as La@ C_{82}. These results were also confirmed by the NRL group [14]. In addition, they also reported the production of mixed di-metallofullerenes such as (LaY)@C_{80}. McElvany [15] reported the production of a series of yttrium fullerenes, Y$_m$@C_n including Y@C_{82}, by direct laser vaporization of samples containing graphite, yttrium oxide, and fullerenes in the gas phase.

Scandium metallofullerenes were also produced in macroscopic quantity and solvent-extracted by Shinohara et al. [16] and Yannoni et al. [17]. Interestingly, the Sc fullerenes exist in extracts as a variety of species (mono-, di-, tri-, and even tetra-scandium fullerenes), typically as Sc@C_{82}, Sc$_2$@C_{74}, Sc$_2$@C_{82}, Sc$_2$@C_{84}, Sc$_3$@C_{82}, and Sc$_4$@C_{82}. It was found that Sc$_3$@C_{82} was also an ESR-active species whereas di- and tetra-scandium fullerenes such as Sc$_2$@C_{84} and Sc$_4$@C_{82} were ESR-silent. (See Section 3.2.2 for the present correct assignment for some of the di- and tri-scandium metallofullerenes.) A detailed discussion on the electronic structures of the scandium fullerenes accrued from these ESR experiments is given in Sections 5.1 and 5.2.

The formation of lanthanide metallofullerenes R@C_{82} (R = Ce, Pr, Nd, Sm, Eu, Gd, Tb, Dy, Ho, Er, Yb, and Lu) was also reported by the UCLA [18] and SRI international [19] groups. These metallofullerenes were also found to be based on the C_{82} fullerene.

In addition to group 3 (Sc, Y, La) and the lanthanide metallofullerenes, group 2 metal atoms (Ca, Sr, Ba) also form endohedral metallofullerenes, and have been produced and isolated in milligram quantity [20–26]. These metal atoms have been encaged not only by C_{82} and C_{84} but also by smaller fullerenes such as C_{72}, C_{74}, and C_{80}. Furthermore, group 4 metallofullerenes (Ti, Zr, Hf) were synthesized and isolated [27,28].

C_{60}-based endohedral fullerenes which were produced but not isolated in the early days are Ca@C_{60} [29–33] and U@C_{60} [29]. The Ca@C_{60} and U@C_{60} fullerenes are unique metallofullerenes in which Ca and U atoms are encaged by C_{60}, and are quite different from group 3 and lanthanide, R@C_{82} type, metallofullerenes. An *ab initio* self-consistent field (SCF)

Table 1.1 The "bucky periodic table" showing the elements which have been reported to form endohedral metallofullerenes and isolated as purified forms. The black elements form endohedral metallofullerenes which have been purified, whereas the gray elements form endohedral non-metallofullerenes

Bucky periodic table

1												13	14	15	16	17	18
H	2																He
Li	Be											B	C	N	O	F	Ne
Na	Mg	3	4	5	6	7	8	9	10	11	12	Al	Si	P	S	Cl	Ar
K	Ca	Sc	Ti	V	Cr	Mn	Fe	Co	Ni	Cu	Zn	Ga	Ge	As	Se	Br	Kr
Rb	Sr	Y	Zr	Nb	Mo	Te	Ru	Rh	Pd	Ag	Cd	In	Sn	Sb	Te	I	Xe
Cs	Ba		Hf	Ta	W	Re	Os	Ir	Pt	Au	Hg	Tl	Pb	Bi	Po	At	Rn
Fr	Ra		Rf	Db	Sg	Bh	Hs	Mt									

La	Ce	Pr	Nd	Pm	Sm	Eu	Gd	Tb	Dy	Ho	Er	Tm	Yb	Lu
Ac	Th	Pa	U	Np	Pu	Am	Cm	Bk	Cf	Es	Fm	Md	No	Lr

Hartree–Fock calculation indicates that the Ca ion in $Ca@C_{60}$ is displaced by 0.7 Å from the center and that the electronic charge of Ca is 2^+ [30,34]. A similar theoretical prediction has been made on $Sc@C_{60}$ by Scuseria and co-workers [35]. Metallofullerenes based on C_{60} are known to be unstable in air and in normal fullerene solvents such as toluene and carbon disulfide. We will discuss the stability and properties of C_{60}-based metallofullerenes together with an inability to extract and purify these in Chapter 14. The metal atoms which have been reported to form endohedral metallofullerenes are shown in Table 1.1.

1.3 CONVENTIONAL AND IUPAC NOMENCLATURE FOR METALLOFULLERENES

The symbol @ is conventionally used to indicate that atoms listed to the left of the @ symbol are encaged in the fullerenes. For example, a C_{60}-encaged metal species (M) is then written as $M@C_{60}$ [7]. The corresponding IUPAC nomenclature is different from this conventional $M@C_{60}$ representation. It is recommended by IUPAC that $La@C_{82}$ should be called [82] fullerene-incar-lanthanum and be written as $iLaC_{82}$ [36]. However, throughout this book the conventional $M@C_{2n}$ form is used for endohedral metallofullerenes for brevity, unless otherwise noted.

REFERENCES

[1] Kroto H W, Heath J R, O'Brien S C *et al*. 1985 C_{60}: Buckminsterfullerene *Nature* 318 162

[2] Heath J R, O'Brien S C, Zhang Q *et al*. 1985 Lanthanum complexes of spheroidal carbon shells *J. Am. Chem. Soc.* 107 7779

[3] Weiss F D, Elkind J L, O'Brien S C *et al*. 1988 Photophysics of metal complexes of spheroidal carbon shells *J. Am. Chem. Soc.* 110 4464

[4] Kräetschmer W, Fostiropoulos K and Huffman D R 1990 The infrared and ultraviolet absorption spectra of laboratory-produced carbon dust: evidence for the presence of the C_{60} molecule *Chem. Phys. Lett.* 170 167

[5] Kräetschmer W, Fostiropoulos K, Lamb L D and Huffman D R 1990 Solid C_{60}: a new form of carbon *Nature* 347 354

[6] Haufler R E, Chai Y, Chibante L P F *et al*. 1991 *Cluster-Assembled Materials*, Vol. 206, eds R S Averback, J Bernhoc and D L Nelson (Pittsburgh, PA: Materials Research Society), pp. 627–637

[7] Chai Y, Guo T, Jin C *et al*. 1991 Fullerenes with metals inside *J. Phys. Chem.* 95 7564

[8] Alvarez M M, Gillan E G, Holczer K *et al* 1991 La_2C_{80}: a soluble dimetallofullerene *J. Phys. Chem.* 95 10561

[9] Yeretzian C, Hansen K, Alvarez M M *et al*. 1992 Collisional probes and possible structures of La_2C_{80} *Chem. Phys. Lett.* 196 337

[10] Johnson R D, de Vries M S, Salem J *et al*. 1992 Electron-paramagnetic resonance studies of lanthanum-containing C_{82} *Nature* 355 239

[11] Bethune D S, Johnson R D, Salem J R *et al*. 1993 Atoms in carbon cages: the structure and properties of endohedral fullerenes *Nature* 366 123

[12] Weaver J H, Chai Y, Kroll G H *et al*. 1992 XPS probes of carbon-caged metals *Chem. Phys. Lett.* 190 460

[13] Shinohara H, Sato H, Saito Y *et al*. 1992 Mass spectroscopic and ESR characterization of soluble yttrium-containing metallofullerenes YC_{82} and Y_2C_{82} *J. Phys. Chem.* 96 3571

[14] Ross M M, Nelson H H, Callahan J H and McElvany S W 1992 Production and characterization of metallofullerenes *J. Phys. Chem.* 96 5231

[15] McElvany S W 1992 Production of endohedral yttrium-fullerene cations by direct laser vaporization *J. Phys. Chem.* 96 4935

[16] Shinohara H, Sato H, Ohchochi M *et al*. 1992 Encapsulation of a scandium trimer in C_{82} *Nature* 357 52

[17] Yannoni C S, Hoinkis M, de Vries M S *et al*. 1992 Scandium clusters in fullerene cages *Science* 256 1191

[18] Gillan E, Yeretzian C, Min K S *et al* 1992 Endohedral rare-earth fullerene complexes *J. Phys. Chem.* 96 6869

[19] Moro L, Ruoff R S, Becker C H *et al*. 1993 Studies of metallofullerene primary soots by laser and thermal desorption mass spectrometry *J. Phys. Chem.* 97 6801

[20] Xu Z, Nakane T and Shinohara H 1996 Production and isolation of $Ca@C_{82}$ (I-IV) and $Ca@C_{82}$ (I, II) metallofullerenes *J. Am. Chem. Soc.* 118 11309

[21] Dennis T J S and Shinohara H 1997 Production and isolation of endohedral strontium- and barium-based mono-metallofullerenes: $Sr/Ba@C_{82}$ and $Sr/Ba@C_{84}$ *Chem. Phys. Lett.* 278 107

[22] Dennis T J S and Shinohara H 1997 *Fullerenes: Recent Advances in the Chemistry and Physics of Fullerenes and Related Materials*, Vol. 4, eds K Kadish and R Ruoff (Pennington, NJ: Electrochemical Society), pp. 182–190

[23] Dennis T J S and Shinohara H 1998 Production, isolation, and characterization of group-2 metal-containing endohedral metallofullerenes *Appl. Phys. A* **66** 243

[24] Dennis T J S and Shinohara H 1998 Isolation and characterisation of the two major isomers of [84]fullerene (C_{84}) *J. Chem. Soc., Chem. Commun.* 619

[25] Wan S M, Zhang H-W, Tso TSC *et al.* 1997 *Fullerenes: Recent Advances in the Chemistry and Physics of Fullerenes and Related Materials*, Vol. 4, eds K Kadish and R Ruoff (Pennington, NJ: Electrochemical Society), pp. 490–506

[26] Wan S M, Zhang H-W, Nakane T *et al.* 1998 Production, isolation, and electronic properties of missing fullerenes: $Ca@C_{72}$ and $Ca@C_{74}$ *J. Am. Chem. Soc.* **120** 6806

[27] Cao B P, Hasegawa M, Okada K *et al.* 2001 EELS and ^{13}C NMR characterization of pure $Ti_2@C_{80}$ metallofullerene *J. Am. Chem. Soc.* **123** 9679

[28] Cao B P, Suenaga K, Okazaki T and Shinohara H 2002 Production, isolation, and EELS characterization of $Ti_2@C_{84}$ dititanium metallofullerenes *J. Phys. Chem. B* **106** 9295

[29] Guo T, Diener M D, Chai Y *et al* 1992 Uranium stabilization of C_{28}: a tetravalent fullerene *Science* **257** 1661

[30] Wang L S, Alford J M, Chai Y *et al.* 1993 The electronic structure of $Ca@C_{60}$ *Chem. Phys. Lett.* **207** 354

[31] Wang L S, Alford J M, Chai Y *et al.* 1993 Photoelectron spectroscopy and electronic structure of $Ca@C_{60}$ *Z. Phys. D* **26** S297

[32] Wang Y, Tomanek D and Ruoff R 1993 Stability of $M@C_{60}$ endohedral complexes *Chem. Phys. Lett.* **208** 79

[33] Kubozono Y, Ohta T, Hayashibara T *et al.* 1995 Preparation and extraction of $Ca@C_{60}$ *Chem. Lett.* 457

[34] Scuseria G E 1992 Comparison of coupled-cluster results with a hybrid of Hartree-Fock and density functional theory *J. Chem. Phys.* **97** 7528

[35] Guo T, Odom G K and Scuseria G E 1994 Electronic structure of $Sc@C_{60}$: an *ab initio* theoretical study *J. Phys. Chem.* **98** 7745

[36] Godly E W and Taylor R 1997 Nomenclature and terminology of fullerenes: a preliminary study *Pure Appl. Chem.* **69** 1411

2

Synthesis, Extraction, and Purification

2.1 SYNTHESIS OF ENDOHEDRAL METALLOFULLERENES

2.1.1 Laser-Furnace Synthesis

Metallofullerenes can be synthesized typically in two ways similar to the synthesis of empty fullerenes, which involves the generation of a carbon-rich vapor or plasma in He or Ar gas atmosphere. The two methods have been routinely used to date for preparing macroscopic amounts of metallofullerenes: the high-temperature laser vaporization or the "laser-furnace" method [1–3] and the DC arc-discharge method [4]. Both methods simultaneously generate a mixture of hollow fullerenes (C_{60}, C_{70}, C_{76}, C_{78}, C_{84}, etc.) together with metallofullerenes. The production of metallofullerenes can be followed by procedures to extract from soot and to separate/purify the metallofullerenes from the hollow fullerenes (see Section 2.2).

There is a less common but important method called the metal ion implantation technique for metallofullerene synthesis. This technique has been used to produce some alkaline metallofullerenes [5–7] and N atom endofullerenes [8–10] such as $Li@C_{60}$ and $N@C_{60}$, respectively. Also, Hirata et al. [11] reported the production of $K@C_{60}$ by introducing negatively charged C_{60} into a low-temperature (about 0.2 eV) potassium plasma column by a strong axial magnetic field. However, up until recently, isolation and structural characterization of the $Li@C_{60}$ metallofullerene prepared by the above methods have been difficult due

Endohedral Metallofullerenes: Fullerenes with Metal Inside, First Edition.
Hisanori Shinohara and Nikos Tagmatarchis.
© 2015 John Wiley & Sons, Ltd. Published 2015 by John Wiley & Sons, Ltd.

to the insufficiency of the materials produced by the implantation and the plasma techniques, although partial high-performance liquid chromatography (HPLC) separation on Li@C_{60} was reported [12]. The complete purification and molecular/crystal X-ray structures of Li@C_{60} are described in section 3.2.3. In Chapter 14, the separation, purification, and structural characterization of some M@C_{60} (M = metal atom) are described. In the following, we briefly review the laser-furnace and arc-discharge methods for the synthesis of endohedral metallofullerenes.

In the laser-furnace method (Figure 2.1), a target composite rod or disk for laser vaporization, which is composed of metal oxide/graphite with a high-strength pitch binder, is placed in a furnace at about 1200 °C [1]. A frequency-doubled Nd:YAG laser at 532 nm is focused onto the target rod, which is normally rotating/translating to ensure a fresh surface under an Ar gas flow (100–200 Torr) condition. Metallofullerenes and empty fullerenes are produced by the laser vaporization and then flow down the tube with the Ar gas carrier, and are finally trapped on the quartz tube wall near the end of the furnace.

To produce fullerenes and metallofullerenes a temperature above 800°C was found to be necessary, and below this critical temperature no fullerenes were produced [1,13,14], suggesting that relatively slow thermal annealing processes are required to form fullerenes and metallofullerenes. The laser-furnace method is suited to the study of growth mechanism of fullerenes and metallofullerenes [1,3,15–18]. The laser-furnace method is also known to be an efficient production method for single-wall carbon nanotubes when Ni/Co or Ni/Fe binary metal is stuffed with graphite powder for target composite rods [19].

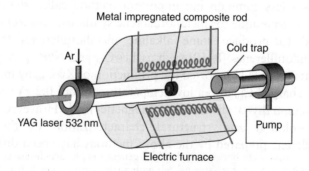

Figure 2.1   Schematic diagram of the high-temperature laser-furnace apparatus to produce fullerenes and metallofullerenes by laser vaporization of a rotating metal-impregnated graphite target in an electric furnace with flowing argon carrier gas

2.1.2   DC Arc-Discharge Synthesis

Figure 2.2 represents a third-generation large-scale DC arc-discharge apparatus for the production of metallofullerenes developed and installed at Nagoya [20–24]. The arc generator consists of a production chamber and a collection chamber, equipped with an anaerobic sampling and

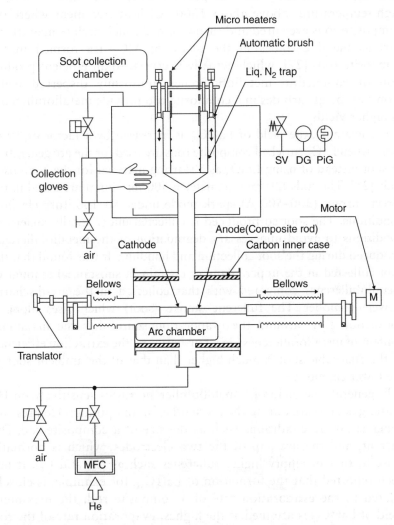

Figure 2.2   A cross-sectional view of the third-generation DC arc-discharge apparatus (the so-called Nagoya arc-discharge model) with an anaerobic collection and sampling mechanism. The produced metallofullerene-containing soot is effectively trapped by the liquid N_2 trap installed in the center of the collection chamber. Typical arc-discharge conditions: 40–100 Torr He flow, 300–500 A, and 25–30 V

collection mechanism of raw soot containing metallofullerenes [25,26]. Anaerobic sampling of the soot is preferred to conventional collection under ambient conditions because many of the metallofullerenes in primary soot are air (moisture)-sensitive and may be subjected to degradation during the soot handling.

Metal oxide/graphite composite rods, for example, La_2O_3 to prepare $La@C_{82}$, are normally used as positive electrodes (anodes) after a high-temperature (above about 1600 °C) heat treatment where the composite rods are cured and carbonized. At such high temperatures, various metal carbides in the phase of MC_2 are formed in the composite rods [27], which actually is crucial to an efficient production of endohedral metallofullerenes: uniformly dispersed metal atoms as metal carbides in a composite rod provide metallofullerenes in higher yields.

For example, the yield of $La@C_{82}$ is increased by a factor of 10 or more when LaC_2-enriched composite rods are used for the arc generation of soot instead of using La_2O_3 as a starting material for the composite rods [25]. The rods (20 mm diameter × 500 mm long) are arced in the direct current (300–500 A) spark mode under 50–100 Torr He flow conditions. The soot so produced is collected under totally anaerobic conditions to avoid unnecessary degradation of the metallofullerenes produced during the soot collection and handling. It was found that the soot collected in the upper chamber contains a substantial amount of metallofullerenes compared with that collected in the arc-discharge (lower) chamber. The fullerene smoke (soot) which rises along a convection flow around the evaporation source has the maximum content of metallofullerenes [28]. Furthermore, the extraction efficiency of the anaerobic soot is much higher than that of the ambient soot in the lower chamber.

In general, the yield of a metallofullerene varies sensitively on He buffer gas pressure during the arc synthesis. An optimum He pressure depends on arc conditions such as the size of a composite rod, DC current, and the arc gap of the two electrodes, which is normally close to that of empty higher fullerenes such as C_{82} and C_{84}. It has been reported that the formation of $La@C_{82}$, for example, is closely related to the evaporation rate of a composite rod; the maximum yield of $La@C_{82}$ is attained at the highest evaporation rate of the rod [28]. Mieno [29] reported that the production of endohedral metallofullerenes can be much enhanced under gravitation-free arc-discharge conditions as compared with the normal gravitational condition. This is due to the fact that the gravitation-free conditions suppress thermal convection of hot gas in the arc region and thus enable long-duration hot reaction of carbon clusters suited to metallofullerene production.

The production of scandium fullerenes, $Sc_n@C_{82}$ and/or $Sc_nC_2@C_{80}$ ($n = 1$–4), is especially interesting, because scandium fullerenes exist in solvent extract as mono-, di-, and tri-scandium fullerenes [30,31], which is quite unique compared with the other rare earth metallofullerenes. Even a tetra-scandium fullerene, $Sc_4C_2@C_{80}$, has been produced and isolated (K. Kuroki, T. Kuriyama, M. Inakuma, and H. Shinohara, 1999, unpublished results) [32]. The synthesis of the mono-, di-, tri-, and tetra-scandium fullerenes was found to be sensitive to the mixing ratio of Sc and C atoms in the composite rods; the relative abundance of di-, tri-, and tetra-scandium fullerenes increases as the carbon/scandium ratio decreases. For example, with a carbon/scandium (atomic) ratio of 86.2, the formation of mono- and di-scandium fullerenes such as $Sc@C_{82}$ and $Sc_2C_2@C_{82}$ was dominant, and the production of $Sc_3C_2@C_{80}$ and $Sc_4C_2@C_{80}$ was almost negligible. It was observed that the major scandium fullerene produced was $Sc_2C_2@C_{82}$ over a wide range of the carbon/scandium mixing ratios (10–100) [30,33].

2.1.3 Ion Implantation Technique

Weidinger and co-workers developed an ion implantation (bombardment) technique to produce nitrogen-containing C_{60}, $N@C_{60}$, with nitrogen ions from a conventional plasma discharge ion source which were implanted onto C_{60} films [8]. The yield of $N@C_{60}$ after atomic nitrogen ion implantation is about 10^{-5}–$10^{-4}\%$ yield. $N_2@C_{60}$ was also detected at a similar yield.

Campbell and co-workers applied a similar technique to implant alkali metal ions (Li^+, Na^+, K^+, Rb^+ ions) into C_{60} [5,7]. Figure 2.3 shows a schematic experimental setup of the alkali metal ion implantation apparatus. The synthesis of $Li@C_{60}$ was achieved. However, subsequent trial of separation of $Li@C_{60}$ had not been achieved due to difficulties in the extraction of this metallofullerene.

In 2011, $Li@C_{60}$ produced by ion bombardment was finally isolated and structurally characterized by Aoyagi et al. [34] in a form of its cationic salt, $[Li@C_{60}]^+(SbCl_6)^-$. The details of this characterization are described in Chapter 3. The ion implantation technique using an ultrafine-particle plasma including large negative fullerene ions was employed in producing $Li@C_{60}$ [11].

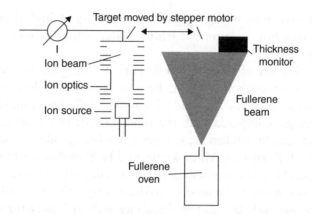

Figure 2.3 Schematic experimental setup for alkali metal ions implantation into C_{60} vapor to produce $M@C_{60}$ (M = Li, Na, K, Rb) metallofullerenes. Reproduced from Ref. 7] with permission from Elsevier

2.2 SOLVENT EXTRACTION OF METALLOFULLERENES FROM PRIMARY SOOT

The so-called solvent extraction method by toluene, o-xylene, or CS_2 is the most common and frequently used extraction method, in which metallofullerenes and hollow fullerenes are preferentially dissolved in solvents. The so-called Soxhlet extraction (a continuous and hot solvent extraction) or ultrasonic extraction is normally employed to increase the solvent extraction efficiency [35]. Insolubles in soot are easily separated from this solution by filtration. However, in many cases, the toluene or CS_2 extraction is not sufficient, since nearly half of the metallofullerene still remains in the residual soot even after the extensive CS_2 extraction. It has been found that metallofullerenes are further extracted from the residual soot by such solvents as pyridine [36] and 1,2,4-trichlorobenzene [37,38]. The metallofullerenes were found to be concentrated in this pyridine- or trichlorobenzene-extracted fraction. When necessary, the metallofullerene extracts can be stored in CS_2 solution for an extended period of time, up to a year.

In a sublimation method [2,39,40], as in the case of empty fullerenes [41–46], the raw soot containing metallofullerenes is heated in He gas or in vacuum up to 400 °C where metallofullerenes such as $La@C_{82}$ and $Y@C_{82}$ start to sublime. The metallofullerenes then condense in a cold trap, leaving the soot and other nonvolatiles behind in the sample holder. However, a complete separation of metallofullerenes has not been achieved to date by sublimation. Extraction by sublimation has the advantage over solvent extraction for obtaining "solvent-free" extracts, whereas the latter method is suited to large-scale extraction of metallofullerenes.

2.3 PURIFICATION AND ISOLATION BY HPLC

As in the case of hollow fullerenes [42,47,48], liquid chromatography (LC) is the main purification technique for metallofullerenes. LC has been frequently and traditionally used in separation chemistry. One of the most powerful LC techniques is HPLC (high-performance LC) which allows separation of fullerenes according to their molecular weight, size, shape, or other parameters [49–52]. The HPLC technique can even allow the structural isomers of various metallofullerenes to be separated [20].

The purification of endohedral metallofullerenes via HPLC had been difficult, mainly because the content of metallofullerenes in raw soot is normally very limited and, furthermore, the solubility in normal HPLC solvents is generally lower than that of various empty higher fullerenes. It took almost 2 years for metallofullerenes to be completely isolated by the HPLC method [33,53] after the first extraction of La@C_{82} by the Rice group [2]. Following these initial isolations of metallofullerenes, isolation with different HPLC columns was also reported [37,54]. The success of the purification/isolation was a real breakthrough for further characterization of the endohedral metallofullerenes.

Scandium metallofullerenes are, in particular, interesting in terms of separation and purification because, as described in Section 2.2, scandium fullerenes appear as mono-, di-, tri-, and even tetra-scandium fullerenes with several structural isomers which can be separated completely by HPLC. As an example, the HPLC separation of scandium fullerenes is briefly described in the following.

The scandium fullerenes, such as Sc@C_{82}, Sc$_2$C$_2$@C_{82}, and Sc$_3$C$_2$@C_{80}, were separated and isolated from various hollow (C_{60}–C_{110}) fullerenes by the so-called two-stage HPLC method [20,26,33,53]. The two-stage HPLC method uses two complementary HPLC columns which have different types of fullerene adsorption mechanisms and are suited for a complete separation of the metallofullerenes. The two-stage HPLC method was first successfully applied to the isolation of several di-scandium fullerenes including Sc$_2$@C_{74}, Sc$_2$@C_{82}, and Sc$_2$C$_2$@C_{82} [33] as shown in Figure 2.4. Kikuchi et al. [53] employed a similar method for the first isolation of La@C_{82}. To simplify the separation, an automated HPLC separation on some endohedral metallofullerenes has been also reported by Stevenson et al. [55].

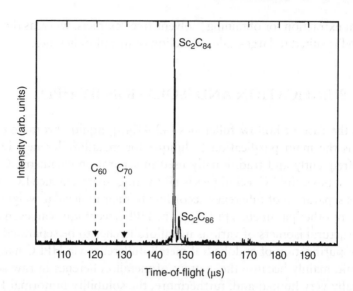

Figure 2.4 The first mass spectral evidence of the isolation/purification of metallofullerenes: Sc_2C_{84} (which is currently identified as $Sc_2C_2@C_{82}$). Reproduced with permission from Ref. [33]. Copyright 1993 American Chemical Society

In the first HPLC stage, the toluene solution of the extracts was separated by a preparative recycling HPLC system (Japan Analytical Industry LC-908-C_{60}) with a Trident-Tri-DNP column (Buckyclutcher I, 21 mm × 500 mm: Regis Chemical) or a 5-PBB (pentabromobenzyl) column (20 mm × 250 mm, Nacalai Tesque) with CS_2 eluent. In this HPLC process, the scandium fullerene-containing fractions were separated from other fractions including C_{60}, C_{70}, and higher fullerenes (C_{76}–C_{110}). The complete purification and isolation of various scandium fullerenes were performed in the second HPLC stage by using a Cosmosil Buckyprep column (20 mm × 250 mm, Nacalai Tesque) with a 100% toluene eluent. Figure 2.5 shows the first and the second HPLC stages of the purification of $Sc@C_{82}$ as an example. It has been found [36–38] that most of the mono-metallofullerenes, $M@C_{82}$, have at least two types of structural isomers (conventionally called isomers I and II), which can be separated by the two-stage HPLC technique.

The retention times of isomer I are normally shorter than those of isomer II and, in general, isomer I is much more stable than isomer II in air and in various solvents. In some cases, several isomers have been found for a metallofullerene. For example, a mono-calcium fullerene, $Ca@C_{82}$, has four isomers (I–IV) which have been produced, isolated, and characterized [56]. Unless otherwise noted, we discuss the structural and electronic properties of "major isomers" (usually isomer I) in this book.

The isolation of various metallofullerenes is normally confirmed by laser-desorption time-of-flight mass spectrometry. For ESR (electron spin resonance)-active metallofullerenes, the observation of the corresponding hyperfine structures can further confirm the identification and isolation.

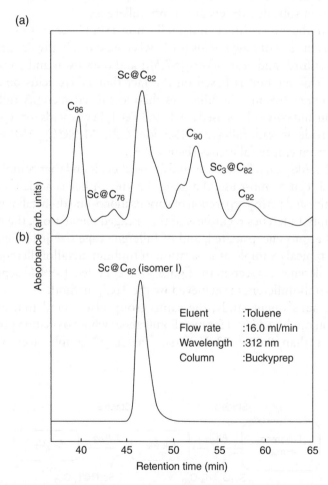

Figure 2.5 (a) An overall HPLC spectrum for the fraction which contains various scandium metallofullerenes in the first HPLC stage. (b) An isolated HPLC chromatogram for Sc@C$_{82}$ (I) after the second HPLC stage. The experimental conditions are presented in the lower right of the figure. Reproduced from Ref. [20]. Copyright 2000 IOP Publishing

2.4 FAST SEPARATION AND PURIFICATION WITH LEWIS ACIDS

Although the HPLC separation/purification described above is powerful and straightforward, the cost performance of the method is low because of limited quantities of separation (normally milligram quantity) together with the inherent time-consuming multi-step HPLC processes involved. The most time-consuming part of the HPLC separation is the separation of metallofullerenes from empty fullerenes, because the majority of the fullerenes in solvent extracts are empty fullerenes.

One of the most direct and facile non-HPLC separations of this process is the use of Lewis acids (or Lewis bases) originally developed by Bolskar, Alford, and co-workers [57,58] and Stevenson and co-workers [59,60]. This method is based on the fact that Lewis acids bind more selectively reactive metallofullerenes than less reactive empty fullerenes. Stevenson and co-workers used $AlCl_3$ and $FeCl_3$ Lewis acids for separating some nitride metallofullerenes (Section 6.2), $M_3N@C_{2n}$ (M = metal atoms), from empty fullerenes (Figure 2.6).

In 2012, Akiyama, Shinohara, and co-workers found that complexation with $TiCl_4$ can be most useful among the known Lewis acids for almost quantitative and perfect separation of various mono- and di-metallofullerenes from empty fullerenes regardless of the encaged metal atom, the number of metal atoms encapsulated, and of fullerene cage size [61]. Figure 2.7 shows a typical example of separation of thulium metallofullerenes from empty fullerenes. As seen in the figure, almost perfect (>99%) separation of the metallofullerenes is achieved by the $TiCl_4$ method.

In the subsequent study, the same group also found that efficient separation can be achieved for metallofullerenes whose oxidation potentials are lower than 0.62–0.72 V versus $Fe(Cp)_2^{+/0}$ couple [62]. Vis-NIR

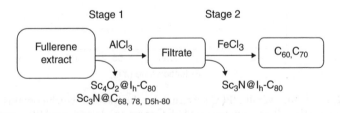

Figure 2.6 Schematic flowchart describing the separation of the metallofullerenes from empty fullerenes using $AlCl_3$ and $FeCl_3$ Lewis acids. Reproduced with permission from Ref. [60]. Copyright 2009 American Chemical Society

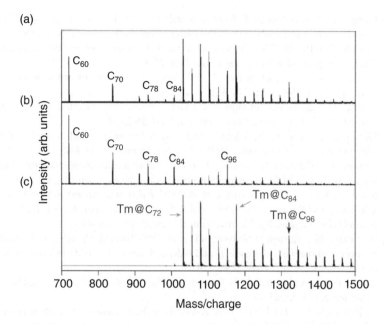

Figure 2.7 Laser-desorption mass spectra of (a) crude extract, (b) filtered solution, and (c) various thulium metallofullerenes after the $TiCl_4$ separation. Reproduced with permission from Ref. [61]. Copyright 2012 American Chemical Society

(visible-near infrared) absorption spectroscopy showed that electron transfer took place from metallofullerenes to $TiCl_4$, which provides efficient separation of metallofullerenes from empty fullerenes.

REFERENCES

[1] Haufler R E, Chai Y, Chibante L P F *et al*. 1991 *Cluster-Assembled Materials*, Vol. **206**, eds R S Averback, J Bernhoc and D L Nelson (Pittsburgh, PA: Materials Research Society), pp. 627–637

[2] Chai Y, Guo T, Jin C *et al*. 1991 Fullerenes with metals inside *J. Phys. Chem.* **95** 7564

[3] Ying Z C, Jin C, Hettich R L *et al*. 1994 *Fullerenes: Recent Advances in the Chemistry and Physics of Fullerenes and Related Materials*, Vol. **1**, eds K Kadish and R Ruoff (Pennington, NJ: Electrochemical Society), pp. 1402–1412

[4] Haufler R E, Conceicao J, Chibante L P F. *et al*. 1990 Efficient production of C_{60} (Buckminsterfullerene), $CH_{60}H_{36}$, and the solvated buckide ion *J. Phys. Chem.* **94**, 8634.

[5] Tellgmann R, Krawez N, Lin S-H *et al*. 1996 Endohedral fullerene production *Nature* **382** 407

[6] Tellgmann R, Krawez N, Hertel I V and Campbell E E B 1996 *Fullerenes and Fullerene Nanostructures*, eds H Kuzmany, J Fink, M Mehring and S Roth (London: World Scientific), pp. 168–172

[7] Campbell E E B, Tellgmann R, Krawez N and Hertel I V 1997 Production and LDMS characterisation of endohedral alkali-fullerene films *J. Phys. Chem. Solids* **58** 1763

[8] Murphy T A, Pawlik T, Weidinger A *et al.* 1996 Observation of atomlike nitrogen in nitrogen-implanted solid C_{60} *Phys. Rev. Lett.* **77** 1075

[9] Knapp C, Dinse K-P, Pietzak B *et al.* 1997 Fourier-transform EPR study of $N@C_{60}$ in solution *Chem. Phys. Lett.* **272** 433

[10] Mauser H, van Eikema Hommes N, Clark T *et al.* 1997 Stabilization of atomic nitrogen inside C_{60} *Angew. Chem., Int. Ed. Engl.* **36** 2835

[11] Hirata T, Hatakeyama R, Mieno T and Sato N Production and control of $K-C_{60}$ plasma for material processing 1996 *J. Vac. Sci. Technol., A* **14** 615

[12] Krawez N, Tellgmann R, Gromov A *et al.* 1998 *Molecular Nanostructures*, eds H Kuzmany, J Fink, M Mehring and S Roth (London: World Scientific), pp. 184–188

[13] Suzuki S, Kasuya D, Suganuma T *et al.* 1997 *Fullerenes: Recent Advances in the Chemistry and Physics of Fullerenes and Related Materials*, Vol. 4, eds K Kadish and R Ruoff (Pennington, NJ: Electrochemical Society), pp. 485–489

[14] Wakabayashi T, Kasuya D, Shiromaru H *et al.* 1997 Towards the selective formation of specific isomers of fullerenes: *T*- and *p*-dependence in the yield of various isomers of fullerenes $C_{60}-C_{84}$ *Z.Phys. D* **40** 414

[15] Curl R F and Smalley R E 1991 Formation of fullerides and fullerene-based hetero-structures *Sci. Am.* **265** 54

[16] Wakabayashi T and Achiba Y 1992 A model for the C_{60} and C_{70} growth mechansim *Chem. Phys. Lett.* **190** 465

[17] Smalley R E 1992 Self-assembly of the fullerenes *Acc. Chem. Res.* **25** 97

[18] Wakabayashi T, Kikuchi K, Shiromaru H *et al.* 1993 Ring-stacking consideration on higher fullerene growth *Z. Phys. D* **26** S258

[19] Thess A, Lee, R., Nikolaev, P. *et al.* 1996 Crystalline ropes of metallic carbon nanotubes *Science* **273** 483

[20] Shinohara H 2000 Endohedral metallofullerenes *Rep. Prog. Phys.* **63** 843

[21] Shinohara H, Takata M, Sakata M *et al.* 1996 Metallofullerenes: their formation and characterization *Mater. Sci. Forum* **232** 207

[22] Nakane T, Xu Z, Yamamoto E *et al.* 1997 A review on endohedral metallofullerenes - structures and properties *Fullerene Sci. Technol.* **5** 829

[23] Dennis T J S and Shinohara, H. 1998 Production, isolation, and characterization of group-2 metal-containing endohedral metallofullerenes *Appl. Phys. A* **66** 243

[24] Shinohara H 1998 *Advances in Metal and Semiconductor Clusters*, Vol. 4, ed. M Duncan (New York, NY: JAI Press), pp. 205–226

[25] Bandow S, Shinohara H, Saito Y *et al.* 1993 High yield synthesis of lanthanofullerenes via lanthanum carbide *J. Phys. Chem.* **97** 6101

[26] Shinohara H, Inakuma M, Hayashi N *et al.* 1994 Spectroscopic properties of isolated $Sc_3@C_{82}$ metallofullerene *J. Phys. Chem.* **98** 8597

[27] Adachi G, Imanaka N and Fuzhong Z 1991 *Handbook on the Physics and Chemistry of Rare Earths*, Vol. 15, eds K A Gschneider Jr and L Eyring (Amsterdam: Elsevier), p. 61

[28] Saito Y, Yokoyama S, Inakuma M and Shinohara H 1996 An ESR study of the formation of $La@C_{82}$ isomers in arc synthesis *Chem. Phys. Lett.* **250** 80

[29] Mieno T 1998 Efficient production of endohedral metallofullerenes in repetitive gravitation-free arc discharge using a vertical swing tower *Jpn. J. Appl. Phys.* **37** L761

[30] Shinohara H, Sato H, Ohchochi M *et al.* 1992 Encapsulation of a scandium trimer in C_{82} *Nature* **357** 52

[31] Yannoni C S, Hoinkis M, de Vries M S *et al.* 1992 Scandium clusters in fullerene cages *Science* **256** 1191

[32] Wang T S, Chen N, Xiang J F *et al.* 2009 Russian-doll-type metal carbide endofullerene: synthesis, isolation, and characterization of $Sc_4C_2@C_{80}$ *J. Am. Chem. Soc.* **131**, 16646

[33] Shinohara H, Yamaguchi H, Hayashi N *et al.* 1993 Isolation and spectroscopic properties of $Sc_2@C_{74}$, $Sc_2@C_{82}$, and $Sc_2@C_{84}$ *J. Phys. Chem.* **97** 4259

[34] Aoyagi S, Nishibori E and Sawa H *et al.* 2010 A layered ionic crystal of polar $Li@C_{60}$ superatoms *Nature Chem.* **2** 678

[35] Khemani K C, Prato M and Wudl F 1992 A simple Soxhlet chromatographic method for the isolation of pure C_{60} and C_{70} *J. Organomet. Chem.* **57** 3254

[36] Inakuma M, Ohno M and Shinohara H 1995 *Fullerenes: Recent Advances in the Chemistry and Physics of Fullerenes and Related Materials*, Vol. 2, eds K Kadish and R Ruoff (Pennington, NJ: Electrochemical Society), pp. 330–342

[37] Yamamoto K, Funasaka H, Takahashi T and Akasaka T 1994 Isolation of an ESR-active metallofullerene of $La@C_{82}$ *J. Phys. Chem.* **98** 2008

[38] Yamamoto K, Funasaka H, Takahashi T *et al.* 1994 Isolation and characterization of an ESR-active $La@C_{82}$ isomer *J. Phys. Chem.* **98** 12831

[39] Yeretzian C, Wiley J B, Holczer K *et al.* 1993 Partial separation of fullerenes by gradient sublimation *J. Phys. Chem.* **97** 10097

[40] Diener M D, Smith C A and Veirs D K 1997 Anaerobic preparation and solvent-free separation of uranium endohedral metallofullerenes *Chem. Mater.* **9** 1773

[41] Kräetschmer W, Fostiropoulos K, Lamb L D and Huffman D R 1990 Solid C_{60}: a new form of carbon *Nature* **347** 354

[42] Taylor R, Hare J P, Abdul-Sada A K and Kroto H W 1990 Isolation, separation and characterisation of the fullerenes C_{60} and C_{70}: the third form of carbon *J. Chem. Soc., Chem. Commun.* 1423

[43] Cox D M, Behal S, Disko M *et al.* 1991 Characterization of C_{60} and C_{70} clusters *J. Am. Chem. Soc.* **113** 2940

[44] Abrefah J, Olander D R, Balooch M and Siekhaus W J 1992 Vapor pressure of Buckminsterfullerene *Appl. Phys. Lett.* **60** 1313

[45] Pan C, Sampson M P, Chai Y *et al.* 1991 Heats of sublimation from a polycrystalline mixture of carbon clusters (C_{60} and C_{70}) *J. Phys. Chem.* **95** 2944

[46] Averitt R D, Alford J M and Halas N J 1994 High-purity vapor phase purification of C_{60} *Appl. Phys. Lett.* **65** 374

[47] Ajie H, Alvarez M M, Anz S J *et al.* 1990 Characterization of the soluble all-carbon molecules C_{60} and C_{70} *J. Phys. Chem.* **94** 8630

[48] Scrivens W A, Bedworth P V and Tour M J 1992 Purification of gram quantities of C_{60}. A new inexpensive and facile method *J. Am. Chem. Soc.* **114** 7917

[49] Kikuchi K, Nakahara N, Honda M *et al.* 1991 Separation, detection and UV/visible absorption spectra of C_{76}, C_{78} and C_{84} *Chem. Lett.* 1607

[50] Kikuchi K, Nakahara N, Wakabayashi T *et al.* 1992 Isolation and identification of fullerene family: C_{76}, C_{78}, C_{82}, C_{84}, C_{90} and C_{96} *Chem. Phys. Lett.* **188** 177

[51] Klute R C, Dorn H C and McNair H M 1992 HPLC separation of higher (C_{84}^+) fullerenes *J. Chromatogr. Sci.* **30** 438

[52] Meier M S and Selegue J P 1992 Efficient preparative separation of C_{70} and C_{60} –Gel permeation chromatography of fullerenes using 100 percent toluene as mobile phase *J. Org. Chem.* **57** 1924

[53] Kikuchi K, Suzuki S, Nakao Y *et al.* 1993 Isolation and characterization of the metallofullerenes LaC$_{82}$ *Chem. Phys. Lett.* **216** 67

[54] Savina M, Martin G, Xiao J *et al.* 1994 *Fullerenes: Recent Advances in the Chemistry and Physics of Fullerenes and Related Materials*, Vol. 1, eds K Kadish and R Ruoff (Pennington, NJ: Electrochemical Society), pp. 1309–1319

[55] Stevenson S Dorn H C, Burbank P *et al.* 1994 Automated HPLC separation of endohedral metallofullerene Sc@C$_{2n}$ and Y@C$_{2n}$ fractions *Anal. Chem.* **66** 2675

[56] Xu Z, Nakane T and Shinohara H 1996 Production and Isolation of Ca@C$_{82}$ (I–IV) and Ca@C$_{84}$ (I, II) metallofullerenes *J. Am. Chem. Soc.* **118** 11309

[57] Bolskar R D and Alford J M 2003 Chemical oxidation of endohedral metallofullerenes: identification and separation of distinct classes *Chem. Commun.* **1292**

[58] Raebiger J W and Bolskar R D 2008 Improved production and separation processes for gadolinium metallofullerenes *J. Phys. Chem. C* **112** 6605

[59] Stevenson S, Harich K, Yu H *et al.* 2006 Nonchromatographic "stir and filter approach" (SAFA) for isolating Sc$_3$N@C$_{80}$ metallofullerenes *J. Am. Chem. Soc.* **128** 8829

[60] Stevenson S, Mackey M A, Pickens J E *et al.* 2009 Lewis acids and use as an effective purification method *Inorg. Chem.* **48** 11685

[61] Akiyama K, Hamano T, Nakanishi Y *et al.* 2012 Non-HPLC rapid separation of metallofullerenes and empty cages with TiCl$_4$ Lewis acid *J.Am. Chem. Soc.* **134** 9762

[62] Wang Z, Nakanishi Y, Noda S *et al.* 2012 The origin and mechanism of non-HPLC purification of metallofullerenes with TiCl$_4$ *J. Phys. Chem. C* **116** 25563

3

Molecular and Crystal Structures

3.1 ENDOHEDRAL OR EXOHEDRAL? A BIG CONTROVERSY

Since the first studies on production and solvent extraction of metallofullerenes such as La/Y/Sc@C_{82}, there had been great controversy as to whether or not the metal atom is really trapped inside the fullerene cage [1,2]. In the gas phase, the stability of endohedral metallofullerenes has been studied by laser photofragmentation for La@C_{82} and Sc_2C_2@C_{82} [3,4], collisional fragmentation with atomic and molecular targets for La@C_{82} and Gd@C_{82} [5], and fragmentation induced by surface impact for La$_2$@C_{80} [6], La@C_{82}, La$_2$@C_{100} [7], La@C_{60} [8], Ce@C_{82} and Ce$_2$@C_{100} [9], Y@C_{82} and Ca@C_{84} [10].

Although the most extensive fragmentation was observed in the laser photofragmentation, the general tendency of the fragmentation induced by the three excitations was found to be similar: the main fragments from La@C_{82} were C_2-loss species such as La@C_{80}, La@C_{78}, La@C_{76}, and so on, and the empty C_{82} fragment was not observed. This result was interpreted as being due to the endohedral nature of La@C_{82} since exohedral La(C_{60}) [11] and Fe(C_{60}) [12] prepared by gas phase reactions gave C_{60} as the main product upon collisional fragmentation against rare-gas targets.

Endohedral Metallofullerenes: Fullerenes with Metal Inside, First Edition.
Hisanori Shinohara and Nikos Tagmatarchis.
© 2015 John Wiley & Sons, Ltd. Published 2015 by John Wiley & Sons, Ltd.

However, in the solid state, the evidence for the two independent contradictory results was reported for extended X-ray absorption fine structure (EXAFS) experiments on an unpurified extract of $Y@C_{82}$ (i.e., a mixture of $Y@C_{82}$ and empty fullerenes). Soderholm *et al.* [13] reported that the Y atom is exohedrally attached from the outside to the C_{82} cage, whereas Park *et al.* [14] reported an endohedral nature of $Y@C_{82}$: the nearest-neighbor C—Y distances obtained were 2.53 ± 0.02 and 2.4 Å, respectively. Kikuchi *et al.* [15] performed an EXAFS experiment on a purified $La@C_{82}$ powder material and reported that the nearest- and next-to-nearest-neighbor C—La distances are 2.47 ± 0.02 and 2.94 ± 0.07 Å, respectively.

Most of the major experimental evidence suggested, however, the endohedral nature of the metallofullerenes: the IBM Almaden group reported a high-resolution transmission electron microscopy (HRTEM) experiment on a purified $Sc_2C_2@C_{82}$ (III) material identified as $Sc_2@C_{84}$ at the time (Figure 3.1) which strongly suggests that the two Sc atoms are encapsulated in the C_{82} cage [16].

Similar evidence on the endohedral nature based on HRTEM images was reported on $Gd@C_{82}$ by Tanaka *et al.* [17]. The UCLA group

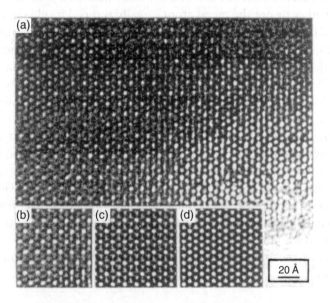

Figure 3.1   (a) HRTEM image of a $Sc_2@C_{84}$ crystal, taken along the [0001] direction. Inset (b) is a Fourier-filtered image that brings out the periodicities present in the original image. Insets (c) and (d) are simulated images of 0.67 nm thick $Sc_2@C_{84}$ and C_{84} crystals, respectively. Reprinted by permission from Macmillan Publishers Ltd: [16]. Copyright 1994 Rights Managed by Nature Publishing Group

reported a high-energy collision experiment on $La_2@C_{80}$ against silicon surfaces and found that no collision fragments such as La atoms and C_{80} were observed, also suggesting an endohedral structure of $La_2@C_{80}$ [18]. Similar surface-induced dissociation experiments were done on $La@C_{82}$ and $La@C_{60}$ [8] against a self-assembled monolayer (SAM) film and on $Y@C_{82}$, $Ca@C_{82}$, and $Ca@C_{84}$ against solid (silicon and gold) surfaces and SAM films [10], all of which indicated the endohedral nature of these metallofullerenes.

The Tohoku–Nagoya University group reported a series of ultra-high vacuum scanning tunneling microscopy (UHV-STM) studies on $Sc_2C_2@C_{82}$ and $Y@C_{82}$ adsorbed on silicon and copper clean surfaces, respectively [19–21]. All of the obtained scanning tunneling microscopy (STM) images showed a spherical shape which strongly suggests that the metal atoms are encapsulated in the fullerene cages. Gimzewski also studied $Sc_2C_2@C_{82}$ molecules deposited from a CS_2 solution onto Au(110) by STM and obtained some internal structure on the top part of the images [22].

Although the above experimental results strongly suggest an endohedral nature of the metallofullerenes, the final confirmation of the endohedral nature and detailed endohedral structures of the metallofullerenes was obtained by synchrotron X-ray diffraction (XRD) measurements on purified powder samples [23].

3.2 STRUCTURAL ANALYSES

A fullerene molecule has structural isomers with different five- and six-membered ring patterns. The most important structural isomers are called isolated pentagon rule (IPR) isomers [24,25]. IPR is considered to be the most important and essential rule governing the geometry of fullerenes, stating that the most stable fullerenes are those in which all pentagons are surrounded by five hexagons (cf. Section 6.1). In fact, all the empty fullerenes so far produced, isolated, and structurally charac- terized have been known to satisfy IPR. IPR can be best understood as a logical consequence of minimizing the number of dangling bonds and steric strain of fullerenes [26]. As a result, the smallest IPR-satisfying fullerene is C_{60}, and C_{70} is the second smallest; there are no IPR fuller- enes between C_{60} and C_{70}. Although IPR has been equally applied for metallofullerenes, IPR isomers of a metallofullerene can be different from those of the corresponding empty fullerene [27,28]. Electron transfers from a caged metal atom to the carbon cage may alter the stability of the fullerene.

3.2.1 Confirmation on Endohedral Structures as Determined by Synchrotron X-Ray Powder Diffraction

3.2.1.1 Y@C_{82} Mono-Metallofullerene

Previous experimental evidence including EXAFS [14,15] and HRTEM [16] suggested that the metal atoms are trapped inside the fullerenes. Theoretical calculations also indicated that such endohedral metallofullerenes are most stable[19,29–40]. However, the first conclusive experimental evidence on the endohedral nature of a metallofullerene, Y@C_{82}, was obtained by a synchrotron XRD study. The result indicated that the Y atom is encapsulated within the C_{82} fullerene and is strongly bound to the carbon cage [23].

The experimental data were analyzed in an iterative way of combination of Rietveld analysis [41] and the maximum entropy method (MEM) [42, 43]. The MEM can produce an election density distribution map from a set of X-ray structure factors without using any structural model. By the MEM analysis [44,45], the reliability factor, R_I, becomes as low as 1.5% for Y@C_{82}.

By using the revised structural model based on the previous MEM map and MEM analysis, a series of iterative steps involving Rietveld analysis were carried out until no significant improvement was obtained. Eventually, the R_I factor improved from 14.4 to 5.9% (R_{WP} = 3.0%). In Figure 3.2, the best fit of the Rietveld analysis of Y@C_{82} is shown. To display the endohedral nature of the Y@C_{82}, the MEM electron density distribution of Y@C_{82} is shown in Figure 3.3. There exists a high-density area just inside the C_{82} cage. The density maximum at the interior of the C_{82} cage corresponds to the Y atom, unequivocally indicating the endohedral structure of the metallofullerene.

It was also found that the cage structure of Y@C_{82} differs from that of the hollow C_{82} fullerene. There are many local maxima along the cage in Y@C_{82}, whereas electron densities of the C_{82} cage are relatively uniform. This suggests that in Y@C_{82} the rotation of the C_{82} cage is very limited around a certain axis even at room temperature, whereas that in C_{82} is almost free.

The MEM election density map further reveals that the Y atom does not reside at the center of the C_{82} cage but is very close to the carbon cage, as suggested theoretically [35,37–40, 46]. The ESR [47,48] and theoretical [40,49] studies suggest the presence of a strong charge transfer interaction between the Y^{3+} ion and the C^{3-} cage which may cause the aspherical electron density distribution of atoms. The Y—C distance

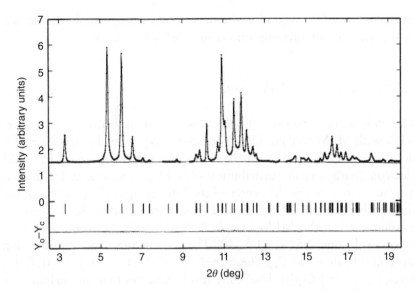

Figure 3.2 Powder X-ray diffraction patterns and the corresponding fitting results of $Y@C_{82}$ based on the calculated intensities from the MEM electron density. Reprinted by permission from Macmillan Publishers Ltd: [23]. Copyright 1995 Rights Managed by Nature Publishing Group

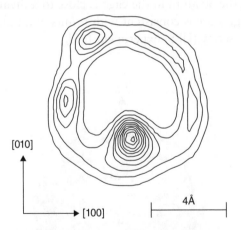

Figure 3.3 The MEM electron density distribution of $Y@C_{82}$ for the (001) section. The density maximum corresponds to the Y atom. Reprinted by permission from Macmillan Publishers Ltd: [23]. Copyright 1995 Rights Managed by Nature Publishing Group

calculated from the MEM map is 2.9(3) Å which is slightly longer than a theoretical prediction of 2.55–2.65 Å [40]. The X-ray study also reveals that the $Y@C_{82}$ molecules are aligned along the [001] direction in a head-to-tail $(...Y@C_{82}... Y@C_{82}...Y@C_{82}...)$ order in the toluene (as a solvent) containing crystal, suggesting the presence of strong dipole–dipole and charge transfer interactions among the $Y@C_{82}$ fullerenes.

3.2.1.2 $Sc@C_{82}$ Metallofullerene

The endohedral structure of $Sc@C_{82}$ was also studied by synchrotron XRD with MEM analysis [28]. The $Sc@C_{82}$ crystal includes solvent toluene molecules and has $P2_1$ space group as in the $Y@C_{82}$ case. The MEM electron charge density distribution of $Sc@C_{82}$ is shown in Figure 3.4. The Sc atom is not at the center of the fullerene but close to one of the six-membered rings of the cage. The nearest-neighbor Sc—C distance estimated from the MEM map is 2.5(8) Å which is very close to a theoretical value, 2.52–2.61 Å [40]. There are in total nine IPR satisfying structural isomers for C_{82}. These are $C_2(a)$, $C_2(b)$, $C_2(c)$, C_{2v}, $C_s(a)$, $C_s(b)$, $C_s(c)$, $C_{3v}(a)$, and $C_{3v}(b)$. The X-ray result indicates that the carbon cage of $Sc@C_{82}$ has C_{2v} symmetry (Figure 3.4).

There has been controversy as to whether the encaged Sc atom has a divalent state or a trivalent state [40,50–52]. The synchrotron X-ray result shows that the number of electrons around the Sc atom is 18.8e, indicating that the Sc atom in the cage is close to a divalent state, $Sc^{2+}@C_{82}^{2-}$. The charge state is consistent with results of an ultraviolet photoelectron spectroscopy (UPS) experiment [53].

C_2

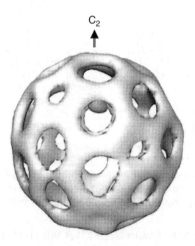

Figure 3.4 Equi-contour density map of the MEM charge density for the side view of a $Sc@C_{82}$ molecule. The Sc atom is drawn in a different contrast. The C_2 axis is indicated. Reprinted with permission from Ref. [28], Elsevier

3.2.1.3 La@C$_{82}$ Metallofullerene

The La@C$_{82}$ metallofullerene is one of the first endohedral metallofullerenes that was macroscopically produced and solvent extracted [1]. Suematsu *et al.* [54] first reported the crystal structure of La@C$_{82}$ precipitated from CS$_2$ solution via synchrotron X-ray powder diffraction. The composition of the microcrystal is expressed by La@C$_{82}$(CS$_2$)$_{1.5}$. The crystal has a cubic structure.

The results suggest a molecular alignment in the unit cell, in which the molecules align in the [111] direction with the molecular axis orienting in the same [111] direction. Watanuki *et al.* [55,56] performed synchrotron XRD measurements on solvent-free powder samples of La@C$_{82}$, and concluded that the major part of the crystal has a face-centered cubic (fcc) lattice. Their results strongly suggest the endohedral nature of La@C$_{82}$ where the La atom is displaced from the center of the C$_{82}$ cage by 1.9 Å.

The detailed endohedral structure of La@C$_{82}$ was revealed experimentally [57]. The electron density distribution of La@C$_{82}$ based on the MEM analysis of the powder XRD data is presented in Figure 3.5. The result shows that the La atom is encapsulated by the C$_{2v}$ isomer of C$_{82}$ as in Sc@C$_{82}$ described above. As is seen from the figure, the La atom is not at rest in the cage but rather is floating along the nearest six-membered ring at room temperature. The result is different from the Sc@C$_{82}$ and Y@C$_{82}$ cases in which Sc and Y atoms are almost at a standstill in the

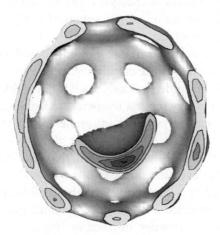

Figure 3.5 The section of the equi-charge density surface of a La@C$_{82}$ molecule. Reprinted with permission from Ref. [57], Elsevier

cage even at room temperature. A light metal atom such as Sc seems to be more strongly bound to the fullerene cage than a heavy La atom.

3.2.2   Dynamics of Metal Atoms within the Fullerene Cage

Intrafullerene metal motions have been theoretically predicted by Andreoni and Curioni [37,38,58,59] on La@C_{60} and La@C_{82} on the basis of molecular dynamics simulations. Experimentally, dynamic motion of metal atoms has been reported on La@C_{82} [57], Sc_2C_2@C_{82} [60], and La_2@C_{80} [61,62]. It is noted that the La atom is moving in the C_{82} cage at room temperature (Figure 3.5).

A particularly interesting case has been found in La_2@C_{80}. La_2@C_{80} metallofullerene was first produced by Whetten and co-workers [18] and was first isolated by Kikuchi *et al.* [15]. The empty C_{80} has seven IPR structures [D_2, D_{5d}, $C_{2v}(3)$, $C_{2v}(5)$, D_3, D_{5h}, and I_h]. A ^{13}C NMR study indicated that the most abundant C_{80} has D_2 symmetry [63]. However, theoretical calculations [64] have shown that encapsulation of two La atoms inside the I_h-C_{80} cage is most favorable. This is due to the fact that the I_h-C_{80} cage has only two electrons in the fourfold degenerate highest occupied molecular orbital (HOMO) level and can accommodate six more electrons to form the stable closed-shell electronic state of $(La^{3+})_2$@C_{80}^{6-} with a large HOMO–LUMO (highest occupied molecular orbital–lowest unoccupied molecular orbital) gap.

On the basis of ^{13}C NMR and ^{139}La NMR results, Akasaka *et al.* [61] reported a circular motion of encaged La atoms in the C_{80} cage. Two La atoms may circuit the inside of the spherical I_h-C_{80} cage. The energy barrier for the circuit of the metal cations is very small (about 5 kcal mol^{-1}). The dynamic behavior of metal atoms should also be reflected in the ^{139}La NMR linewidth, since circulation of two La^{3+} cations produces a new magnetic field inside the cage. Such a linewidth broadening was actually observed with increasing temperature from 32 to 90 °C [62].

A similar but greatly restricted intrafullerene dynamics of encaged metal ions has been reported by Miyake *et al.* [60] on Sc_2C_2@C_{82}. They observed a single ^{45}Sc NMR line, indicating that two Sc atoms in the cage are equivalent. However, in contrast to the La_2@C_{80} case, the internal rotation is hindered by a large barrier of about 50 kcal mol^{-1} [65].

Nishibori *et al.* revealed a detailed La dynamic motion in a C_{80} cage by synchrotron X-ray powder diffraction (Figure 3.6, [66]), where a perfect pentagonal-dodecahedral charge density of La_2 was seen in an icosahedral I_h-C_{80} cage. The characteristic charge density results from a

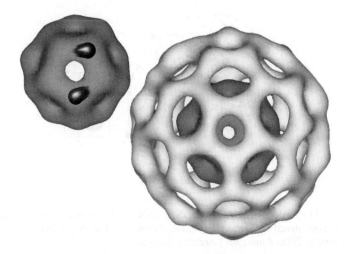

Figure 3.6 MEM charge density of $La_2@C_{80}$ as the equal-density contour surface along the S_{10} axis. The La_2 dodecahedral charge density is shown in dark gray and is also additionally shown beside the fullerene molecule. Reproduced with permission from Ref. [66], John Wiley & Sons

highly selective trajectory of the two La atoms, which hop along the hexagonal rings of the I_h-C_{80} polyhedral network. This highly symmetrical hopping of the two La atoms in a C_{80} cage was supported and analyzed in detail by a quantum chemical study [67].

The intrafullerene dynamics of the Ce atom in a C_{82} cage was also studied by time-differential perturbed angular correlation measurements [68]. The observed angular correlation shows the presence of two different chemical species of $Ce@C_{82}$. The data at low temperatures reveal that Ce stays at a certain site for one of the species, whereas for the other the atom has intramolecular dynamic motion.

3.2.2.1 A Di-metallofullerene: $Sc_2C_2@C_{82}$

Various metallofullerenes supposed to encapsulate two or three metal atoms within fullerene cages, such as $La_2@C_{80}$ [15,18,61,62,66,67,69, 70], $La_2@C_{72}$ [71–73], $Y_2@C_{82}$ [47,48], $Sc_2@C_{74}$ ([19,74), $Sc_2@C_{82}$ [48,75], $Sc_2C_2@C_{82}$ [16,48,50,75–78], $Sc_3C_2@C_{80}$ [50,64,71,75, 79–83], and $Er_2@C_{82}$ [84–86], have been successfully synthesized and purified. Among them the scandium di-metallofullerenes, $Sc_2C_2@C_{82}$, are especially interesting, because three structural isomers were found and isolated [77].

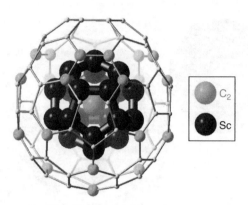

Figure 3.7 The molecular structure of $(Sc_2C_2)@C_{82}$ (III) carbide di-metallofullerene [originally considered to be $Sc_2@C_{84}$ (III)]. Reproduced with permission from Ref. [90]. Copyright 2006 American Chemical Society

Ab initio theoretical studies [87,88] and the experimental results on $Sc_2C_2@C_{82}$ including STM ([19,20]), transmission electron microscopy (TEM) [16] and ^{13}C-NMR [77] have suggested an endohedral nature. Similar to the mono-metallofullerenes, a synchrotron powder X-ray study is reported on $Sc_2C_2@C_{82}$ (isomer III) based on the Rietveld/MEM analysis [78]. Figure 3.7 shows a structural model based on a three-dimensional MEM electron density distribution of $Sc_2@C_{84}$ (III), which is now identified as $Sc_2C_2@C_{82}$ (III) carbide metallofullerene as described in detail in Chapter 5 [89,90].

The number of electrons around each maximum charge density inside the cage is 18.8 which is close to that of a divalent scandium ion Sc^{2+} (19.0). A theoretical study predicted that the formal electronic structure of $Sc_2C_2@C_{82}$ is well represented by $(Sc_2C_2)^{2+}{}_2@C_{82}{}^{4-}$, where two 4s electrons of each Sc atom transfer to the C_{82} cage [87]. The positive charge of the Sc atom from the MEM charge density is +2.2 which is in good agreement with the theoretical value. Furthermore, Pichler *et al.* reported that comparison of the Sc $2p$-$3d$ X-ray absorption spectrum with calculated ionic multiplet spectra shows a formal charge transfer to the fullerene cage of 2.6 [91].

3.2.2.2   A Tri-metallofullerene: $Sc_3C_2@C_{80}$

A tri-scandium fullerene, $Sc_3@C_{82}$ (currently identified as $Sc_3C_2@C_{80}$, cf. Section 4.2), has been produced [50,75] and characterized by ESR (see Section 4.2). A synchrotron X-ray structural study on $Sc_3@C_{82}$ has been reported based on Rietveld/MEM analysis [92]. The result revealed an

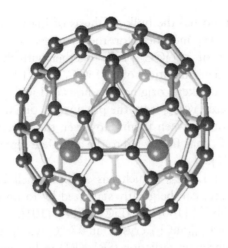

Figure 3.8 The molecular structure model of $Sc_3C_2@C_{80}$ along the S_{10} axis determined by the MEM/Rietveld method from the synchrotron X-ray powder diffraction data. Reprinted with permission from Ref. [95], Elsevier

intriguing feature of this metallofullerene: three Sc atoms are encapsulated in the form of a triangle Sc_3 cluster inside the C_{3v}-C_{82} fullerene cage. Furthermore, the charge state of the encaged Sc_3 cluster is 3^+ leading to a formal molecular charge state of $(Sc_3)^{3+}@C_{82}{}^{3-}$ as a result of an intrafullerene electron transfer. This was the first example in which a metal cluster is encaged by a fullerene. The presence of a Sc_3 trimer in the cage is consistent with an extended Hückel calculation [93].

The detailed XRD studies currently indicate that $Sc_3@C_{82}$ should be $Sc_3C_2@C_{80}$ carbide metallofullerene [94,95], where a C_2 molecule is encapsulated in a C_{80}-I_h cage. The encapsulated three Sc atoms form a triangle. A spherical charge distribution originating from the C_2 molecule is located at the center of the triangle. Intra-atomic distances between Sc and Sc are 3.61(3) Å in the triangle. The distance between Sc and the center of the C_2 molecule is 2.07(1) Å. The molecular structure is shown in Figure 3.8.

3.2.3 C_{60}-Based Metallofullerene: Li@C_{60}

A series of pioneering experimental studies on the production of Li@C_{60} was reported by Campbell and co-workers using the ion implantation technique (Section 2.1.3) [96–98]. They found that the polymerization of Li@C_{60} with C_{60} prevents the accurate determination of the structure and physical properties of the polymer.

Aoyagi *et al.* reported the bulk synthesis of Li@C$_{60}$ using a reported plasma method [99] modified for the synthesis of Li@C$_{60}$ and the complete isolation of a [Li@C$_{60}$] salt, together with the molecular and crystal structure determination of a [Li@C$_{60}$](SbCl$_6$) [100]. Ever since Campbell and co-workers' original synthesis of Li@C$_{60}$, it had not been possible to separate Li@C$_{60}$ from the pristine C$_{60}$ due to a strong charge transfer interaction between them. However, the selective one-electron oxidation of Li@C$_{60}$ has finally led to the complete separation of Li@C$_{60}$ from C$_{60}$ [100,101].

They found that Li@C$_{60}$ (which has an open-shell Li$^+$@C$_{60}$$^-$ electron transferred structure) can be greatly stabilized by forming salt structures such as [Li@C$_{60}$](SbCl$_6$) [100] and [Li@C$_{60}$](PF$_6$) [101]. Structural determination of the [Li@C$_{60}$](SbCl$_6$) crystal at 97 °C was performed using a single-crystal synchrotron radiation (SR) XRD technique at SPring-8. An idealized molecular structure of a Li@C$_{60}$ cation is depicted in Figure 3.9a,b. A Li cation lies in the vicinity of one of the six-membered rings, suggesting the presence of an attractive force exerted between the ring and the Li cation as in M@C$_{82}$-type metallofullerenes.

One of the most striking features of Li@C$_{60}$ revealed by the study (which has never been observed in higher-fullerene-based metallofullerenes) is an extremely high tendency to form ion-pair states (species) such as [Li@C$_{60}$]$^+$ (SbCl$_6$)$^-$ and [Li@C$_{60}$]$^+$(PF$_6$)$^-$. Li@C$_{60}$ can only be stabilized significantly in

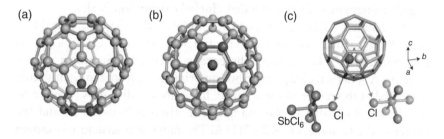

Figure 3.9 Structure of Li@C$_{60}$. (a) Structure of one orientation of the Li@C$_{60}$ cation extracted from a disordered structure. An encapsulated Li cation (**dark ball**) is located in the vicinity of a six-membered ring (**dark color**). The distance from the center of the carbon cage to the Li cation is 1.34 Å. (b) View from the center of the six-membered ring close to the Li cation. The distances between the Li cation and the six carbon atoms are in the range 2.24–2.45 Å. (c) Geometrical relationships between a Li@C$_{60}$ cation and two equivalent adjacent SbCl$_6$ anions. Two disordered Li sites are shown as magenta and white spheres. The encapsulated Li cation is ratchet-type disordered in the cage. The two arrows indicate the molecular dipole moment that the Li@C$_{60}$ cation directs to the Cl atoms. The Li–Cl distance is 5.58 Å. Reprinted by permission from Macmillan Publishers Ltd: [100]. Copyright 2010 Rights Managed by Nature Publishing Group

ambient conditions only when it coexists with an appropriate counter-anion. In fact, this is one of the main reasons why $M@C_{60}$-type metallofullerenes had never been purified at a level of more than 99% in previous decades (cf. Chapter 14).

3.2.4   Crystal Structures

Microcrystals of a metallofullerene grown from solution and those from solvent-free conditions generally have different crystal structures because of the presence of solvent molecules in the former case. For example, solvent-free $La@C_{82}$ crystals prepared by sublimation have fcc structure [55,56], whereas $La@C_{82}$ crystals grown from toluene and CS_2 exhibit monoclinic [57] and cubic [54] structures, respectively. Crystals of $Y@C_{82}$ [23], $Sc@C_{82}$ [78], $Sc_2C_2@C_{82}$ [28], and $Sc_3C_2@C_{80}$ [92] grown from toluene solutions also have monoclinic structures. Crystals of empty higher fullerenes such as C_{76} and C_{82} grown from toluene solutions also show monoclinic structures [102]. The observed crystal data and structural parameters of $La@C_{82}$, $Y@C_{82}$, and $Sc@C_{82}$ are listed in Table 3.1.

Table 3.1   X-ray structural data for $Sc@C_{82}$, $Y@C_{82}$, and $La@C_{82}$ determined by synchrotron powder X-ray diffraction

	Lattice type[a]	Space group	Lattice parameters	Nearest metal–carbon distance (Å)
$Sc@C_{82}$	Monoclinic	$P2_1$	a = 18.362(1) Å b = 11.2490(6) Å c = 11.2441(7) Å β = 107.996(9)°	2.53(8)
$Y@C_{82}$	Monoclinic	$P2_1$	a = 18.401(2) Å b = 11.281(1) Å c = 11.265(1) Å β = 108.07(1)°	2.40(15)[d]
$La@C_{82}$	Monoclinic	$P2_1$	a = 18.3345(8) Å b = 11.2446(3) Å c = 11.2320(3) Å β = 107.970(6)°	2.55(7)
$La@C_{82}$	Cubic[b]	$I\bar{4}3d$	a_0 = 25.72 ± 0.007 Å	2.55(7)
$La@C_{82}$	fcc[b]	$Fm3m^c$	a_0 = 15.78 Å	2.55(7)

[a] All crystals as grown from toluene solution have monoclinic lattice.
[b] Crystals as grown from CS_2 solution have cubic lattice, whereas solvent-free crystals have fcc lattice.
[c] Patterson symmetry.
[d] The structure with five positional disorder of Y atom.

One of the important findings in the crystal structures of the mono-metallofullerenes is the alignment of molecules in a certain direction in the crystal. For example, the Y@C_{82} molecules are aligned along the [001] direction in a head-to-tail (...Y@C_{82}...Y@C_{82}...Y@ C_{82}...) configuration in the crystal, indicating the presence of strong dipole–dipole and charge transfer interactions among the Y@C_{82} fullerenes [23]. The crystal data in Table 3.1 also support this idea since the c parameter of Y@C_{82} (11.265 Å) is shorter than that of C_{82} (11.383 Å), whereas the a parameter of Y@C_{82} is much longer than that of C_{82}, indicating a shrinkage in crystal packing in the [001] direction.

3.2.5   Single-Crystal Structural Characterization

On account of the inherent spherical shape of fullerenes, it had been difficult to obtain high quality single crystals [103]. However, the random rotation of the spherical fullerene cage can be suppressed by chemical modification or co-crystallization with suitable compounds.

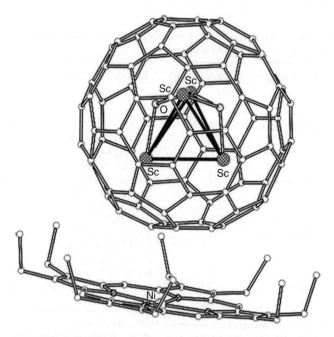

Figure 3.10   Single crystal X-ray structure of the metallofullerene $Sc_4(\mu_3\text{-}O)_2$@ $C_{80}[I_h] \cdot Ni^{II}\text{-}(OEP) \cdot 2C_6H_6$. Reproduced with permission from Ref. [108]. Copyright 2008 American Chemical Society

One of the frequently used and common techniques to obtain single crystals of metallofullerenes is to employ co-crystallization with metal porphyrin compounds, which was originally developed by Balch, Dorn, and co-workers [104,105]. The technique has extensively been applied to single crystal X-ray structural determination of metal nitride metallofullerenes (cf. Chapter 5) in particular [106,107]. The co-crystallization procedure produces samples with sufficient order due to the non-covalent interactions between the metal porphyrin and the fullerene cage [106]. Figure 3.10 shows the single crystal X-ray structure of the metallofullerene $Sc_4(m_3\text{-}O)_2@C_{80}[I_h]$ co-crystallized with $Ni^{II}\text{-}(OEP)\cdot 2C_6H_6$ (where OEP is octaethylporphyrin).

REFERENCES

[1] Chai Y, Guo T, Jin C, Haufler R E et al. 1991 Fullerenes with metals inside J. Phys. Chem. **95** 7564

[2] Bethune D S, Johnson R D, Salem J R et al. 1993 Atoms in carbon cages: the structure and properties of endohedral fullerenes Nature **366** 123

[3] Wakabayashi T, Shiromaru H, Suzuki S et al. Kikuchi K and Achiba Y 1996 C_2-loss fragmentation of higher fullerenes and metallofullerenes Surf. Rev. Lett. **3** 793

[4] Suzuki S, Kojima Y, Shiromaru H et al. 1997 Photoionization/fragmentation of endohedral fullerenes Z. Phys. D **40** 410

[5] Lorents D C, Yu D H, Brink C et al. 1995 Collisional fragmentation of endohedral fullerenes Chem. Phys. Lett. **236** 141

[6] Yeretzian C, Hansen K, Alvarez M M et al. 1992 Collisional probes and possible structures of La_2C_{80} Chem. Phys. Lett. **196** 337

[7] Beck R D, Weis P, Rockenberger J et al. 1996 Fragmentation of fullerenes and metallofullerenes by surface impact Surf. Sci. Lett. **3** 881

[8] Kimura T, Sugai T and Shinohara H 1999 Surface-induced dissociation of lanthanum metallofullerenes on a fluorinated self-assembled monolayer film Chem. Phys. Lett. **304** 211

[9] Beck R D, Weis P, Rockenberger J and Kappes M M 1996 Delayed ionization of fullerenes and fullerene derivatives upon laser desorption and surface collision Surf. Sci. Lett. **3** 771

[10] Kimura T, Sugai T and Shinohara H 1999 Surface-induced fragmentation of higher fullerenes and endohedral metallofullerenes J. Chem. Phys. **110** 9681

[11] Huang Y and Freiser B S 1991 Externally bound metal ion complexes of buckminsterfullerene, MC_{60}^+, in the gas phase J. Am. Chem. Soc. **113** 9418

[12] Roth L M, Huang Y, Schwedler J T et al. 1991 Evidence for an externally bound Fe^+–Buckminsterfullerene complex, FeC_{60}^+, in the gas phase J. Am. Chem. Soc. **113** 6298

[13] Soderholm L, Wurz P, Lykke K R et al. 1992 An EXAFS study of the metallofullerene YC_{82}: is the yttrium inside the cage? J. Phys. Chem. **96** 7153

[14] Park C H, Wells B O, DiCarlo J et al. 1993 Structural information on yttrium ions in fullerene C_{82} from EXAFS experiments Chem. Phys. Lett. **213** 196

[15] Kikuchi K, Nakao Y, Achiba Y and Nomura M 1994 *Fullerenes: Recent Advances in the Chemistry and Physics of Fullerenes and Related Materials*, Vol. 1, eds K Kadish and R Ruoff (Pennington, NJ: Electrochemical Society), pp. 1300–1308

[16] Beyers R, Kiang C-H, Johnson R D *et al.* 1994 Preparation and structure of crystals of the metallofullerene $Sc_2@C_{84}$ *Nature* **370** 196

[17] Tanaka N, Honda Y, Kawahara M *et al.* 1996 High-resolution electron microscopy of $Gd@C_{82}$ metal fullerenes grown on $MgO(001)$ surfaces *Thin Solid Films* **282** 613

[18] Yeretzian C, Hansen K, Alvarez MM *et al.* 1992 Collisional probes and possible structures of La_2C_{80} *Chem,Phys. Lett.* **196** 337

[19] Wang X D, Hashizume T, Xue Q *et al.* 1993 Geometry of metallofullerenes adsorbed on the Si (100) 2×1 surface studied by scanning tunneling microscopy *Chem. Phys. Lett.* **216** 409

[20] Shinohara H, Yamaguchi H, Hayashi N *et al.* 1993 Isolation and spectroscopic properties of scandium fullerenes ($Sc_2@C_{74}$, $Sc_2@C_{82}$, and $Sc_2@C_{84}$) *J. Phys. Chem.* **97** 4259

[21] Sakurai T, Wang X D, Xue Q K *et al.* 1996 Scanning tunneling microscopy study of fullerenes *Prog. Surf. Sci.* **51** 263

[22] Gimzewski J K 1996 *The Chemical Physics of Fullerenes 10 (and 5) Years Later*, Vol. **316**, ed. W Andreoni (Dordrecht: Kluwer), pp. 117–136

[23] Takata M, Umeda B, Nishibori E *et al* 1995 Confirmation by X-ray diffraction of the endohedral nature of the matallofullerene $Y@C_{82}$ *Nature* **377** 46

[24] Kroto H W 1987 The stability of the fullerenes C_n, with n = 24, 28, 32, 50, 60 and 70 *Nature* **329** 529

[25] Schmalz T G, Seits W A, Klein D J and Hite G E 1988 Elemental carbon cages *J. Am. Chem. Soc.* **110** 1113

[26] Fowler P W and Manolopoulos D E 1995 *An Atlas of Fullerenes* (Oxford: Clarendon)

[27] Dennis T J S, Kai T, Tomiyama T and Shinohara H 1998 Isolation and characterisation of the two major isomers of [84]fullerene (C_{84}) *J. Chem. Soc., Chem. Commun.* 619

[28] Nishibori E, Takata M, Sakata M *et al.* 1998 Determination of the cage structure of $Sc@C_{82}$ by synchrotron powder diffraction *Chem. Phys. Lett.* **298** 79

[29] Rosén A and Waestberg B 1989 Electronic structure of spheroidal metal containing carbon shells: study of the lanthanum-carbon and sixty-atom carbon (LaC_{60} and C_{60}) clusters and their ions within the local density approximation *Z. Phys. D* **12** 387

[30] Rosén, A. and Waestberg, B., 1998 1st-Principle calculations of the ionization-potentials and electron affinities of the spheroidal molecules C_{60} and LaC_{60} *J. Am. Chem. Soc.* **110** 8701.

[31] Cioslowski J and Fleishcmann E D 1991 Endohedral complexes: Atoms and ions inside the C_{60} cage *J. Chem. Phys.* **94** 3730

[32] Chang A H H, Ermler W C and Pitzer R M 1991 The ground and excited states of $C_{60}M$ and $C_{60}M^+$ (M=O, F, K, Ca, Mn, Cs, Ba, La, Eu, U) *J. Chem. Phys.* **94** 5004

[33] Manolopoulos D E and Fowler P W 1991 Structural proposals for endohedral metal-fullerene complexes *Chem. Phys. Lett.* **187** 1

[34] Manolopoulos D E and Fowler P W 1992 Hypothetical isomerisations of LaC_{82} *J. Chem. Soc., Faraday Trans.* **88** 1225

[35] Laasonen K, Andreoni W and Parrinello M 1992 Structural and electronic properties of $La@ C_{82}$ *Science* **258** 1916

[36] Saito S and Sawada S 1992 Growth mechanism and geometry of LaC_{82} *Chem. Phys. Lett.* **198** 466

[37] Andreoni W and Curioni A 1996 Freedom and constraints of a metal atom encapsulated in fullerene cages *Phys. Rev. Lett.* **77** 834

[38] Andreoni W and Curioni A 1996 *The Chemical Physics of Fullerenes 10 (and 5) Years Later*, ed. W Andreoni (Dordrecht: Kluwer), pp. 183–196

[39] Guo T, Odom G K and Scuseria G E 1994 Electronic structure of $Sc@C_{60}$: an *ab initio* theoretical study *J. Phys. Chem.* **98** 7745

[40] Nagase S and Kobayashi K 1993 Metallofullerenes MC_{82} (M = Sc, Y, and La). A theoretical study of the electronic and structural aspects *Chem. Phys. Lett.* **214** 57

[41] Rietveld H M 1969 A profile refinement method for nuclear and magnetic structures *J. Appl. Crystallogr.* **2** 65

[42] Collins D M 1982 Electron density images from imperfect data by iterative entropy maximization *Nature* **298** 49

[43] Bricogne G 1988 A Bayesian statistical theory of the phase problem. I. A multichannel maximum-entropy formalism for constructing generalized joint probability distributions of structure factors *Acta Crystallogr., A* **44** 517

[44] Sakata M and Sato M 1990 Accurate structure analysis by the maximum-entropy method *Acta Crystallogr., A* **46** 263

[45] Kumazawa S, Kubota Y, Tanaka M *et al.* 1993 MEED: a program package for electron-density-distribution calculation by the maximum-entropy method *J. Appl. Crystallogr.* **26** 453

[46] Nagase S and Kobayashi K 1994 Theoretical study of the lanthanide fullerene CeC_{82}. Comparison with ScC_{82}, YC_{82} and LaC_{82} *Chem. Phys. Lett.* **228** 106

[47] Weaver J H, Chai Y, Kroll G H *et al.* 1992 XPS probes of carbon-caged metals *Chem. Phys. Lett.* **190** 460

[48] Shinohara H, Sato H, Saito Y *et al.* 1992 Mass spectroscopic and ESR characterization of soluble yttrium-containing metallofullerenes YC_{82} and Y_2C_{82} *J. Phys. Chem.* **96** 3571

[49] Schulte J, Boehm M C and Dinse K P 1996 Electronic-structure of endohedral $Y@C_{82}$ – an ab-initio Hartree-Fock investigation *Chem. Phys. Lett.* **259** 48

[50] Shinohara H, Sato H, Ohchochi M *et al.* 1992 Encapsulation of a scandium trimer in C_{82} *Nature* **357** 52

[51] Ruebsam M, Knapp C P, Schweitzer P and Dinse K P 1996 *Fullerenes: Recent Advances in the Chemistry and Physics of Fullerenes and Related Materials*, Vol. 3, eds K Kadish and R Ruoff (Pennington, NJ: Electrochemical Society), pp. 602–608

[52] Schulte J, Boehm M C and Dinse K P 1998 *Molecular Nanostructures*, eds Kuzmany H, Fink J, Mehring M *et al.* (London: World Scientific), pp. 189–192

[53] Hino S, Umishita K, Iwasaki K *et al.* 1998 Ultraviolet photoelectron spectra of $Sc@C_{82}$ *Chem. Phys. Lett.* **300** 145

[54] Suematsu H, Murakami Y, Kawata H *et al.* 1994 Crystal structure of endohedral-metallofullerene $La@C_{82}$ *Proc. Mater. Res. Soc. Symp.* **349** 213

[55] Watanuki T, Fujiwara A, Ishii K *et al.* 1995 Crystal structure of endohedral metallofullerene $La@C_{82}$ *Photon Factory Report* **13** 333

[56] Watanuki T, Fujiwara A, Ishii K *et al.* 1996 Crystal structure and molecular structure of endohedral metallofullerene $La@C_{82}$ *Photon Factory Report* **14** 403

[57] Nishibori E, Sakata M, Takata M *et al.* 2000 Giant motion of La atom inside C_{82} cage *Chem. Phys. Lett.* **330** 497

[58] Andreoni W and Curioni A 1997 *Fullerenes: Recent Advances in the Chemistry and Physics of Fullerenes and Related Materials*, Vol. 4, eds K Kadish and R Ruoff (Pennington, NJ: Electrochemical Society), pp. 516–522

[59] Andreoni W and Curioni A 1998 Ab initio approach to the structure and dynamics of metallofullerenes *Appl. Phys. A* 66 299

[60] Miyake Y, Suzuki S, Kojima Y *et al.* 1996 Motion of scandium ions in Sc_2C_{84} observed by Sc-45 solution NMR *J. Phys. Chem.* 100 9579

[61] Akasaka T, Nagase S, Kobayashi K *et al.* 1995 Synthesis of the first adducts of the dimetallofullerenes $La_2@C_{80}$ and $Sc_2@C_{84}$ by addition of a disilirane *Angew. Chem., Int. Ed. Engl.* 34 2139

[62] Akasaka T, Nagase S, Kobayashi K *et al.* 1997 ^{13}C and ^{139}La NMR studies of $La_2@C_{80}$: first evidence for circular motion of metal atoms in endohedral dimetallofullerenes *Angew. Chem., Int. Ed. Engl.* 36 1643

[63] Hennrich F H, Michel R H, Fisher A *et al.* 1996 Isolation and characterization of C_{80} *Angew. Chem., Int. Ed. Engl.* 35 1732

[64] Kobayashi K, Nagase S and Akasaka T 1995 A theoretical study of C_{80} and $La_2@C_{80}$ *Chem. Phys. Lett.* 245 230

[65] Nagase S and Kobayashi K 1994 Theoretical study of the dimetallofullerene $Sc_2@C_{84}$ *Chem. Phys. Lett.* 231 319

[66] Nishibori E, Takata M, Sakata M *et al.* 2001 Pentagonal-dodecahedral La_2 charge density in $[80-I_h]$fullerene: $La_2@C_{80}$ *Angew. Chem. Int. Ed.* 40 2998

[67] Shimotani H, Ito T, Iwasa Y *et al.* 2004 Quantum chemical study on the configurations of encapsulated metal ions and the molecular vibration modes in endohedral dimetallofullerene $La_2@C_{80}$ *J. Am. Chem. Soc.* 126 364

[68] Sato W, Sueki K, Kikuchi K *et al.* 1998 Novel dynamic behavior of CeC_{82} at low temperature *Phys. Rev. Lett.* 80 133

[69] Alvarez M M, Gillan E G, Holczer K *et al.* 1991 La_2C_{80}: A soluble dimetallofullerene *J. Phys.Chem.* 95 10561

[70] Suzuki T, Maruyama Y, Kato T *et al.* 1995 Electrochemistry and ab initio study of the dimetallofullerene $La_2@C_{80}$ *Angew. Chem., Int. Ed. Engl.* 34 1094

[71] van Loosdrecht P H M, Johnson R D, Beyers R *et al.* 1994 *Recent Advances in the Chemistry and Physics of Fullerenes and Related Materials*, eds K Kadish and R Ruoff (Pennington, NJ: Electrochemical Society), pp. 1320–1330

[72] Bethune D S 1996 *The Chemical Physics of Fullerenes 10 (and 5) Years Later*, ed. W Andreoni (Dordrecht: Kluwer), pp. 165–181

[73] Stevenson S, Burbank P, Harich K *et al.* 1998 $La_2@C_{72}$: metal-mediated stabilization of a carbon cage *J. Phys. Chem.* 102 2833

[74] Shinohara H, Yamaguchi H, Hayashi N *et al.* 1993 Isolation and spectroscopic properties of $Sc_2@C_{74}$, $Sc_2@C_{82}$, and $Sc_2@C_{84}$ *J. Phys. Chem.* 97 4259

[75] Yannoni C S, Hoinkis M, de Vries M S *et al.* 1992 Scandium clusters in fullerene cages *Science* 256 1191

[76] Takahashi T, Ito A, Inakuma M and Shinohara H 1995 Divalent scandium atoms in the cage of C_{84} *Phys. Rev. B* 52 13812

[77] Yamamoto E, Tansho M, Tomiyama T *et al.* 1996 ^{13}C-NMR study on the structure of isolated $Sc_2@C_{84}$ metallofullerene *J. Am. Chem. Soc.* 118 2293

[78] Takata M, Nishibori E, Umeda B *et al.* 1997 Structure of endohedral dimetallofullerene $Sc_2@C_{84}$ *Phys. Rev. Lett.* 78 3330

[79] van Loosdrecht P H M, Johnson R D, Bethune D S *et al.* 1994 Orientational dynamics of the Sc_3 trimer in $Sc_3@C_{82}$: an EPR study *Phys. Rev. Lett.* **73** 3415

[80] Shinohara H, Inakuma M, Hayashi N *et al.* 1994 Spectroscopic properties of isolated $Sc_3@C_{82}$ metallofullerene *J. Phys. Chem.* **98** 8597

[81] Stevenson S, Dorn H C, Burbank P *et al.* 1994 Isolation and monitoring of the endohedral metallofullerenes $Y@C_{82}$ *Anal. Chem.* **66** 2675

[82] Stevenson S, Dorn H C, Burbank P *et al.* 1994 Automated HPLC separation of endohedral metallofullerene $Sc@C_{2n}$ and $Y@C_{2n}$ fractions *Anal. Chem.* **66** 2680

[83] Anderson M R, Dorn H C, Stevenson S *et al.* 1997 The voltammetry of $Sc_3@C_{82}$ *J. Am. Chem. Soc.* **119** 437

[84] Dorn H C, Stevenson S, Burbank P *et al.* 1995 *Science and Technology of Fullerenes*, eds P Bernier, D S Bethune, L Y Chiang *et al.* (Pittsburgh, PA: Materials Research Society), pp. 123–135

[85] Macfarlane R M, Wittmann G, van Loosdrecht P H M *et al.* 1997 Measurement of pair interactions and 1.5 μm emission from Er3+ ions in a C_{82} fullerene cage *Phys. Rev. Lett.* **79** 1397

[86] Ding X, Alford J M and Wright J C 1997 Lanthanide fluorescence from erbium endohedral fullerenes *Chem. Phys. Lett.* **269** 72

[87] Nagase S and Kobayashi K 1997 Structural study of endohedral metallofullerenes $Sc_2@C_{84}$ and $Sc_2@C_{74}$ *Chem. Phys. Lett.* **276** 55

[88] Nagase S, Kobayashi K and Akasaka T 1996 Endohedral metallofullerences: new spherical properties *Bull. Chem. Soc. Jpn.* **69** 2131

[89] Iiduka Y, Wakahara T, Nakajima K *et al.* 2006 ^{13}C NMR spectroscopic study of scandium dimetallofullerene, $Sc_2@C_{84}$ vs. $Sc_2C_2@C_{82}$ *Chem. Commun.* 2057

[90] Nishibori E, Ishihara M, Takata M *et al.* 2006 Bent $(metal)_2C_2$ clusters encapsulated in $(Sc_2C_2)@C_{82}(III)$ and $(Y_2C_2)@C_{82}(III)$ metallofullerenes *Chem. Phys. Lett.* **433** 120

[91] Pichler T, Hu Z, Grazioli C *et al.* 2000 Proof for trivalent Sc ions in $Sc_2@C_{84}$ from high-energy spectroscopy *Phys. Rev. B* **62** 13196

[92] Takata M, Nishibori E, Sakata M *et al.* 1999 Triangle scandium cluster imprisoned in a fullerene cage *Phys. Rev. Lett.* **83** 2214

[93] Ungerer J R and Hughbanks T 1993 The electronic structure of $Sc_3@C_{82}$ *J. Am. Chem. Soc.* **115** 2054

[94] Iiduka Y, Wakahara T, Nakahodo T *et al.* 2005 Structural determination of metallofullerene Sc_3C_{82} revisited: a surprising finding *J. Am. Chem. Soc.* **127** 12500

[95] Nishibori E, Terauchi I, Sakata M *et al.* 2006 High-resolution analysis of $(Sc_3C_2)@C_{80}$ metallofullerene by third generation synchrotron radiation X-ray powder diffraction *J. Phys. Chem. B* **110**, 19215

[96] Tellgmann R, Krawez N, Lin S-H *et al.* 1996 Endohedral fullerene production *Nature* **382** 407

[97] Campbell E E B, Tellgmann R, Krawez N and Hertel I V 1997 Production and LDMS characterisation of endohedral alkali-fullerene films *J. Phys. Chem. Solids* **58** 1763

[98] Gromov A, Ostrovskii D, Lassesson A *et al.* 2003 FTIR and Raman spectroscopical study of chromatographically isolated $Li@C_{60}$ and $Li@C_{70}$ *J. Phys. Chem. B* **107** 11290

[99] Hirata T, Hatakeyama R, Mieno T and Sato N 1996 Production and control of K-C_{60} plasma for material processing *J. Vac. Sci. Technol., A* **14** 615

[100] Aoyagi S, Nishibori E, Sawa H *et al.* 2010 A layered ionic crystal of polar Li@C$_{60}$ superatoms *Nat. Chem.* **2** 678

[101] Aoyagi S, Sado Y, Nishibori E *et al.* 2012 Rock-salt-type crystal of thermally contracted C$_{60}$ with encapsulated lithium cation *Angew. Chem. Int. Ed.* **51** 3377

[102] Kawata H, Fujii Y, Nakao H *et al.* 1995 Structural aspects of C$_{82}$ and C$_{76}$ crystals studied by x-ray diffraction *Phys. Rev. B* **51** 8723

[103] Shinohara H 2000 Endohedral metallofullerenes *Rep. Prog. Phys.* **63** 843

[104] Olmstead M M, Costa D A, Maitra K *et al.* 1999 Interaction of curved and flat molecular surfaces. The structures of crystalline compounds composed of fullerene (C$_{60}$, C$_{60}$O, C$_{70}$, and C$_{120}$O) and metal octaethylporphyrin units *J. Am. Chem. Soc.* **121** 7090

[105] Stevenson S, Rice G, Glass T *et al.* 1999 Small-bandgap endohedral metallofullerenes in high yield and purity *Nature* **401** 55

[106] Rodriguez-Fortea A, Balch A L and Poblet J M 2011 Endohedral metallofullerenes: a unique host–guest association *Chem. Soc. Rev.* **40** 3551

[107] Zhang J, Stevenson S and Harry D 2013 Trimetallic nitride template endohedral metallofullerenes: discovery, structural characterization, reactivity, and applications. *Acc. Chem. Res.* **46** 1548

[108] Stevenson S, Mackey M A, Stuart M A *et al.* 2008 A distorted tetrahedral metal oxide cluster inside an icosahedral carbon cage. Synthesis, isolation, and structural characterization of Sc$_4$(μ_3-O)$_2$@I$_h$-C$_{80}$ *J. Am. Chem. Soc.* **130** 11844

4

Electronic States and Structures

4.1 ELECTRON TRANSFER IN METALLOFULLERENES

Group 3 (Sc, Y, La) metallofullerenes exhibit ESR (electron spin resonance) hfs (hyperfine splitting), which provides important information on the electronic structures of the metallofullerenes. Typical ESR-active monometallofullerenes are $La@C_{82}$, $Y@C_{82}$, and $Sc@C_{82}$. The ESR hfs of a metallofullerene was first observed in $La@C_{82}$ by the IBM Almaden group [1] (Figure 4.1) and was discussed within the framework of an intrafullerene electron transfer. The observation of eight equally spaced lines provides evidence of isotropic electron–nuclear hyperfine coupling (hfc) to ^{139}La with a nuclear spin quantum number $I = 7 / 2$. The observed electron g-value of 2.0010, close to that measured for the C_{60} radical anion [2,3], indicates that a single unpaired electron resides in the LUMO (lowest unoccupied molecular orbital) of the carbon cage. They also observed ^{13}C hyperfine satellites, provided that the unpaired electron couples to both the ^{139}La and the C atoms. The observed hfc (1.2 G) is very small compared with that (50 G) measured for La^{2+} substituted in CaF_2 [4]. Therefore, it was concluded that the La atom in the C_{82} cage must be in the 3+ state, which gives a formal charge state of $La^{3+}@C_{82}^{3-}$. Ultraviolet photoelectron spectroscopy (UPS) [5,6] and recent X-ray diffraction results [7] strongly support this conclusion.

ESR hfs obtained for $Y@C_{82}$ indicated that the yttrium in C_{82} is also in the 3+ oxidation state [8,9]. However, there has been controversy as to whether $Sc@C_{82}$ has 3+ or 2+ charge state, as described in Section 4.3.2 [10–12]. Recent X-ray diffraction [13] and UPS [14] results indicate a 2+ state leading

Endohedral Metallofullerenes: Fullerenes with Metal Inside, First Edition.
Hisanori Shinohara and Nikos Tagmatarchis.
© 2015 John Wiley & Sons, Ltd. Published 2015 by John Wiley & Sons, Ltd.

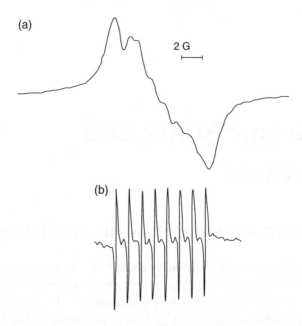

Figure 4.1   The first ESR hyperfine spectra (9.112 GHz) of La@C$_{82}$ at room temperature of (a) a solid degassed toluene extract and (b) a degassed solution of the dried extract in 1,1,2,2-tetrachloroethane. Reprinted by permission from Macmillan Publishers Ltd: [1]. Copyright 2010 Rights Managed by Nature Publishing Group

to Sc^{2+}@C$_{82}$$^{2-}$. Theoretical calculations suggest that the electronic structure of Sc@C$_{82}$ is well represented by Sc^{2+}@C$_{82}$$^{2-}$ [11,15–17]. Unlike La@C$_{82}$ and Y@C$_{82}$, Sc@C$_{82}$ forms a divalent state and an electron residing in the Sc(3d) orbital, owing to a large energy separation between the 3d and 4s orbitals, might be responsible for the observed ESR hfs.

The temperature-dependent linewidths of the ESR hfs of La@C$_{82}$, Sc@C$_{82}$, and Gd@C$_{82}$ have been discussed by Kato et $al.$ [18–21] in terms of the spin–rotation coupling interaction. Dinse and co-workers [12,22,23] investigated temperature dependence of ESR linewidths of La@C$_{82}$, La@C$_{90}$, and Sc@C$_{82}$ in different solvents and obtained information on the nuclear quadrupole interactions in these metallofullerenes. Dunsch and co-workers [24–26] studied ^{13}C satellite structures of M@C$_{82}$ (M = Sc, Y, La) in detail and reported that the manifold of ^{13}C hfc constants could be interpreted by the calculated spin density distributions.

ESR spectra of La@C$_{82}$, Y@C$_{82}$, Ho@C$_{82}$, and Tm@C$_{82}$ taken from the solid soot extract were reported by Bartl et $al.$ [24,27–29] and showed low resolved but split hyperfine structure, indicating that the metal atoms exist in ionic form in the fullerene cage also in the solid state. The research group also reported [30] the principal values of the hyperfine tensor A and the relative orientation of g and A tensors of $M@C_{82}$ ($M = Sc, Y, La$) applying three- and four-pulse electron spin–echo envelope modulation (ESEEM) techniques.

4.2 ESR EVIDENCE ON THE EXISTENCE OF STRUCTURAL ISOMERS

Suzuki *et al.* [31] reported the presence of structural isomers of $M@C_{82}$ ($M = Sc, Y, La$) via ESR hfs measurements and also hfs to ^{13}C in natural abundance on the fullerene cage. Figure 4.2 shows ESR spectra of Sc@C_{82}, Y@C_{82}, and La@C_{82} which exhibit ^{13}C hfs. Hoinkis *et al.* [32] similarly reported the presence of isomers of La@C_{82} and Y@C_{82}. Bandow *et al.* [33] also found the structural isomers of La@C_{82} as well as the presence of other lanthanum fullerenes such as La@C_{76} and La@C_{84} by ESR hfs measurements only when anaerobic sampling of soot containing lanthanum fullerenes was employed.

These structural isomers of $M@C_{82}$ have been separated and isolated by high-performance liquid chromatography (HPLC) (Section 2.3). The minor isomer of La@C_{82}, that is, La@C_{82} (II), was isolated by Yamamoto *et al.* [34]. Similar to the major isomer, La@C_{82} (I), La@C_{82} (II) exhibits an equally spaced octet line but with a smaller hfs constant. Sc@C_{82} also has two structural isomers (I and II) which have been separated and isolated [35]. The two Sc@C_{82} isomers show eight equally spaced, narrow ESR hfs owing to the hfc to the scandium nucleus [10,35,36]. The hfs of Sc@C_{82} is much bigger than that of La@C_{82}.

Two isomers of Y@C_{82} (I, II) have been separated and isolated by the two-stage HPLC method [35,37–39]. Both isomers of Y@C_{82} show distinct ESR hyperfine doublets due to the $I = 1/2$ yttrium nucleus. The overall ESR spectral patterns of Y@C_{82} (I, II) in CS_2 solution at room temperature are similar, but the hfs values differ. Moreover, the appearance of the small satellite peaks due to ^{13}C, adjacent to the main doublets, is much less clear in Y@C_{82} (II). Obviously, the electronic structures of the two isomers are different. It has been found that isomer II is less stable in air and much more reactive toward various solvents than isomer I. Yamamoto *et al.* [40] produced and isolated three isomers of La@C_{90} (I–III) and observed the hfs. The observed hfc values are much smaller than those of La@C_{82} (I, II).

Other mono-metallofullerenes are either simply ESR-silent or show fine structures (instead of hfs) only at low temperature. For example, the Gd@C_{82} metallofullerene exhibits a fine structure at low temperature

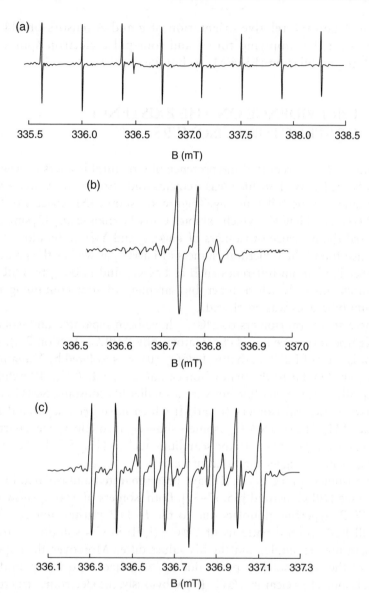

Figure 4.2   ESR spectra of: (a) Sc@C_{82} in CS_2 (9.4325 GHz, 0.1 mW); (b) Y@C_{82} in toluene (9.4296 GHz, 0.1 mW); and (c) La@C_{82} in toluene (9.4307 GHz, 0.1 mW). Reproduced with permission from Ref. [31]. Copyright 1992 American Chemical Society

[20,21]. Di-metallofullerenes, such as La$_2$@C_{80}, Y$_2$@C_{82}, and Sc$_2$C$_2$@C_{82}, are known to be ESR-silent, indicating that these species are diamagnetic. Knapp *et al.* [41] observed ESR signals of Lu@C_{82} arising from unresolved hyperfine interaction with $I = 7/2$ nuclear spin of ^{175}Lu, whereas no ESR

signals were detected for Ho@C_{82}. The unobservability of solution and frozen matrix ESR for Ho@C_{82} can be attributed to a high spin state (a $^5I_8$ ground state for Ho^{3+}) leading to strong spin lattice relaxation and rigid limit spectral broadening [41]. The general inability to observe well-resolved ESR hfs for the lanthanide metallofullerenes Ln@C_{82} (Ln = Ce,...,Lu) as in group 3 metallofullerenes might also be due to this strong nuclear spin relaxation.

The tri-scandium fullerene, Sc_3C_2@C_{80} (formerly assigned as Sc_3@C_{82}), has so far been the only ESR-active metallofullerene identified other than the mono-metallofullerenes of the type M@C_{82} (M = Sc, Y, La). Figure 4.3 shows 22 perfectly symmetric, equally spaced, narrow ESR hfs, which is a manifestation of the isotropic hfc of three scandium nuclei with $I = 7/2$ in the C_{80} cage [10,36]. The presence of the 22 perfectly symmetric hfs lines suggests the geometrical equivalence of the three Sc atoms in the C_{80} cage [10,19,36,43–47]. Based on the appearance of the 22 perfectly symmetric hfs and the results of theoretical calculations [48], three Sc atoms form an equilateral Sc_3 trimer within the C_{80} cage so as to retain a threefold axis as an entire Sc_3C_2@C_{80} molecule. The result is consistent with a recent X-ray structural study on Sc_3C_2@C_{80} as described in Section 3.2.2. The molecular structure of Sc_3C_2@C_{80} is shown in Figure 3.8.

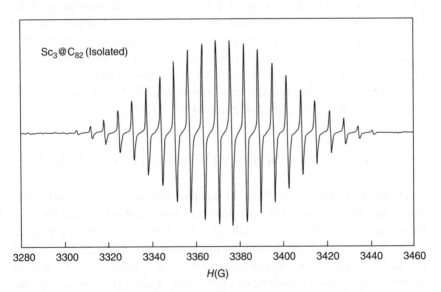

Sc_3@C_{82} (Isolated)

3280	3300	3320	3340	3360	3380	3400	3420	3440	3460

H(G)

Figure 4.3 ESR spectrum (X-band, 9.4360 GHz) for Sc_3C_2@C_{80} in CS_2 solution at 220 K, showing 22 well resolved, equally spaced, and perfectly symmetric hfs ($g = 1.9985$, hfs = 6.51, $\Delta H_{pp} = 0.770$ G). Reproduced with permission from Ref. [42]. Copyright 1994 American Chemical Society

Table 4.1   ESR parameters for some isolated metallofullerenes

Species	Hyperfine coupling (G)	g-value	H_{pp} (G)[a]
Sc@C$_{82}$ (I)	3.82	1.9999	0.036 (300 K)
Sc@C$_{82}$ (II)	1.16	2.0002	0.019 (300 K)
Sc@C$_{84}$	3.78	1.9993	0.017 (300 K)
Sc$_3$C$_2$@C$_{82}$	6.45	1.999	0.4 (220 K)
Y@C$_{82}$ (I)	0.48	2.0004	0.087 (300 K)
Y@C$_{82}$ (II)	0.34	2.0002	0.12 (300 K)
La@C$_{82}$ (I)	1.20	2.0008	0.049 (300 K)
La@C$_{82}$ (II)	0.83	2.0002	0.052 (220 K)
La@C$_{90}$ (I)	0.60	2.0010	—
La@C$_{90}$ (II)	0.53	2.0011	—
La@C$_{90}$ (III)	0.12	2.0022	—

[a] hfs linewidth

The temperature dependence of the 22 hfs lines can provide us with further structural information on Sc$_3$C$_2$@C$_{80}$ [20,44]. The H (magnetic field) value has a minimum at 220 K above which the hfs linewidth increases as temperature increases. A similar temperature dependence has been reported for La@C$_{82}$ [18,33]. Even at this temperature the hfs linewidth of Sc$_3$C$_2$@C$_{80}$ (H_{pp} = 0.4 G at 220 K) is about 10 times as broad as that of Sc@C$_{82}$ (0.036 G at 300 K). Such a large linewidth could be due to incomplete motional averaging of local field variations due to strong magnetic anisotropy of the entire molecule [44]. The intramolecular dynamics is the inherent nature of the Sc$_3$ trimer encapsulated in the C$_{80}$ cage. In addition to Sc$_3$C$_2$@C$_{80}$ described above, Suzuki et al. [49] observed a series of ESR hfs due to non-equivalent Sc trimers encaged in fullerene cages other than C$_{80}$.

The observed ESR hyperfine parameters for M@C$_{82}$ (M = Sc, Y, La), La@C$_{90}$, and Sc$_3$C$_2$@C$_{80}$ are summarized in Table 4.1.

4.3   ELECTROCHEMISTRY OF METALLOFULLERENES

Electronic properties of endohedral metallofullerenes based on reduction/oxidation (redox) properties have also been investigated electrochemically by using cyclic voltammetry (CV). Electrochemical techniques provide valuable information on the interaction between the encaged species and the fullerene cage. The electrochemical method is particularly important in the study of metallofullerenes, where only microgram (to milligram) quantities are necessary to perform measurements.

Suzuki et al. [50] measured cyclic voltammograms of La@C$_{82}$ and found unusual redox properties of the metallofullerene which differ

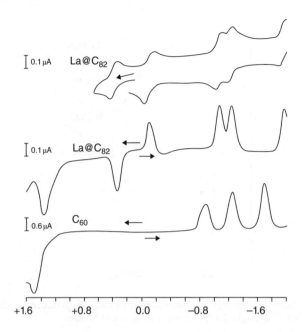

Figure 4.4 Cyclic voltammogram of La@C_{82} and differential pulse voltammograms of La@C_{82} and C_{60} in o-dichlorobenzene. Reproduced with permission from Ref. [50]. Copyright 1993 American Chemical Society

significantly from those of empty fullerenes. Figure 4.4 shows the cyclic voltammogram of La@C_{82}. The first reversible oxidation potential is approximately equal to that of ferrocene, indicating that La@C_{82} is a moderate electron donor. The first reduction and oxidation potentials indicate that it should form both cationic and anionic charge transfer complexes. In addition, La@C_{82} is a stronger electron acceptor than empty fullerenes such as C_{60}, C_{70}, and C_{84}. A schematic energy level diagram of La@C_{82} is presented in Figure 4.5. The CV results of La@C_{82} revealed that La@C_{82} is a good electron donor as well as a good electron acceptor and that at least five electrons can be transferable to the C_{82} cage while maintaining the 3+ charge state of the encaged La atom, that is, $(La^{3+}@C_{82}^{3-})^{5-}$.

The CV measurements indicate that the oxidation state of the Y atom in Y@C_{82} [51] was close to that of La@C_{82}, probably +3. The electrochemistry of Y@C_{82} is almost identical to that of La@C_{82} [51,52]. Other mono-metallofullerenes such as Y@C_{82}, Ce@C_{82}, and Gd@C_{82} have a similar tendency in their redox properties [51]. Anderson et $al.$ [47] reported the CV data of Sc_3C_2@C_{80}. CV measurements were also taken for other lanthanide fullerenes such as Pr@C_{82}, Nd@C_{82}, Tb@C_{82},

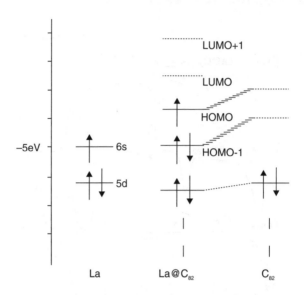

Figure 4.5   Schematic energy level diagram of La@C$_{82}$. Reproduced with permission from Ref. [50]. Copyright 1993 American Chemical Society

Dy@C$_{82}$, Ho@C$_{82}$, Er@C$_{82}$, and Lu@C$_{82}$ by Wang *et al.* [53]. The first reduction potentials of all seven M@C$_{82}$ were found to locate within a close vicinity to each other, suggesting that all the entrapped metal atoms adopt a similar valence state, presumably a trivalent cation state.

As seen in Figure 4.5, trivalent M@C$_{82}$ metallofullerenes provide the paramagnetic state of the fullerene cage, in which an open-shell electronic structure leads to small electrochemical gaps (the difference between first oxidation and first reduction potentials). This is due to the fact that the first oxidation potential is much more negative than that of the corresponding empty fullerenes. Dunsch and co-workers [29,54–56] studied electron transfers in metallofullerenes by CV coupled with *in situ* ESR experiments. The electron transfer to the endohedral La@C$_{82}$ molecule studied by this method gives evidence of a charge in the electronic state of the fullerene; the electrochemical reaction in the anodic scan causes the formation of La^{3+}@C$_{82}$$^{4-}$, and during the cathodic scan the spin concentration decreases as the La^{3+}@C$_{82}$$^{2-}$ structure formed by reduction is not paramagnetic.

Figure 4.6 shows a cyclic voltammogram of Sc$_3$N@C$_{80}$-I_h(7), where, importantly, reductions of this metallofullerene are electrochemically irreversible at moderate voltammetric scan rates in contrast to mono- and di-metallofullerenes [57]. Furthermore, the electrochemical gap of Sc$_3$N@C$_{80}$-I_h (7), 1.85 V, is significantly larger than the gaps of

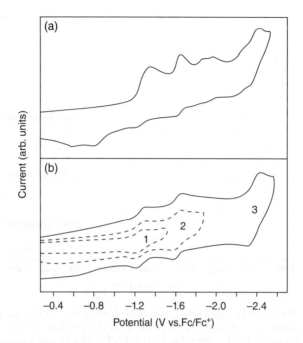

Figure 4.6 Cyclic voltammogram of $Sc_3N@C_{80}$-I_h(7) at scan rates of (a) 100 mV/s and (b) 6 V/s (1), 10 V/s (2), and 20 V/s (3). The different scan rates for each reduction reflect the scan rates necessary to achieve full electrochemical reversibility. Reproduced with permission from Ref. [57]. Copyright 2005 American Chemical Society

mono- and di-metallofullerenes and is approaching the values of empty fullerenes. CV studies of $Sc_3N@C_{80}$-I_h (7) at different scan rates showed that the electrochemical reversibility of the first reduction step can be achieved at a scan rate of 6 V/s, the second reduction step requires 10 V/s to be reversible, while the third reduction step is reversible at 20 V/s (Figure 4.6). Comprehensive treatments of the electrochemistry of metallofullerenes can be found in recent reviews [58,59].

4.4 SIMILARITY IN THE UV-VIS-NIR ABSORPTION SPECTRA

Absorption spectra of endohedral metallofullerenes in the UV-Vis-NIR (ultraviolet-visible-near infrared) region are unique as compared with those of empty fullerenes. Normally, the absorption spectra of metallofullerenes have long tails to the red down to 1500 nm or more. The absorption spectra of the major isomers of mono-metallofullerenes $M@C_{82}$ (M = Y, La, Ce, Pr, Nd, Gd, Tb, Dy, Ho, Er, Lu) are similar to

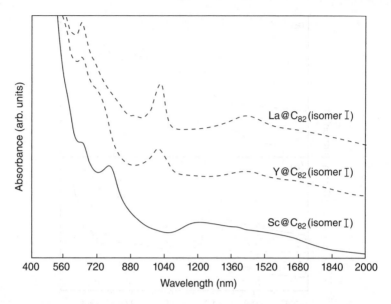

Figure 4.7   UV-Vis-NIR absorption spectra of La@C$_{82}$ (I), Y@C$_{82}$ (I), and Sc@C$_{82}$ (I). The spectral feature of Sc@C$_{82}$ is different from those of Y@C$_{82}$ and La@C$_{82}$. The absorption spectra of M@C$_{82}$ (M = Ce, Pr, Gd, Tb, Dy, Ho, Er, and Lu) are essentially the same as that of La@C$_{82}$. Reproduced with permission from Ref. [35]. Copyright 1995 The Electrochemical Society

each other and well represented by a sharp peak around 1000 nm and a broad peak around 1400 nm. These absorption peaks are related to the intrafullerene electron transfers from the encaged metal atom to the carbon cage.

Figure 4.7 shows UV-Vis-NIR absorption spectra for the isolated group 3 metallofullerenes: La@C$_{82}$ [60,61], Y@C$_{82}$ [37], and Sc@C$_{82}$ [35] in CS$_2$ solution. The spectrum of Sc@C$_{82}$ is very different from those of Y@C$_{82}$ and La@C$_{82}$, indicating that the electronic structure of Sc@C$_{82}$ is different from those of Y@C$_{82}$ and La@C$_{82}$. As described in Section 4.2.1, the charge state of Sc^{2+}@C$_{82}^{2-}$ (divalency) is different from those of La^{3+}@C$_{82}^{3-}$ and Y^{3+}@C$_{82}^{3-}$ (trivalency) [5,11,62]. The UV-Vis-NIR absorption spectra of many mono-metallofullerenes encapsulating lanthanide elements, M@ C$_{82}$ (M = Ce–Nd, Gd–Er, Lu) [51,63], Pr [64,65], Nd [66], Gd [64,67], Tb [68], Dy [69,70], Ho [41,53], Er [69,71], Lu [41], are similar to those of La@C$_{82}$ and Y@C$_{82}$. However, the absorption spectra of the C$_{82}$-based metallofullerenes containing the divalent lanthanide elements (i.e., Sm, Eu, Tm, and Yb) [72–74] are different from those of La@C$_{82}$, Y@C$_{82}$, and Ln@C$_{82}$ but are similar to that of Sc@C$_{82}$ in that the sharp absorption peak at 1000 nm are missing in these divalent metallofullerenes.

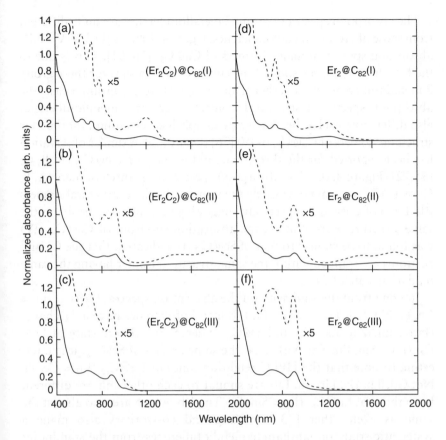

Figure 4.8 UV-Vis-NIR absorption spectra of (a)–(c): $(Er_2C_2)@C_{82}$ (I, II, III) and (d)–(f): $Er_2@C_{82}$ (I, II, III) in CS_2 solvent at room temperature. These spectra are normalized with the absorbance at 400 nm. Reproduced with permission from Ref. [71]. Copyright 2007 American Chemical Society

UV-Vis-NIR absorption spectra of structural isomers for a metallofullerene generally differ from each other. Such differences stem from the fullerene cage (isomer) structure as well as from the difference in charge state of the fullerene. For example, isomers I and II of $La@C_{82}$ [34], $Y@C_{82}$ [35], and $Sc@C_{82}$ [75] have different spectral features in their respective absorption spectra. In contrast, absorption spectra of carbide metallofullerenes are, in many cases, similar to those of the corresponding pure metallofullerenes. For example, the absorption spectra of $Er_2C_2@C_{82}$ (I, II, III) are almost exactly the same as those of $Er_2@C_{82}$ (I, II, III) as shown in Figure 4.8, suggesting that the charge states and isomer structures of the corresponding three isomers are the same as each other.

The absorption spectra of group 2 metallofullerenes are quite different from those of group 3 metallofullerenes. Figure 4.9 shows the UV-Vis-NIR absorption spectra of major isomers of $Ca@C_{82}$ [76–81]. This is due to the fact that the charge state of the encaged metal atoms for these group 2 metallofullerenes is 2+ rather than 3+. Similar to the group 3 case, the absorption spectra of structural isomers of group 2 metallofullerenes also differ from each other. Such an example has been found for the four structural isomers of $Ca@C_{82}$ (I–IV) [76] (Figure 4.9). A similar observation has been reported for the three structural isomers of $Tm@C_{82}$ (A, B, C) [55,72] (Figure 4.10). The absorption spectra of the three isomers (A, B, C) of $Tm@C_{82}$ are almost exactly the same as the three structural isomers (III, I, IV) of $Ca@C_{82}$, respectively, suggesting that each isomer shares the same structure. Furthermore, the absorption spectrum of $Ca@C_{82}$ (IV) was found to be similar to that of $Sc@C_{82}$ (I), indicating that the Ca and Sc atoms are trapped within the same structural isomer having the same oxidation state of 2+.

Judging from the similarity on the absorption spectra, $Ln@C_{82}$ (Ln = Ce, Pr, Nd, Gd, Tb, Dy, Ho, Er, Lu) metallofullerenes have a 3+ charge state similar to $La@C_{82}$ and $Y@C_{82}$, whereas the charge state of $Ln@C_{82}$ (Ln = Sm, Eu, Tm, Yb) is 2+, the same as that of $Sc@C_{82}$. It is interesting to note that the HPLC retention times of $Ln@C_{82}$ (Ln = Ce, Pr, Nd, Gd, Tb, Dy, Ho, Er, Lu) are similar to each other but are different from those of $Ln@C_{82}$ (Ln = Sm, Eu, Tm, Yb) which are also almost the same as each other [73,82]. Yang and co-workers also made a systematic study on lanthanide metallofullerenes from the standpoint of Vis-NIR absorption spectra [83] together with their HPLC elution behavior [84]. They found that there is a fairly good correlation between the relative HPLC retention time and the charge state (i.e., divalent or trivalent) of encapsulated metal atoms. According to their results, the four lanthanide atoms Sm, Eu, Tm, and Yb have divalent state whereas the other lanthanide elements (La, Ce, Pr, Nd, Gd, Tb, Dy, Ho, Er, Lu) form a trivalent state in the C_{82} cage which is consistent with the former results.

Here we are able to derive an empirical rule regarding the relationship between absorption features and the isomer structure of a metallofullerene: *a UV-Vis-NIR absorption spectrum of a metallofullerene* ($M@C_{82}$, *where* M *is a metal atom) is very similar to that of another metallofullerene irrespective of the kind of encaged metal atom when the cage (isomer) structure and the charge state of the atoms are the same.* This empirical rule can be best understood by molecular excitations of the isomer cage which should not be changed as long as the cage and the filling of the molecular levels are the same.

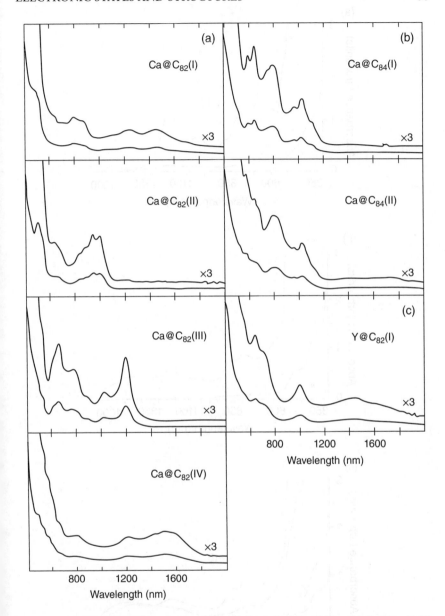

Figure 4.9 UV-Vis-NIR absorption spectra of (a) Ca@C$_{82}$ (I–IV), (b) Ca@C$_{84}$ (I, II), and (c) Y@C$_{82}$ (I) for comparison. Reproduced with permission from Ref. [76]. Copyright 1996 American Chemical Society

Fluorescence experiments of endohedral metallofullerenes, on the other hand, have been limited because of their weak emission feature. IR emissions from Er^{3+} for Er$_2$@C$_{82}$ around 1.52 μm have been observed [71,85–88]. The emission was ascribed to the characteristic

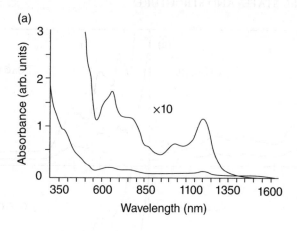

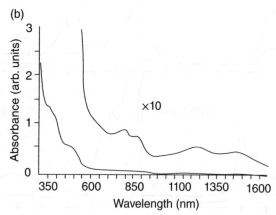

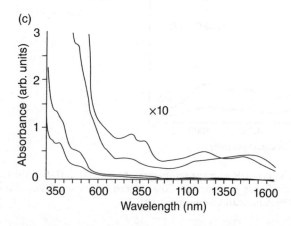

Figure 4.10 UV-Vis-NIR absorption spectra for three structural isomers of Tm@C$_{82}$: (a) isomer A; (b) isomer B; and (c) isomer C. Spectroscopic features of Tm@C$_{82}$ (A, B, C) are almost identical to those of Ca@C$_{82}$ (III, I, IV) in Figure 4.9, respectively. Reproduced with permission from Ref. [72]. Copyright Wiley-VCH Verlag GmbH & Cpo. KGaA

intraconfigurational $4f^{11}$ fluorescence of the trivalent $\text{Er}\;{}^4I_{15/2} \rightarrow {}^4I_{15/2}$ transition. The 1.52 µm emission was observed to be enhanced in di-erbium-carbide metallofullerenes $(Er_2C_2)@C_{82}$ (I, II, III) as compared with those observed in $Er_2@C_{82}$ (I, II, III) [71]. The photoluminescence properties of $(Er_2C_2)@C_{82}$ (I, II, III) and $Er_2@C_{82}$ (I, II, III) correlate strongly with the absorbance at 1.52 µm.

4.5 FERMI LEVELS AND THE ELECTRONIC STRUCTURES

The electronic properties of several metallofullerenes in the solid state have been studied by UPS [5,6,89–91]. Figure 4.11 is a UPS spectrum of $La@C_{82}$ [6]. The first (small) peak is 0.64 eV (number 1, and a in the

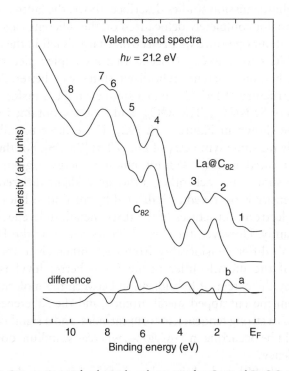

Figure 4.11 Comparison of valence band spectra for C_{82} and $La@C_{82}$ measured at 21.2 eV. The observed onset at 0.35 eV indicates insulating character for solid La@C_{82}. Peaks 1–3 and 4–8 are assigned as due to the π and σ electrons, respectively. The difference curve at the bottom is obtained after shifting the C_{82} spectrum by –0.18 eV for the best overall alignment. Reproduced with permission from Ref. [6]. Copyright 1994 The American Physical Society

difference plot) below the Fermi level corresponding to the singly occupied molecular orbital (SOMO) level of La@C$_{82}$ where the spectral onset is 0.35 eV from the Fermi level. Hino *et al.* [5] also observed peaks at 0.9 and 1.6 eV below the Fermi level in their UPS measurements on La@C$_{82}$ which are absent in the corresponding empty C$_{82}$ spectrum. The peaks at 1.6 and 0.9 eV correspond to electron transfers from the La atom to the LUMO and (LUMO + 1) levels of La@C$_{82}$ which are now occupied, respectively (see Figure 4.5). Since the observed intensity ratio of the two peaks is about 2:1, they concluded that three electrons of the La atom are transferred to the fullerene cage, that is, La^{3+}@C$_{82}^{3-}$, which is consistent with the results obtained by ESR hfs (see Section 4.1). Eberhardt and co-workers also studied the valency of La@C$_{82}$ by UPS measurements on the sublimed layers. They observed a resonant enhancement of the lanthanum-derived valence states via the La 3d to 4f transition [91]. They concluded that, in contrast to the interpretations of the lanthanum core-level photoemission studies described above, the lanthanum valence electrons are not completely delocalized on the fullerene cage. They estimated that about one-third of an electron charge is left in the lanthanum-valence orbitals for La@C$_{82}$. It seems that a complex picture involving several lanthanum-fullerene hybridized states can better describe the electronic structure of La@C$_{82}$ than a simple charge transfer.

The UPS of Sc$_3$N@C$_{78}$, Ti$_2$C$_2$@C$_{78}$, and La$_2$@C$_{78}$ obtained with 40 eV photons are shown in Figure 4.12 [92]. These three metallofullerenes have exactly the same symmetry of D_{3h} (78:5) [93–96]. For the electronic structures located in the 5–11 eV binding energy region, these UPS spectra are similar to each other, although a slight difference in their relative intensity is observed as indicated by the dotted lines, indicating that the σ-electronic structures of the three metallofullerenes do not differ significantly. However, the three UPS spectra near the Fermi levels (below 1 eV) differ considerably from each other, since the electronic structure of the metallofullerene near the Fermi level is generally governed by the extent of hybridization between the molecular orbitals derived from the entrapped metal atom(s) and the fullerene cage [94]. Titanium has a strong tendency to form carbides, and a significant hybridization might be occurring in Ti$_2$C$_2$@C$_{78}$, whereas scandium does not have such a tendency.

Pichler *et al.* [89] studied the valency of the Tm ion in the endohedral Tm@C$_{82}$ fullerene by UPS and XPS. The resemblance of the Tm 4d core level photoemission spectrum to that calculated for Yb^{3+} suggests a 4f^{13} ground state configuration of the Tm ion. The real part of the optical conductivity of the C$_{3v}$ and C$_s$ isomers is shown in Figure 4.13 [90].

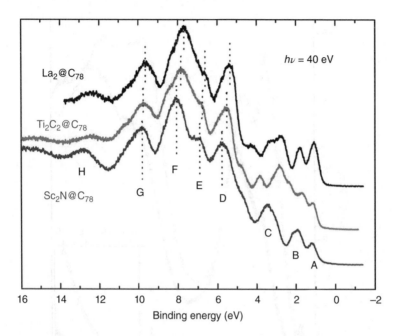

Figure 4.12 UPS spectra of $Sc_3N@C_{78}$, $Ti_2C_2@C_{78}$, and $La_2@C_{78}$ obtained with 40 eV photons. The cage symmetry of these metallofullerenes is the same D_{3h}, but their electronic structures near the Fermi levels differ significantly. Reproduced with permission from Ref. [92]. Copyright 2012 American Chemical Society

The energy gap transitions of the two isomers are different, where the onsets of the isomers are 0.6 (C_s) and 0.8 (C_{3v}) eV. The UPS measurements on $Sc_2C_2@C_{82}$ (III) [97], $Gd@C_{82}$ [98], $Sc@C_{82}$ (I) [14], and $Ca@C_{82}$ (I, III) [99] were also reported in which band gaps of these metallofullerenes were obtained.

4.6 METAL–CAGE VIBRATION WITHIN METALLOFULLERENES

Vibrational structures of several metallofullerenes have been studied by IR and Raman spectroscopy [60,100–104]. Some of the vibrational absorption lines of $Sc_2C_2@C_{82}$ (III) are strongly enhanced if compared with the spectrum of the empty cage [100,101,105]. With decreasing temperature, a dramatic narrowing of the lines was observed. The linewidth shows an Arrhenius-like behavior between −73 °C and 27 °C provided that the main contribution to it comes from rotational diffusion.

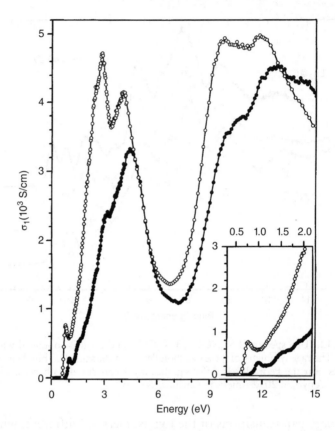

Figure 4.13   Real part of the optical conductivity of the C_{3v} isomer (●) and C_s isomer (○) of Tm@C_{82}. The inset shows the region of the energy gap on an extended scale. Reproduced with permission from Ref. [90]. With kind permission from Springer Science and Business Media

Lebedkin *et al.* [102,103] reported vibrations due to the encapsulated metal ions in the cage for M@C_{82} (M = La, Y, Ce, Gd) based on IR and Raman measurements. Figure 4.14 shows Raman spectra of M@C_{82} (M = La, Y, Ce, Gd) [103]. The peaks around 150 cm^{-1} can be attributed to internal vibrational modes, most probably metal-to-cage vibrations. Almost all peaks are observed at similar positions. These peaks were strongly broadened when the samples were exposed to air. This result is in agreement with a near-edge X-ray absorption fine-structure study [106] where the pronounced effect of air on the spectra of La@C_{82} films could be reduced only by heating them to 600 °C. The far infrared (FIR) spectra of M@C_{82} (M = La, Y, Ce) [103] support the picture derived from the Raman measurements. The metal-dependent FIR peaks between 150 cm^{-1} and 200 cm^{-1} correspond well to their Raman counterparts.

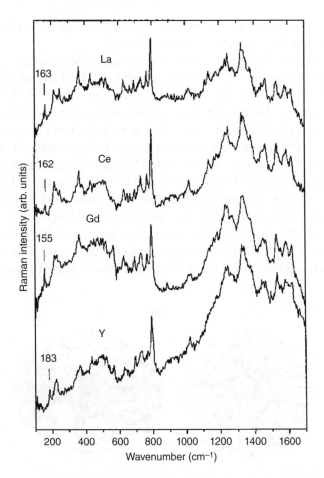

Figure 4.14 Raman spectra of M@C$_{82}$ (M = La, Ce, Gd, Y) metallofullerenes. The spectra are shifted for clarity. Reproduced with permission from Ref. [103]. With kind permission from Springer Science and Business Media

More detailed IR and Raman measurements were reported by Dunsch and co-workers [107] on the three isomers of Tm@C$_{82}$ [55,108]. The frequency of the metal-to-cage vibration is only slightly affected by the fullerene cage isomerization, but strongly depends on the kind of metal ion inside the fullerene. Generally, the metal-to-cage peaks around 100 cm^{-1} are almost invariant among the isomers but are much smaller than those of M@C$_{82}$ (M = La, Y, Ce, Gd) described above. It was proposed that the metal-to-cage vibration is sensitive toward the charge state of the encaged metal atom; the frequencies of this vibration for the trivalent metallofullerenes, M@C$_{82}$ (M = La, Y, Ce, Gd), are much higher than those of the divalent metallofullerenes such as Tm@C$_{82}$ and Eu@C$_{74}$ [104].

This is expected if the metal–cage bonding is basically electrostatic [62,109] and the metals bear the same charge.

A systematic Raman and IR study on di-metallofullerenes, $Sc_2@C_{84}$ (I, II, III), was reported by Krause *et al.* [105,110]. As in the mono-metallofullerenes they observed metal (Sc)–cage vibrations below 300 cm^{-1}. They concluded that the amount of metal-to-fullerene charge transfer and the distance between the oppositely charged centers determines the carbon cage–metal bond strength. Inelastic neutron scattering results for $La@C_{82}$ and $Y@C_{82}$ [102,103] also presented evidence of metal-to-cage vibrations at 180,150, and 85 cm^{-1}. The interval from 100 to 200 cm^{-1} in the Raman and FIR spectra of $M@C_{82}$ (M = divalent and trivalent metal atoms) can be regarded as a "metal-fingerprint" range.

Although the ionic (electrostatic) interaction exerted between encaged metal atoms and the fullerene cage due to the charge transfer definitely plays a crucial role in the metal–cage interaction as shown theoretically

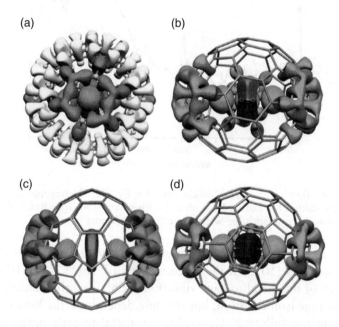

Figure 4.15 Illustration of metal–cage and metal–metal bonding in endohedral metallofullereness with the use of ELF: (a–d) ELF (isovalue 0.72) in $Y_3@C_{80}$ (a,b), $Y_2@$ C_{82} (c), and $Y_3N@C_{80}$ (d). Key for basins: white part in (a) = valence V(C,C), central part in (b) and (c) = core C(C), polar parts in (a)–(d) = valence V(Y,C,C), balls inside the cages = core C(Y), black parts in (b) and (d) = valence V(Y,Y,Y), V(Y,Y), and V(N,Y,Y). In (b–d), valence V(C,C) and core C(C) basins are not shown. Valence trisynaptic V(Y,C,C) basins are real-space representations of the yttrium–cage bonding. Reproduced with permission from Ref. [112]. Copyright 2010 American Chemical Society

[109,111], the electronic state of the cage cannot be considered as purely ionic, and the covalent contribution to the metal–cage bonding must be also taken into account. For example, according to Dunsch and co-workers (e.g., [59,112]), chemical bonds in terms of electron localization function (ELF) can be visualized as polysynaptic valence basins (Figure 4.15), whereas in the empty fullerene molecule only the cage bond is seen in ELF as a spatial extension of some valence $V(C,C)$ basins in the vicinity of the metal atoms and their transformation into *trisynaptic* $V(M,C,C)$ basins (Figure 4.15).

As well summarized by Popov and Dunsch [59], for describing metal–cage bonding in metallofullerenes, distinguishing two types of the atomic charges in metallofullerenes, a formal charge, and an effective charge, is necessary. The formal charge, which is the same as an oxidation state or a valence state, is an integer and implies that the metal–cage bonding is basically ionic. However, formal charges do not describe the actual electron distribution in the metallofullerene, but rather they are useful for understanding the spectroscopic and structural properties of metallofullerenes.

REFERENCES

[1] Johnson R D, de Vries M S, Salem J *et al.* 1992 Electron-paramagnetic resonance studies of lanthanum-containing C_{82}. *Nature* **355** 239
[2] Allemand P M, Srdanov G, Koch A *et al.* 1991 The unusual electron spin resonance of fullerene C_{60} anion radical *J. Am. Chem. Soc.* **113** 2780
[3] Krusic P J, Wasserman E, Parkinson B A *et al.* 1991 Electron spin resonance study of the radical reactivity of C_{60} *J. Am. Chem. Soc.* **113** 6274
[4] Pilla O and Bill H 1984 Study of the intermediate Jahn-Teller system La^{2+} in CaF_2 Raman and EPR spectroscopy *J. Phys. C: Solid State Phys.* **17** 3263
[5] Hino S, Takahashi H, Iwasaki K *et al.* 1993 Electronic structure of metallofullerene LaC_{82}: electron transfer from lanthanum to C_{82} *Phys. Rev. Lett.* **71** 4261
[6] Poirier D M, Knupfer M, Weaver J H *et al.* 1994 Electronic and geometric structure of La@C_{82} and C_{82}: theory and experiment *Phys. Rev. B* **49** 17403
[7] Nishibori E, Sakata M, Takata M *et al.* 2000 Giant motion of La atom inside C_{82} cage *Chem. Phys. Lett.* **330** 497
[8] Weaver J H Chai Y, Kroll G H *et al.* 1992 XPS probes of carbon-caged metals *Chem. Phys. Lett.* **190** 460
[9] Shinohara H, Sato H, Saito Y *et al.* 1992 Mass spectroscopic and ESR characterization of soluble yttrium-containing metallofullerenes YC_{82} and Y_2C_{82} *J. Phys. Chem.* **96** 3571
[10] Shinohara H, Sato H, Ohchochi M *et al.* 1992 Encapsulation of a scandium trimer in C_{82} *Nature* **357** 52.
[11] Nagase S and Kobayashi K 1993 Metallofullerenes MC_{82} (M = Sc, Y, and La). A theoretical study of the electronic and structural aspects *Chem. Phys. Lett.* **214** 57

[12] Ruebsam M, Schweitzer P, Dinse K.P and Dinse K P 1996 *Fullerenes and Fullerene Nanostructures*, eds H Kuzmany, J Fink, M Mehring *et al.* (London: World Scientific), pp. 173–177

[13] Nishibori E, Takata M, Sakata M *et al.* 1998 Determination of the cage structure of Sc@C$_{82}$ by synchrotron powder diffraction *Chem. Phys. Lett.* **298** 79

[14] Hino S, Umishita K, Iwasaki K *et al.* 1998 Ultraviolet photoelectron spectra of Sc@ C$_{82}$ *Chem. Phys. Lett.* **300** 145

[15] Nagase S and Kobayashi K 1994 Theoretical study of the lanthanide fullerene CeC$_{82}$. Comparison with ScC$_{82}$, YC$_{82}$ and LaC$_{82}$ *Chem. Phys. Lett.* **228** 106

[16] Nagase S and Kobayashi K 1994 The ionization energies and electron affinities of endohedral metallofullerenes MC$_{82}$ (M = Sc, Y, La): density functional calculations *J. Chem. Soc., Chem. Commun.* 1837

[17] Nagase S, Kobayashi K and Akasaka T 1996 Endohedral metallofullerenes: new spherical cage molecules with interesting properties *Bull. Chem. Soc. Jpn.* **69** 2131

[18] Kato T, Suzuki S, Kichuchi K and Achiba Y 1993 ESR study of the electronic structures of metallofullerenes: a comparison between lanthanum fullerene (La@C$_{82}$) and scandium fullerene (Sc@C$_{82}$) *J. Phys. Chem.* **97** 13425

[19] Kato T, Bandow S, Inakuma M and Shinohara H 1995 ESR study on structures and dynamics of Sc$_3$@C$_{82}$ *J.Phys. Chem.* **99** 856

[20] Kato T, Suzuki T, Yamamoto K *et al.* 1995 *Fullerenes: Recent Advances in the Chemistry and Physics of Fullerenes and Related Materials*, Vol. 2, eds K Kadish and R Ruoff (Pennington, NJ: Electrochemical Society), pp. 733–739

[21] Kato T, Akasaka T, Kobayashi K *et al.* 1996 ESR study on the reactivity of two isomers of LaC$_{82}$ with the disilirane. *Appl. Magn. Reson.* **11** 293

[22] Ruebsam M, Plueschau M, Schweitzer P *et al.* 1995 *Physics and Chemistry of Fullerenes and Derivatives*, eds H Kuzmany, J Fink, M Mehring *et al.* (London: World Scientific), pp. 117–121

[23] Ruebsam M, Knapp C P, Schweitzer P and Dinse K P 1996 *Fullerenes: Recent Advances in the Chemistry and Physics of Fullerenes and Related Materials*, Vol. 3, eds H Kuzmany and R Ruoff (Pennington, NJ: Electrochemical Society), pp. 602–608

[24] Bartl A, Dunsch L, Froehner J and Kirbach U 1994 New electron spin resonance and mass spectrometric studies of metallofullerenes *Chem. Phys. Lett.* **229** 115

[25] Bartl A, Dunsch L, Kirbach U and Seifert G 1997 Paramagnetic states of metals and ^{13}C in isolated endohedral fullerenes *Synth. Met.* **86** 2395

[26] Seifert G, Bartl A, Dunsch L *et al.* 1998 Electron spin resonance spectra: geometrical and electronic structure of endohedral fullerenes. *Appl . Phys.* A **66** 265

[27] Bartl A, Dunsch L, Kirbach U and Schandert B 1995 Paramagnetic states in pristine and metallofullerenes *Synth. Met.* **70** 1365

[28] Bartl A, Dunsch L and Kirbach U 1995 Preparation, mass spectrometry and *solid state* ESR spectroscopy of endohedral fullerenes *Solid State Commun.* **94** 827

[29] Bartl A, Dunsch L and Kirbach U 1996 Electron transfer at lanthanum endohedral fullerenes *Appl. Magn. Reson.* **11** 301

[30] Knorr S, Grupp A, Mehring M *et al.* 1998 Pulsed ESR investigations of anisotropic interactions in M@C$_{82}$ (M=Sc, Y, La) *Appl. Phys.* A **66** 257

[31] Suzuki S, Kawata S, Shiromaru H *et al.* 1992 Isomers and carbon-13 hyperfine structures of metal-encapsulated fullerenes M@C$_{82}$ (M = Sc, Y, and La) *J. Phys. Chem.* **96** 7159

[32] Hoinkis M, Yannoni C S, Bethune D S *et al.* 1992 Multiple species of La@C$_{82}$ and Y@C$_{82}$. Mass spectroscopic and solution EPR studies *Chem. Phys. Lett.* **198** 461

[33] Bandow S, Kitagawa H, Mitani T *et al.* 1992 Anaerobic sampling and characterization of lanthanofullerenes: extraction of LaC_{76} and other LaC_{2n} *J. Phys. Chem.* **96** 9609

[34] Yamamoto K, Funasaka H, Takahashi T *et al.* 1994 Isolation and characterization of an ESR-active $La@C_{82}$ isomer *J. Phys. Chem.* **98** 12831

[35] Inakuma M, Ohno M and Shinohara H 1995 *Fullerenes: Recent Advances in the Chemistry and Physics of Fullerenes and Related Materials*, Vol. 2, eds K Kadish and R Ruoff (Pennington, NJ: Electrochemical Society), pp. 330–342

[36] Yannoni C S, Hoinkis M, de Vries M S *et al.* 1992 Scandium clusters in fullerene cages *Science* **256** 1191

[37] Kikuchi K, Nakao Y, Suzuki S *et al.* 1994 Characterization of the isolated $Y@C_{82}$ *J. Am. Chem. Soc.* **116** 9367

[38] Shinohara H, Inakuma M, Kishida M *et al.* 1995 An oriented cluster formation of endohedral $Y@C_{82}$ metallofullerenes on clean surfaces *J. Phys. Chem.* **99** 13769

[39] Shinohara H, Ohno M, Kishida M *et al.* 1995 *Fullerenes: Recent Advances in the Chemistry and Physics of Fullerenes and Related Materials*, Vol. 2, eds K Kadish and R Ruoff (Pennington, NJ: Electrochemical Society), pp. 763–773

[40] Yamamoto Y, Ishiguro T, Sakurai K and Funasaka H 1997 *Fullerenes: Recent Advances in the Chemistry and Physics of Fullerenes and Related Materials*, Vol. 4, eds K Kadish and R Ruoff (Pennington, NJ: Electrochemical Society), pp. 375–389

[41] Knapp C, Weiden N and Dinse K P 1998 EPR investigation of endofullerenes in solution *Appl. Phys. A* **66** 249

[42] Shinohara H, Inakuma M, Hayashi N *et al.* 1994 Spectroscopic properties of isolated $Sc_3@C_{82}$ metallofullerene *J. Phys. Chem.* **98** 8597

[43] Shinohara H, Yamaguchi H, Hayashi N *et al.* 1993 A new characterization of lanthanum- and scandium-endohedral metallofullerenes *Mater. Sci. Eng. B* **19** 25

[44] van Loosdrecht P H M, Johnson R D, Bethune D S *et al.* 1994 Orientational dynamics of the Sc_3 trimer in C_{82}: an EPR study *Phys. Rev. Lett.* **73** 3415

[45] Stevenson S, Dorn H C, Burbank P *et al.* 1994 Isolation and monitoring of the endohedral metallofullerenes $Y@C_{82}$ and $Sc_3@C_{82}$: on-line chromatographic separation with EPR detection *Anal. Chem.* **66** 2680

[46] Kato T, Bandow S, Inakuma M *et al.* 1994 *Fullerenes: Recent Advances in the Chemistry and Physics of Fullerenes and Related Materials*, Vol. 1, eds K Kadish and R Ruoff (Pennington, NJ: Electrochemical Society), pp. 1361–1381

[47] Anderson M R, Dorn H C, Stevenson S *et al.* 1997 The voltammetry of $Sc_3@C_{82}$ *J. Am. Chem. Soc.* **119** 437

[48] Ungerer J R and Hughbanks T 1993 The electronic structure of $Sc_3@C_{82}$ *J. Am. Chem. Soc.* **115** 2054

[49] Suzuki S, Kojima Y, Nakao Y *et al.* 1994 ESR detection of non-equivalent scandium trimer *Chem. Phys. Lett.* **229** 512

[50] Suzuki T, Maruyama Y, Kato T *et al.* 1993 Electrochemical properties of $La@C_{82}$ *J. Am. Chem. Soc.* **115** 11006

[51] Suzuki T, Kikuchi K, Oguri F *et al.* 1996 Electrochemical properties of fullerenolanthanides *Tetrahedron* **52** 4973

[52] Anderson M R, Dorn H C, Burbank P M and Gibson J R 1997 *Fullerenes: Recent Advances in the Chemistry and Physics of Fullerenes and Related Materials*, Vol. 4, eds K Kadish and R Ruoff (Pennington, NJ: Electrochemical Society), pp. 448–456

[53] Wang W, Ding J, Yang S and Li X Y 1997 *Fullerenes: Recent Advances in the Chemistry and Physics of Fullerenes and Related Materials*, Vol. 4, eds K Kadish and R Ruoff (Pennington, NJ: Electrochemical Society), pp. 417–428

[54] Dunsch L, Bartl A, Kirbach U and Froehner J 1995 *Fullerenes: Recent Advances in the Chemistry and Physics of Fullerenes and Related Materials*, Vol. 2, eds K Kadish and R Ruoff (Pennington, NJ: Electrochemical Society), pp. 182–190

[55] Dunsch L, Kuran P, Kirbach U and Scheller D 1997 *Fullerenes: Recent Advances in the Chemistry and Physics of Fullerenes and Related Materials*, Vol. 4, eds K Kadish and R Ruoff (Pennington, NJ: Electrochemical Society), pp. 523–536

[56] Petra A, Dunsch L and Neudeck A 1996 In situ UV-Vis ESR spectroelectrochemistry *J. Electroanal. Chem.* **412** 153

[57] Elliott B, Yu L and Echegoyen L 2005 A simple isomeric separation of D_{5h} and I_h $Sc_3N@C_{80}$ by selective chemical oxidation *J. Am. Chem. Soc.* **127** 10885

[58] Chaur M N, Melin F, Ortiz A L and Echegoyen L 2009 Chemical, electrochemical, and structural properties of endohedral metallofullerenes *Angew. Chem. Int. Ed.* **48**, 7514

[59] Popov A A, Yang S and Dunsch L 2013 Endohedral fullerenes *Chem. Rev.* **113** 5989

[60] Kikuchi K, Suzuki S, Nakao Y et al. 1993 Isolation and characterization of the metallofullerene LaC_{82} *Chem. Phys. Lett.* **216** 67

[61] Yamamoto K, Funasaka H, Takahashi T and Akasaka T 1994 Isolation of an ESR-active metallofullerene of LaC_{82} *J. Phys. Chem.* **98** 2008

[62] Takata M, Umeda B, Nishibori E et al. 1995 Confirmation by X-ray diffraction of the endohedral nature of the metallofullerene $Y@C_{82}$ *Nature* **377** 46

[63] Ding J, Weng L and Yang S 1996 Electronic structure of $Ce@C_{82}$: an experimentnal study *J. Phys. Chem.* **100** 11120

[64] Shinohara H, Kishida M, Nakane T et al. 1994 *Fullerenes: Recent Advances in the Chemistry and Physics of Fullerenes and Related Materials*, Vol. 1, eds K Kadish and R Ruoff (Pennington, NJ: Electrochemical Society), pp. 1361–1381

[65] Ding J and Yang S 1996 Isolation and characterization of $Pr@C_{82}$ and $Pr_2@C_{80}$ *J. Am. Chem. Soc.* **118** 11254

[66] Ding J, Lin N, Weng L et al. 1996 Isolation and characterization of a new metallofullerene $Nd@C_{82}$ *Chem. Phys. Lett.* **261** 92

[67] Kikuchi K, Kobayashi K, Sueki K et al. 1994 Encapsulation of radioactive ^{159}Gd and ^{161}Tb atoms in fullerene cages *J. Am. Chem. Soc.* **116** 9775

[68] Shi Z J, Okazaki T, Shimada T et al. 2003 Selective high-yield catalytic synthesis of terbium metallofullerenes and single-wall carbon nanotubes *J. Phys. Chem. B* **107** 2485

[69] Kikuchi K, Nakao Y, Suzuki S et al. 1994 Characterization of the isolated $Y@C_{82}$ *J. Am. Chem. Soc.* **116** 9367

[70] Tagmatarchis N and Shinohara H 2000 Production, separation, isolation and spectroscopic study of dysprosium endohedral metallofullerenes *Chem. Mater.* **12**, 3222

[71] Ito Y, Okazaki T, Okuba S et al. 2007 Enhanced 1520 nm photoluminescence from Er^{3+} ions in di-erbium-carbide metallofullerenes $(Er_2C_2)@C_{82}$ (isomers I, II, and III) *ACS Nano* **1** 456

[72] Kirbach U and Dunsch L 1996 The existence of stable $Tm@C_{82}$ isomers *Angew. Chem., Int. Ed. Engl.* **35** 2380

[73] Kikuchi K, Sueki K, Akiyama K et al. 1997 *Fullerenes: Recent Advances in the Chemistry and Physics of Fullerenes and Related Materials*, Vol. 4, eds K Kadish and R Ruoff (Pennington, NJ: Electrochemical Society), pp. 408–416

[74] Okazaki T, Lian Y, Gu Z N *et al.* 2000 Isolation and spectroscopic characterization of Sm-containing metallofullerenes *Chem.Phys. Lett.* **320** 435

[75] Inakuma M and Shinohara H 2000 Temperature-dependent EPR studies on isolated scandium metallofullerenes: Sc@C$_{82}$(I, II) and Sc@C$_{84}$ *J. Phys. Chem.* **104** 7595

[76] Xu Z, Nakane T and Shinohara H 1996 Production and isolation of Ca@C$_{82}$ (I–IV) and Ca@C$_{84}$ (I,II) metallofullerenes *J. Am. Chem. Soc.* **118** 11309

[77] Dennis T J S and Shinohara H 1997 Production and isolation of endohedral strontium- and barium-based mono-metallofullerenes: Sr/Ba@C$_{82}$ and Sr/Ba@C$_{84}$ *Chem. Phys. Lett.* **278** 107

[78] Dennis T J S and Shinohara H 1997 *Fullerenes: Recent Advances in the Chemistry and Physics of Fullerenes and Related Materials*, Vol. 4, eds K Kadish and R Ruoff (Pennington, NJ: Electrochemical Society), pp. 182–190

[79] Dennis T J S and Shinohara H 1998 Production, isolation, and characterization of group-2 metal-containing endohedral metallofullerenes *Appl. Phys. A* **66** 243

[80] Dennis T J S, Kai T, Tomiyama T and Shinohara H 1998 Isolation and characterisation of the two major isomers of [84]fullerene *J. Chem. Soc., Chem. Commun.* **619**

[81] Nakane T, Xu Z, Yamamoto E *et al.* 1998 *Fullerenes and Fullerene Nanostructures*, eds H Kuzmany, J Fink, M Mehring and S Roth (London: World Scientific), pp. 193–281

[82] Sueki K, Akiyama K, Yamauchi T *et al.* 1997 New lanthanoid metallofullerenes and their HPLC elution behavior *Fullerene Sci. Technol.* **5** 1435

[83] Ding J and Yang S 1997 Systematic isolation of endohedral fullerenes containing lanthanide atoms and their characterization *J. Phys. Chem. Solids* **11** 1661

[84] Huang H and Yang S 1998 Relative yields of endohedral lanthanide metallofullerenes by arc synthesis and their correlation with the elution behavior *J. Phys. Chem.* **102** 10196

[85] Macfarlane R M, Wittmann G, van Loosdrecht P H M *et al.* 1997 Measurement of pair interaction and 1.5 μm emission from Er^{3+} ions in a C$_{82}$ fullerene cage *Phys. Rev. Lett.* **79** 1397

[86] Ding X, Alford J M and Wright J C 1997 Lanthanide fluorescence from Er^{3+} þ in Er$_2$@C$_{82}$. *Chem. Phys. Lett.* **269** 72

[87] Hoffman K R, Norris B J and Merle R B 1997 *Fullerenes: Recent Advances in the Chemistry and Physics of Fullerenes and Related Materials*, Vol. 4, eds K Kadish and R Ruoff (Pennington, NJ: Electrochemical Society), pp. 475–484

[88] Plant S R, Dantelle G, Ito Y *et al.* 2009 Acuminated fluorescence of Er^{3+} centres in endohedral fullerenes through the incarceration of a carbide cluster *Chem. Phys. Lett.* **476** 41

[89] Pichler T, Golden M S, Knupfer M *et al.* 1997 Monometallofullerene Tm@C$_{82}$: Proof of an encapsuulated divalent Tm ion by high-energy spectroscopy *Phys. Rev. Lett.* **79** 3026

[90] Pichler T, Knupfer M, Golden M S *et al.* 1998 The metallofullerene Tm@C$_{82}$: isomer-selective electronic structure *Appl. Phys. A* **66** 281

[91] Kessler B, Bringer A, Cramm S *et al.* 1997 Evidence for incomplete charge transfer and La-derived states in the valence bands of endohedrally doped La@C$_{82}$ *Phys. Rev. Lett.* **79** 2289

[92] Hino S, Zenki M, Zaima T *et al.* 2012 Photoelectron spectroscopy of Sc$_3$N@C$_{78}$ *J. Phys. Chem. C* **116** 165

[93] Olmstead M M, de Bettencourt-Dias A, Duchamp J C *et al.* 2001 Isolation and structural characterization of the endohedral fullerene Sc$_3$N@C$_{78}$ *Angew. Chem. Int. Ed.* **40** 1223

[94] Hino S, Kato M, Yoshimura D *et al.* 2007 Effect of encapsulated atoms on the electronic structure of the fullerene cage: A case study on La$_2$@C$_{78}$ and Ti$_2$C$_2$@C$_{78}$ via ultraviolet photoelectron spectroscopy *Phys. Rev. B* **75**, 125418

[95] Tan K and Lu X 2005 Ti_2C_{80} is more likely a titanium carbide endohedral metallofullerene $(Ti_2C_2)@C_{78}$ *Chem. Commun.* **4444**

[96] Cao B, Wakahara T, Tsuchiya T *et al.* 2004 Isolation, characterisation, and theoretical study of $La_2@C_{78}$ *J. Am. Chem. Soc.* **126** 9164

[97] Takahashi T, Ito A, Inakuma M and Shinohara H 1995 Divalent scandium atoms in the cage of C_{84} *Phys. Rev. B* **52** 13812

[98] Hino S, Umishita K, Iwasaki K *et al.* 1997 Photoelectron spectra of metallofullerenes, GdC_{82} and La_2C_{80}: Electron transfer from the metal to the cage *Chem. Phys. Lett.* **281** 115

[99] Hino S, Umishita K, Iwasaki K *et al.* 2001 Ultraviolet photoelectron spectra of metallofullerenes, two $Ca@C_{82}$ isomers *Chem. Phys. Lett.* **337** 65

[100] Pichler T, Kuzmany H, Yamamoto E and Shinohara H 1996 *Fullerenes and Fullerene Nanostructures*, eds H Kuzmany, J Fink, M Mehring *et al.* (London: World Scientific), pp. 178–181

[101] Hulman M, Pichler T, Kuzmany H *et al.* 1997 Vibrational structure of C_{84} and $Sc_2@C_{84}$ analyzed by IR spectroscopy *J. Mol. Struct.* **408/409** 359

[102] Lebedkin S, Renker B, Heid R *et al.* 1998 *Molecular Nanostructures*, eds H Kuzmany, J Fink, M Mehring and S Roth (London: World Scientific), pp. 203–206

[103] Lebedkin S, Renker B, Heid R *et al.* 1998 A spectroscopic study of $M@C_{82}$ metallofullerenes: Raman, far-infrared, and neutron scattering results *Appl. Phys. A* **66** 273

[104] Dunsch L, Kuran P and Krause M 1998 *Fullerenes: Recent Advances in the Chemistry and Physics of Fullerenes and Related Materials*, Vol. 6, eds K Kadish and R Ruoff (Pennington, NJ: Electrochemical Society), pp. 1031–1038

[105] Krause M, Hulman M, Kuzmany H *et al.* 1999 Diatomic metal encapsulates in fullerene cages: A Raman and infrared analysis of C_{84} and $Sc_2@C_{84}$ with D_{2d} symmetry *J. Chem. Phys.* **111** 7976

[106] Buerk M, Schmidt M, Cummins T R *et al.* 1996 *Fullerenes and Fullerene Nanostructures*, eds H Kuzmany, J Fink, M Mehring and S Roth (London: World Scientific), pp. 196–199

[107] Dunsch L, Eckert D, Froehner J *et al.* 1998 *Fullerenes: Recent Advances in the Chemistry and Physics of Fullerenes and Related Materials*, Vol. 6, eds K Kadish and R Ruoff (Pennington, NJ: Electrochemical Society), pp. 955–966

[108] Kirbach U and Dunsch L 1996 The existence of stable $Tm@C_{82}$ isomers *Angew. Chem. Int. Ed.* **35** 2380

[109] Kobayashi K and Nagase S 1998 Structures and electronic states of $M@C_{82}$ (M=Sc, Y, La and lanthanides) *Chem. Phys. Lett.* **282** 325

[110] Krause M, Hulman M, Kuzmany H *et al.* 2000 Low-energy vibrations in $Sc_2@C_{84}$ and $Tm@C_{82}$ metallofullerenes with different carbon cages *J. Mol. Struct.* **521** 325

[111] Kobayashi K, Nagase S and Akasaka T 1996 Endohedral dimetallofullerenes $Sc_2@C_{84}$ and $La_2@C_{80}$. Are the metal atoms still inside the fullerene cages? *Chem. Phys. Lett.* **261** 502

[112] Popov A A, Zhang L and Dunsch L 2010 A pseudoatom in a cage: trimetallofullerene $Y_3@C_{80}$ mimics $Y_3N@C_{80}$ with nitrogen substituted by a pseudoatom *ACS Nano* **4** 785

5

Carbide and Nitride Metallofullerenes

5.1 DISCOVERY OF CARBIDE METALLOFULLERENES

The production and isolation of the first carbide metallofullerene was reported for $Sc_2@C_{86}$ ($=Sc_2C_2@C_{84}$) in 2001 [1]. Very interestingly, this carbide metallofullerene was an accidental discovery when Wang $et\ al.$ were trying to isolate and spectroscopically characterize $Sc_2@C_{86}$ ($= Sc_2C_2@C_{84}$) and $Sc_2@C_{88}$ ($= Sc_2C_2@C_{86}$) in 1999 [2]. Up until these studies, no one had come up with the idea that $Sc_2@C_{86}$ is actually a carbide $Sc_2C_2@C_{84}$ metallofullerene, because the normal identification technique for metallofullerenes is mass spectrometry in which only mass numbers can be identified.

In the ensuing years, it has been revealed that the major part of the di-scandium metallofullerenes may have $Sc_2C_2@C_{2n-2}$ carbide structure rather than pure scandium $Sc_2@C_{2n}$ metallofullerenes including $Sc_2@C_{84}$ (isomer III) [3,4].

The schematic molecular structure of $Sc_2C_2@C_{84}$ is shown in Figure 5.1. It was found that the C_2 molecule in the C_{84} cage is rotating like a rigid-rotor at low temperatures below 60 K [5]. Furthermore, as described above, it has been found that $Sc_3@C_{82}$ was actually a scandium carbide fullerene, $Sc_3C_2@C_{80}$, as revealed by X-ray diffraction [6,7]. The possibility of such a carbide structure of $Sc_3C_2@C_{80}$ instead of $Sc_3@C_{82}$ was reported by gas-phase ion mobility measurements [8], where the ion mobility of $Sc_3@C_{82}$ was close to that of C_{80} rather than C_{82}.

$Endohedral\ Metallofullerenes:\ Fullerenes\ with\ Metal\ Inside$, First Edition.
Hisanori Shinohara and Nikos Tagmatarchis.
© 2015 John Wiley & Sons, Ltd. Published 2015 by John Wiley & Sons, Ltd.

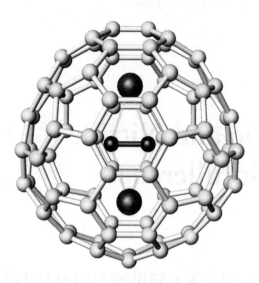

Figure 5.1   Schematic molecular structure of the $Sc_2C_2@C_{84}$ carbide metallofullerene based on the synchrotron X-ray powder diffraction and ^{13}C NMR measurements. Reproduced with permission from Ref. [1]. Copyright Wiley-VCH Verlag GmbH & Cpo. KGaA

Even in non-IPR (isolated pentagon rule) metallofullerenes [9,10], the presence of a carbide fullerene, $Sc_2C_2@C_{68}$, was reported and structurally characterized [11].

The carbide metallofullerenes are now widely observed for the various kinds of di-metallofullerenes such as $Y_2C_2@C_{82}$ [12,13] and $Er_2C_2@C_{82}$ (isomers I, II, III) [14]. One of the main causes for the prevalence of carbide di- or tri-metallofullerenes is that two or three positively charged metal atoms in fullerene cages can be bound tightly together around a negatively charged C_2 molecule [13], where the presence of C_2 molecules provides additional stability of otherwise (mutually repulsive) positively charged metal atoms in fullerene cages. In fact, so far no carbide metallofullerenes have been reported in mono-metallofullerenes such as $MC_2@C_{82}$ (M = metal atom). In the mono-metallofullerenes, basically there is no need to entrap a C_2 moiety to obtain endohedral stabilization of positively charged metal atoms. Interestingly, even a very large fullerene cage such as C_{92} encages Gd_2C_2 metal carbide as depicted in Figure 5.2 [15]. Molecular structures of various carbide metallofullerenes encapsulating Ti, Sc, Y, Gd, and Lu are depicted in Figure 5.3.

Comprehensive treatments on the carbide metallofullerenes can be found in recent review articles [16–21]. The carbide metallofullerenes so far synthesized and characterized are summarized in Table 5.1 [22].

(a)

(b)

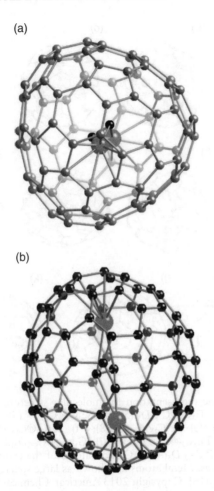

Figure 5.2 A very large carbide metallofullerene $Gd_2C_2@C_{92}$-$D_3(85)$ with cage carbon atoms in gray, the carbide carbon atoms in black, and the Gd atoms shown as large spheres. (a) A view down the non-crystallographic threefold axis of the carbon cage. (b) A side view with the threefold axis in the vertical direction. Only the major Gd sites are shown. Reproduced with permission from Ref. [15]. Copyright 2008 American Chemical Society

In fact, the presence of carbide metallofullerenes was experimentally predicted before Wang *et al.* structurally identified the first carbide metallofullerene, that is, $Sc_2C_2@C_{84}$, as described at the beginning of this chapter [1].

The so-called gas-phase ion mobility measurements [23] have been applied to structural analyses on various species such as clusters and fullerene-related materials. This method has some favorable features in comparison with STM (scanning tunneling miscroscopy), NMR, and X-ray measurements. High sensitivity and high selectivity originating

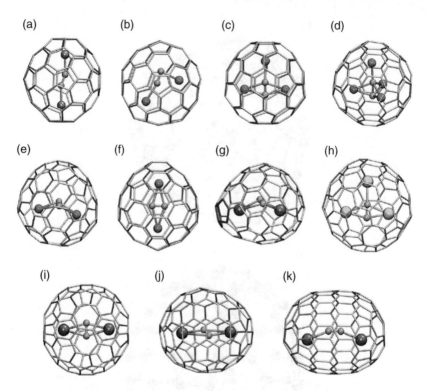

Figure 5.3 Schematic representation for molecular structures of various carbide metallofullerenes: (a) $Ti_2C_2@C_{78}$-$D_{3h}(5)$; (b) $Sc_2C_2@C_{80}$-$C_{2v}(5)$; (c) $Sc_3C_2@C_{80}$-$I_h(7)$; (d) $Sc_4C_2@C_{80}$-$I_h(7)$; (e) $Sc_2C_2@C_{82}$-$C_{3v}(8)$; (f) $Sc_2C_2@C_{84}$-$D_{2d}(23)$; (g) $Y_2C_2@C_{84}$-$C_1(51383)$; (h) $Lu_3C_2@C_{88}$-$D_2(35)$; (i) $Gd_2C_2@C_{88}$-$D_2(35)$; and (j) $Y_2C_2@C_{92}$-$D_3(85)$; (k) $Y_2C_2@C_{100}$-$D_5(450)$. Carbon atoms of the endohedral C_2 units are shown as small spheres. Metal atoms are shown as large spheres. Reproduced with permission from Ref. [16]. Copyright 2013 American Chemical Society

from ion mass detection allow only a very small amount of mixtures of metallofullerenes to be used. Furthermore, high-speed (~ms) gas-phase detection can provide structural information on intermediate species.

Figure 5.4 shows drift time distributions for positive ions of empty fullerenes and metallofullerenes obtained from solvent extractable C_{82} and $Sc_n@C_{82}$ ($n = 1$–3) [8]. The ions of C_{82}^+ and $Sc@C_{82}^+$ show a single peak, and the drift time is exactly the same as that of C_{82}. The di-metallofullerene $Sc@C_{82}^+$ shows two distinct peaks. The main peak with larger drift time corresponds to the peaks measured for C_{82} and $Sc@C_{82}$. The other peak has a substantially smaller drift time, which suggests that it has a smaller drfit time than those of C_{82} and $Sc@C_{82}$. These results strongly suggest that $Sc_2@C_n^+$ ions have a "carbide" structure like $(Sc_2C_2)@C_{n-2}$ as well as $Sc_2@C_n$, simple endohedral structure. The tri-metallofullerene

Table 5.1 Geometry, electronic configurations, and motions of metal carbides in fullerenes. Reproduced with permission from Ref. [22], Elsevier

MCCFs	Cage no.	Shape	Configuration	C_2 bond length (Å)[a]	Metal position[b]	Rotation
$Sc_2C_2@C_{68}$	Non-IPR $C_{2v}(6073)$	Butterfly	$[(Sc^{3+})_2(C_2)^{2-}]^{4+}$	(1.26)	Near [5,5] bond	C_2 rotating and oscillating
$Sc_2C_2@C_{72}$	Non-IPR $C_s(10528)$	Butterfly	$[(Sc^{3+})_2(C_2)^{2-}]^{4+}$	—	Disordered	Metals jumping
$Sc_2C_2@C_{78}$	$C_{2v}(24107)$	Butterfly	$[(Sc^{3+})_2(C_2)^{2-}]^{4+}$	(1.27)	Near [6,6] bond	Fixed
$Ti_2C_2@C_{78}$	$D_{3h}(24109)$	Linear	$[Ti^{4+})_2(C_2)^{2-}]^{6+}$	(1.24)	Under 6 MR	
$Zr_2C_2@C_{78}$	$D_{3h}(24109)$	Butterfly	$[(Zr^{4+})_2(C_2)^{2-}]^{6+}$	(1.28)	Under 6 MR	C_2 rotating
$Hf_2C_2@C_{78}$	$D_{3h}(24109)$	Zigzag	$[(Hf^{4+})_2(C_2)^{2-}]^{6+}$	(1.26)	Under 6 MR	C_2 rotating
$Sc_2C_2@C_{80}$	$C_{2v}(31922)$	Butterfly	$[(Sc^{3+})_2(C_2)^{2-}]^{4+}$	1.20	One under 6 MR, the other near [6,6] bond	Fixed
$Sc_3C_2@C_{80}$	$I_h(31924)$	Planar or trifoliate	$[(Sc^{3+})_3(C_2)^{3-}]^{6+}$	1.11	Disordered	Rotating
$Sc_4C_2@C_{80}$	$I_h(31924)$	Russian doll	$[(Sc^{3+})_4(C_2)^{6-}]^{6+}$	(1.48)	—	Rotating
$Sc_2C_2@C_{82}(I)$	$C_s(39715)$	Butterfly	$[(Sc^{3+})_2(C_2)^{2-}]^{4+}$	1.21	One under 6 MR, the other near [5,6,6] junction	
$Sc_2C_2@C_{82}(II)$	$C_{2v}(39718)$	Butterfly	$[(Sc^{3+})_2(C_2)^{2-}]^{4+}$	1.18	—	—
$Sc_2C_2@C_{82}(III)$	$C_{3v}(39717)$	Butterfly	$[(Sc^{3+})_2(C_2)^{2-}]^{4+}$	1.19–1.20	Disordered	Oscillating
$Y_2C_2@C_{82}(I)$	$C_s(39715)$	Butterfly	$[(Y^{3+})_2(C_2)^{2-}]^{4+}$	—	—	—
$Y_2C_2@C_{82}(II)$	$C_{2v}(39718)$	Butterfly	$[(Y^{3+})_2(C_2)^{2-}]^{4+}$	—	—	—
$Y_2C_2@C_{82}(III)$	$C_{3v}(39717)$	Butterfly	$[(Y^{3+})_2(C_2)^{2-}]^{4+}$	(1.27)	Along C_3 axis	Hopping
$Sc_2C_2@C_{84}$	$D_{2d}(51591)$	Butterfly	$[(Sc^{3+})_2(C_2)^{2-}]^{4+}$	1.20	Disordered	Rotating

(continued overleaf)

Table 5.1 (*continued*)

MCCFs	Cage no.	Shape	Configuration	C$_2$ bond length (Å)[a]	Metal position[b]	Rotation
Y$_2$C$_2$@C$_{84}$	Non-IPR C$_1$ (51383)	Butterfly	$[(Y^{3+})_2(C_2)^{2-}]^{4+}$	—	One under 6 MR, the other near [5,5] bond	—
Gd$_2$C$_2$@C$_{84}$	Non-IPR C$_1$ (51383)	Butterfly	—	1.23	One under 6 MR, the other near [5,5] bond	—
Lu$_3$C$_2$@C$_{88}$	D$_2$(81738)	Planar	$[(Lu^{3+})_3(C_2)^{3-}]^{6+}$	(1.30)	—	Oscillating
Y$_2$C$_2$@C$_{92}$	D$_3$(126408)	Zigzag	$[(Y^{3+})_2(C_2)^{3-}]^{6+}$	(1.27)	—	C$_2$ rotating
Gd$_2$C$_2$@C$_{92}$	D$_3$(126408)	Butterfly	$[(Gd^{3+})_2(C_2)^{2-}]^{4+}$	1.04	Disordered	Metals moving
Y$_2$C$_2$@C$_{100}$	D$_5$(285913)	Bias linear	$[(Y^{3+})_2(C_2)^{2-}]^{4+}$	(1.26)	—	—

[a] Computational results are given in parentheses.
[b] 5 and 6 denote pentagon and hexagon, respectively, 6 MR and 5 MR denote six-membered rings and five-membered rings.

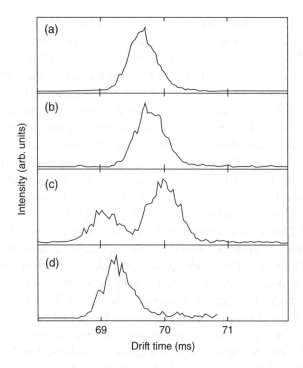

Figure 5.4 High-resolution ion mobility distributions of (a) C_{82}, (b) $Sc@C_{82}$, (c) $Sc_2@C_{82}$, and (d) $Sc_3@C_{82}$ [correct identification: $(Sc_3C_2)@C_{80}$] produced from laser desorption/ionization of these purified metallofullerene materials. The smaller peak in (c) is identified as $Sc_2C_2@C_{80}$. Reproduced with permission from Ref. [8]. Copyright 2001 American Chemical Society

$Sc_3@C_{82}^+$ also shows a peak which corresponds to the shorter peak of the drift time of $Sc_2@C_{82}^+$, indicating that $Sc_3@C_{82}$ may have a carbide $(Sc_3C_2)@C_{80}$ structure. And, in fact, as described in Section 3.2.2, the originally characterized $Sc_3@C_{82}$ in 1992 by Shinohara and co-workers [24] was later found to be a carbide $Sc_3C_2@C_{80}$ metallofullerene [4,6,7], indicating that the gas-phase ion mobility measurement could correctly predict the presence of these carbide metallofullerenes.

5.2 FULLERENE QUANTUM GYROSCOPE: AN IDEAL MOLECULAR ROTOR

In 2004, Kuzmany and co-workers observed the quantized rotational states of a diatomic C_2 unit in solid endohedral fullerene $Sc_2C_2@C_{84}$ [6]. They found that the rotational transitions of C_2 unit in the C_{84} fullerene cage induce a well-resolved periodic line pattern in the low energy Raman spectrum. The rotational constant B and the C—C distance were found to be 1.73 cm^{-1} and 0.127 nm, respectively. Density functional calculations revealed an intrinsic rotational barrier of the order of only a few millielectronvolts for the C_2 unit. The Schrödinger equation involving the potential barrier was solved and the Raman tensor matrix elements were calculated, which provided perfect agreement with the experiment. This carbide metallofullerene can present the first intrinsic rotational spectrum of a diatomic plane molecular rotor, a "quantum gyroscope." Michel *et al.* reported a rigorous study of the superposition of the quantum rotational motion of C_2 unit and classical rotation motions of the surrounding C_{84} carbon cage in $Sc_2C_2@C_{84}$ [25].

Figure 5.5 presents a dynamic molecular structure obtained from molecular dynamics calculations based on the experimental Raman spectra as shown in Figure 5.6. These calculations provide a first indication that $C_2@Sc_2C_{84}$ is a more realistic formula than $Sc_2C_2@C_{84}$ for this carbide metallofullerene. The density functional theory (DFT) calculations gave 0.444, 0.128, and 0.224 nm for the Sc—Sc, C—C, and Sc–cage distances, respectively, in very good agreement with the results obtained by Wang *et al.* [1]. A total charge of −2*e*, +3*e*, and −4*e* is predicted for the C_2 group, each Sc ion, and the cage, respectively, using both theoretical methods.

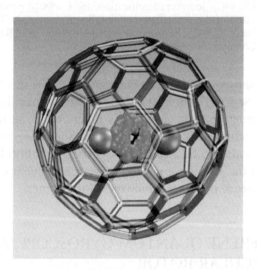

Figure 5.5   Schematic dynamic structure of $C_2@Sc_2C_{84}$. The two spheres represent the Sc ions on the $C_2(z)$ symmetry axis; the *y*-axis lies in the plane of the sheet. The central part of the figure represents a superposition of carbon states as evaluated from DFTB MD (density-functional tight-binding molecular dynamics) calculations at 300 K. Reproduced with permission from Ref. [5]. Copyright 2004 The American Physical Society

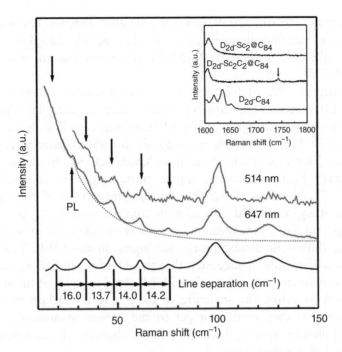

Figure 5.6 Low energy Raman spectra for $Sc_2C_2@C_{84}$ at 80 K. The dotted line is an exponential background and the solid line is the fit using Voigtian lines. PL identifies a plasma line. The arrows indicate positions for equidistant Raman lines. The inset shows the high-energy part of the spectra for $Sc_2@C_{84}$, $Sc_2C_2@C_{84}$, and C_{84} (from top to bottom). The arrow in the inset points to the C—C vibration at 1745 cm^{-1}. Reproduced with permission from Ref. [5]. Copyright 2004 The American Physical Society

A similar quantized rotational motion of the acetylide unit with the tunneling of the C_2 unit through the rotation plane is reported in the Raman study of $Y_2C_2@C_{92}$ [26]. DFT calculations predicted the barrier of the rotation of about 0.1 eV, which explains the slight deviation of the peak Raman values from the theoretical ones computed for the free C_2 rotor. With the help of DFT calculations, a prominent Raman peak at 163.6 cm^{-1} was assigned to a butterfly mode, in which two Y atoms move in a flapping motion around the carbide base.

5.3 NITRIDE METALLOFULLERENES

In the study of metallofullerenes, we sometimes encounter puzzling and unidentified peaks in mass spectra, with mass spectrometry being one of the conventional identification techniques used in fullerene science.

These are not identified from the known metallofullerenes. The discovery of nitride metallofullerenes or trimetallic nitride template (TNT) metallofullerenes of the type $M_3N@C_{2n}$ (M = metal atom) is a typical example [27].

As early as 1995, Achiba and co-workers found an intense peak at $m/z = 1109$ in mass spectra when they were studying laser-induced multiphoton photofragmentation of scandium metallofullerenes [28] (Figure 5.7). They tentatively (erroneously though) assigned this peak as ScO_2C_{86}. They even purified (then assigned) ScO_2C_{86} by the so-called multi-stage high-performance liquid chromatography (HPLC) technique [28]. After discussing the photofragmentation of the purified ScO_2C_{86}, they concluded: "Besides, in contrast with $C_{60}O_n$ ($n = 1$–3), where parent oxide ions cannot be seen in the positive ion mass spectra, ScO_2C_{86} shows the strong parent ion signal in the LD-PI-TOF mass spectrum. Therefore, it suggested that two O atoms in ScO_2C_{86} are not in the epoxide form, as is considered for $C_{60}O_n$. However, at the moment it is hard to define the structure of oxygen in its carbon network." Unfortunately, they could not get to the correct assignment of this metallofullerene species because of the absence of elemental and structural characterization.

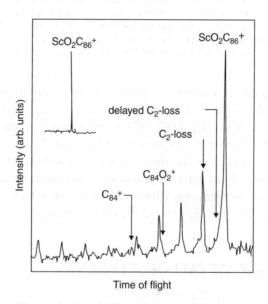

Figure 5.7 Laser-desorption photoionization time-of-flight (LD-PI-TOF) mass spectrum of ScO_2C_{86} having $m/z = 1109$. The current correct assignment of this peak is due to $Sc_3N@C_{80}$. Reproduced from Ref. [28]

Four years later in 1999, Dorn, Balch, and co-workers correctly identified the enhanced mass spectral peak at $m/z = 1109$ as due to $Sc_3N@C_{80}$ [27]. They confirmed the presence of $Sc_3N@C_{80}$ by using ^{13}C NMR and mass spectrometry with ^{13}C enriched samples together with single-crystal X-ray diffraction [27,29]. These experiments unambiguously identified the presence of a four-atom endohedral cluster (Sc_3N) inside an I_h-C_{80} cage (Figure 5.8).

The Sc atoms were found to form an equilateral triangle with a N atom in the center of the planar cluster.

Similar to the $La_2@C_{80}$-I_h [30,31] case, six electrons are transferring from the encapsulated Sc_3N to the C_{80} cage, which stabilizes significantly the C_{80}-I_h cage due to the transformation from an open-shell to a closed-shell electronic structure. This accounts for the fairly high production yield of $Sc_3N@C_{80}$ in arc-discharge syntheses. In fact, irrespective of the fullerene cage size (C_{2n}), most of the $M_3N@C_{2n}$ (M = metal atoms) metallofullerenes possess relatively large HOMO–LUMO (highest occupied molecular orbital–lowest unoccupied molecular orbital) gaps (0.8–1.5 eV) judging from the onsets of optical absorption spectra. Generally, the energy gap between the highest occupied and the lowest unoccupied electronic states, that is, the HOMO–LUMO gap, is an important factor for predicting the electronic stability of molecular structures.

Figure 5.9 shows UV-Vis-NIR (ultraviolet-visible-near infrared) absorption spectra of toluene solutions of $Dy_3N@C_{2n}$ $(2n = 78 - 88)$ [33].

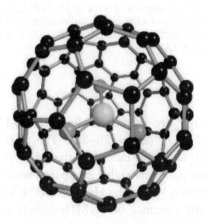

Figure 5.8 The first nitride metallofullerene identified having a structure of $Sc_3N@$ C_{80}-I_h. Reprinted by permission from Macmillan Publishers Ltd: [27]. Copyright 1999 Rights Managed by Nature Publishing Group

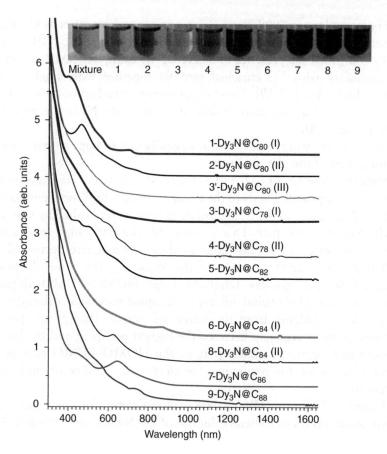

Figure 5.9 UV-Vis-NIR absorption spectra of $Dy_3N@C_{2n}$ ($2n = 78 – 88$) dissolved in toluene. Inset: photographs of a $Dy_3N@C_{2n}$ fullerene extract mixture and dissolved in toluene. Reproduced with permission from Ref. [32]. Copyright Wiley-VCH Verlag GmbH & Cpo. KGaA

As the fullerene cage size increases, the HOMO-LUMO energy gap decreases for $Dy_3N@C_{2n}$ ($2n = 78–88$). The observed large energy gaps are consistent with the fact that the $M_3N@C_{2n}$ type of nitride metallofullerenes are generally stable compared with the $M@C_{82}$ type of mono-metallofullerenes, where the HOMO-LUMO gaps are very small (<0.5 eV) due to the presence of SOMOs (singly occupied molecular orbitals) (cf. Figure 4.5).

Since the successful structural characterization of $Sc_3N@C_{80}$, a number of $M_3N@C_{2n}$ type of nitride metallofullerenes have been synthesized and

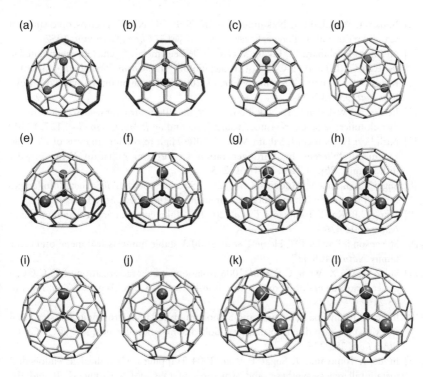

Figure 5.10 Molecular structures of nitride cluster fullerenes: (a) $Sc_3N@$ C_{68}-D_3(6140); (b) $Sc_3N@C_{70}$-C_{2v}(7854); (c) $Sc_3N@C_{78}$-D_{3h}(5); (d) $Sc_3N@C_{80}$-I_h(7); (e) $DySc_2N@C_{76}$-C_s(17490); (f) $M_3N@C_{78}$-C_2(22010); (g) $M_3N@C_{82}$-C_s(39663); (h) $M_3N@C_{84}$-C_s(51365); (i) $M_3N@C_{86}$-D_3(17); (j) $M_3N@C_{88}$-D_2(35); (k) $La_3N@$ C_{92}-T(86); (l) $La_3N@C_{96}$-D_2(186). In (e–j), M denotes Y or lanthanides. Carbon atoms are gray except for the atoms in the adjacent pentagon pairs, which are shown in dark gray; nitrogen is black in the center, metal atoms are gray balls. Reproduced with permission from Ref. [16]. Copyright 2013 American Chemical Society

reported and are both structurally and spectroscopically characterized [17,29,32,34]. The molecular structures of some typical nitride metallofullerenes are presented in Figure 5.10 [17].

REFERENCES

[1] Wang C R, Kai T, Yomiyama T *et al.* 2001 A scandium carbide endohedral metallofullerene: $(Sc_2C_2)@C_{84}$ *Angew. Chem. Int. Ed.* **113** 411

[2] Wang C R, Inakuma M and Shinohara H 1999 Metallofullerenes $Sc_2@C_{82}$(I, II) and $Sc_2@C_{86}$(I, II): isolation and spectroscopic studies *Chem. Phys. Lett.* **300** 379

[3] Iiduka Y, Wakahara T, Nakajima K et al. 2006 ^{13}C NMR spectroscopic study of scandium dimetallofullerene, $Sc_2@C_{84}$ vs. $Sc_2C_2@C_{82}$ Chem. Commun. 2057

[4] Nishibori E, Ishihara M, Takata M et al. 2006 Bent(metal)$_2C_2$ clusters encapsuated in $(Sc_2C_2)@C_{82}(III)$ and $(Y_2C_2)@C_{82}(III)$ metallofullerenes Chem. Phys. Lett. 433 120

[5] Krause M, Hulman M, Kuzmany H et al. 2004 Fullerene quantum gyroscope Phys. Rev. Lett. 93, 137403-1

[6] Iiduka Y, Wakahara T, Nakahodo T et al. 2005 Structural determination of metallofullerene Sc_3C_{82} revisited: a surprising finding J. Am. Chem. Soc. 127, 12500

[7] Nishibori E, Terauchi I, Sakata M et al. 2006 High-resolution analysis of $(Sc_3C_2)@$ C_{80} metallofullerene by third generation synchrotron radiation X-ray powder diffraction J. Phys. Chem. B 110, 19215

[8] Sugai T, Inakuma M, Hudgins R et al. 2001 Structural studies of Sc metallofullerenes by high-resolution ion mobility measurements J. Am. Chem. Soc. 123, 6427

[9] Wang C R, Kai T, Tomiyama T et al. 2000 C_{66} fullerene encaging a scandium dimer Nature 408 426

[10] Stevenson S Fowler PW, Heine T et al. 2000 A stable non-classical metallofullerene family Nature 408 427

[11] Shi Z Q, Wu X, Wang C R et al. 2006 Isolation and characterization of $Sc_2C_2@C_{68}$: a metal-carbide endofullerene with a non-IPR carbon cage Angew. Chem. Int. Ed. 45 2107

[12] Inoue T, Tomiyama T, Sugai T and Shinohara H 2003 Spectroscopic and structural study of Y_2C_2 carbide encapsulating endohedral metallofullerene: $(Y_2C_2)@C_{82}$ Chem. Phys. Lett. 382 226

[13] Inoue T, Tomiyama T, Sugai T et al. 2004 Trapping a C_2 radical in endohedral metallofullerenes: synthesis and structures of $(Y_2C_2)@C_{82}$ (isomer I, II and III) J. Phys. Chem. B 108 7573

[14] Ito Y, Okazaki T, Okuba S et al. 2007 Enhanced 1520 nm photoluminescence from Er^{3+} ions in di-erbium-carbide metallofullerenes $(Er_2C_2)@C_{82}$ (isomers I, II, and III) ACS Nano 1 456

[15] Yang H, Lu C, Liu Z et al. 2008 Detection of a family of gadolinium-containing endohedral fullerenes and the isolation and crystallographic characterization of one member as a metal-carbide encapsulated inside a large fullerene cage J. Am. Chem. Soc. 130 17296

[16] Popov A A, Yang S and Dunsch L 2013 Endohedral fullerenes Chem. Rev. 113 5989

[17] Chaur M N, Melin F, Ortiz A L and Echegoyen L 2009 Chemical, electrochemical, and structural properties of endohedral metallofullerenes Angew. Chem. Int. Ed. 48, 7514

[18] Yang S, Liu F, Chen C et al. 2011 Fullerenes encaging metal clusters – cluster fullerenes Chem. Commun. 47 11822

[19] Lu X, Feng L, Akasaka T and Nagase S 2012 Current status and future developments of endohedral metallofullerenes Chem. Soc. Rev. 41, 7723

[20] Lu X, Akasaka T and Nagase S 2013 Carbide cluster metallofullerenes: structure, properties, and possible origin Acc. Chem. Res. 46 1627

[21] Jin P, Tang C and Chen Z 2014 Carbon atoms trapped in cages: metal carbide cluster fullerenes Coord. Chem. Rev. 270/271 89

[22] Jin P, Tang C and Chen Z 2014 Carbon atoms trapped in cages: metal carbide clusterfullerenes Coord. Chem. Rev. 270/271 89

[23] Dugourd P, Hudgins R R, Clemmer D E and Jarrold M F 1997 High-resolution ion mobility measurements *Rev. Sci. Instrum.* **68** 1122

[24] Shinohara H, Sato H, Ohchochi M *et al.* 1992 Encapsulation of a scandium trimer in C_{82} *Nature* **357** 52

[25] Michel K H, Verberck B, Hulman M *et al.* 2007 Superposition of quantum and classical rotational motions in $Sc_2C_2@C_{84}$ fullerite *J. Chem. Phys.* **126** 064304.

[26] Burke B G, Chan J, Williams K A *et al.* 2011 Vibrational spectrum of the endohedral $Y_2C_2@C_{92}$ fullerene by Raman spectroscopy: evidence for tunneling of the diatomic C_2 molecule *Phys. Rev. B* **83** 115457

[27] Stevenson S, Rice G, Glass T *et al.* 1999 Small-bandgap endohedral metallofullerenes in high yield and purity *Nature* **401** 55

[28] Kojima Y, Suzuki S, Wakabayashi T *et al.* 1995 Isolation and photodissociation of ScO2C86, Proceedings of the 69th Spring Meeting of Chemical Society of Japan, March, p. 411

[29] Zhang J, Stevenson S and Dorn H C 2013 Trimetallic nitride template endohedral metallofullerenes: discovery, structural characterization, reactivity, and applications *Acc.Chem.Res.* **46** 1548

[30] Akasaka T, Nagase S, Kobayashi K *et al.* 1997 ^{13}C and [139]La NMR studies of $La_2@C_{80}$: first evidence for circular motion of metal atoms in endohedral dimetallofullerenes *Angew. Chem., Int. Ed. Engl.* **36** 1643

[31] Nishibori E, Takata M, Sakata M *et al.* 2001 Pentagonal-dodecahedral La_2 charge density in [80-I_h]fullerene: $La_2@C_{80}$ *Angew. Chem. Int. Ed.* **40** 2998

[32] Dunsch L and Yang S 2007 Metal nitride cluster fullerenes: their current state and future prospects *Small* **8** 1298

[33] Yang S F and Dunsch L 2006 Expanding the number of stable isomeric structures of the C_{80} cage: a new fullerene $Dy_3N@C_{80}$ *Chem. Eur. J.* **12** 413

[34] Rodriguez-Fortea A, Balch A L and Poblet J M 2011 Endohedral metallofullerenes: a unique host-guest association *Chem. Soc. Rev.* **40** 3551

6

Non-Isolated Pentagon Rule Metallofullerenes

6.1 ISOLATED PENTAGON RULE

Fullerenes have a novel structure having five- and six-membered ring patterns. In 1987, Kroto, in his *Nature* article entitled "The stability of the fullerenes C_n, with n = 24,28,32,36,50,60, and 70," [1] stated: "... it is clear that a structure in which a pentagon is completely surrounded by hexagons is stable. The simple molecule in which two pentagons are fused has not so far been made, although a stabilized analog exists. These observations suggest that a cage in which all 12 pentagons are completely surrounded by hexagons has optimum stability and one in which pentagons abut is likely to be less stable (Figure 6.1). This argument can be carried further, in that closed structures involving various fused pentagon configurations are likely to exhibit varying degrees of strain-related instability" [1]. This statement, which is based on empirical rules he summarizes in his article, is what is now called the isolated pentagon rule (IPR) [1,2].

IPR can be considered as the most important and essential rule governing the geometry of fullerenes. As Kroto first proposed, IPR simply states that the most stable fullerenes are those in which all pentagons are surrounded by five hexagons. All the empty fullerenes so far produced, isolated, and structurally characterized have been known to satisfy IPR. IPR can be best understood as a logical consequence of minimizing the number of dangling bonds and steric strain of fullerenes [3].

Endohedral Metallofullerenes: Fullerenes with Metal Inside, First Edition.
Hisanori Shinohara and Nikos Tagmatarchis.
© 2015 John Wiley & Sons, Ltd. Published 2015 by John Wiley & Sons, Ltd.

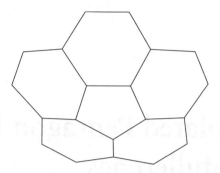

Figure 6.1 Isolated-pentagon unit having a corannulene skeleton. Reprinted by permission from Macmillan Publishers Ltd: [1]. Copyright 1987

As a result, the smallest IPR-satisfying fullerene is C_{60}, and C_{70} is the second smallest; there are no IPR fullerenes between C_{60} and C_{70}.

6.2   NON-IPR METALLOFULLERENES

Although IPR, in principle, has been equally applied for metallofullerenes, IPR isomers of a metallofullerene can be different from those of the corresponding empty fullerene [4,5]. In metallofullerenes, electron transfers from an encaged metal atom to the carbon cage may drastically alter the stability and electronic structure of the fullerene.

Shinohara and co-workers [6] and Dorn and co-workers [7] independently synthesized and structurally characterized IPR-violating (non-IPR) metallofullerenes, $Sc_2@C_{66}$ (Figure 6.2), and $Sc_3N@C_{68}$ (Figure 6.3), respectively. Interestingly, $Sc_2@C_{66}$ has two fused pentagons and $Sc_3N@C_{68}$ has three fused pentagons on the fullerene cage.

Since these reports, many non-IPR metallofullerenes have been produced, isolated, and structurally characterized. Some of these are $La_2@C_{72}$-D_2(10611) [8,9], $Sc_3N@C_{70}$-C_{2v}(7854) [10], $Gd_3N@C_{82}$-C_s(39663), and $M_3N@C_{78}$-C_2(22010) (M = Y, Dy, Tm, Gd) [11,12]. Even a non-IPR carbide metallofullerene, $Sc_2C_2@C_{68}$-C_{2v}, was isolated and structurally characterized [13]. The 21-line ^{13}C NMR spectrum, which is supported by theoretical calculations, indicates that its fullerene cage is C_{68}-(6073) and the encapsulated cluster is a di-scandium acetylide (Figure 6.4). The current state-of-the-art studies of non-IPR metallofullerenes is well reviewed and summarized by Tan *et al.* [14].

The violation of IPR in endohedral metallofullerenes can generally be explained by the changes of the relative stability of the metallofullerenes

(a)   (b)

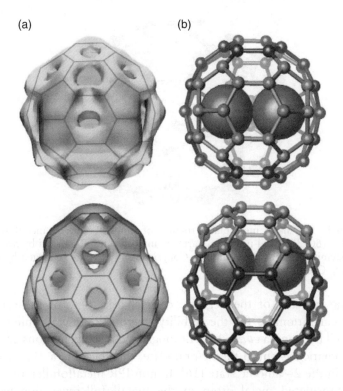

Figure 6.2 (a) X-ray structure of the non-IPR $Sc_2@C_{66}$ fullerene, showing a top view along the C_2 axis and a side view. The equi-contour (1.4 e$Å^{-3}$) surface of the final maximum entropy method (MEM) electron density is shown; the Sc_2 dimer is in black and the two pairs of fused pentagons are evident. (b) Schematic representation of $Sc_2@C_{66}$. Reprinted by permission from Macmillan Publishers Ltd: [6]. Copyright 2000 Rights Managed by Nature Publishing Group

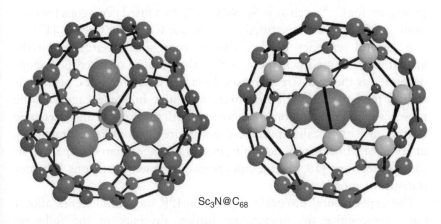

$Sc_3N@C_{68}$

Figure 6.3 Threefold symmetry isomers of C_{68} and proposed $Sc_3N@C_{68}$ non-IPR structure (isomer 6140). Reprinted by permission from Macmillan Publishers Ltd: [7]. Copyright 2000 Rights Managed by Nature Publishing Group

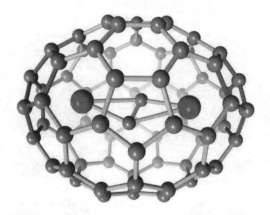

Figure 6.4   Schematic molecular structure of non-IPR carbide metallofullerene $Sc_2C_2@C_{68}$ (6073). Fused-pentagon parts are in both sides. Reprinted by permission from Macmillan Publishers Ltd: [14]. Copyright 2009 Rights Managed by Nature Publishing Group

with the increase of the negative cage charge. One of the probable theoretical rationales for the stability of non-IPR metallofullerenes is presented by Slanina *et al.* [15]. Although two fused pentagons are an 8π anti-aromatic system in the neutral state, they become a 10π aromatic system in the 2– anionic state [16]. In non-IPR metallofullerenes, when the encapsulated metal atom in the metallofullerenes donates two electrons to the pentagon pair, the fused pentagon moiety becomes aromatic.

This is consistent with the experimental findings that the maximum number of fused pentagons found in metallofullerenes is so far equal to half of the negative charge residing on the fullerene cage: three fused pentagons for C_{2n}^{6-} such as in $Sc_3N@C_{68}$ [7], two fused pentagons for C_{2n}^{4-} as in $Sc_2@C_{66}$ [6], and one fused pentagon for C_{2n}^{2-} as in $Ca@C_{72}$ [17].

A general tendency of the fullerene cage size dependence on the appearance of fused pentagons is that the number of fused pentagons decreases as the cage size increases. In larger fullerene cages, pentagons should tend to be distributed over the entire fullerene cage and the presence of fused pentagons, which cause both steric strain and electronic charge density localization, becomes more and more energetically unfavorable. In contrast, normally, in smaller cages than C_{80} apart from some exceptions [18,19], the appearance of fused pentagons becomes statistically more frequent.

The experimental discovery of the non-IPR metallofullerenes allows one to recognize the unexpected further diversity of the fullerene pentagon–hexagon geometry.

REFERENCES

[1] Kroto H W 1987 The stability of the fullerenes C_n, with n = 24, 28, 32, 36, 50, 60 and 70 *Nature* **329** 529

[2] Schmalz T G, Seits W A, Klein D J and Hite G E 1988 Elemental carbon cages *J. Am. Chem. Soc.* **110** 1113

[3] Fowler P W and Manolopoulos D E 1995 *An Atlas of Fullerenes* (Oxford: Clarendon)

[4] Dennis T J S, Kai T, Tomiyama T and Shinohara H 1998 Isolation and characterisation of the two major isomers of [84]fullerene (C_{84}) *J. Chem. Soc., Chem. Commun.* 619

[5] Nishibori E, Takata M, Sakata M *et al.* 1998 Determination of the cage structure of Sc@C_{82} by synchrotron powder diffraction *Chem. Phys. Lett.* **298** 79

[6] Wang C R, Kai T, Tomiyama T *et al.* 2000 C_{66} fullerene encaging a scandium dimer *Nature* **408** 426

[7] Stevenson S, Fowler PW, Heine T *et al.* 2000 A stable non-classical metallofullerene family *Nature* **408** 427

[8] Kato H, Kanazawa Y, Okumura M *et al* 2003 Lanthanoid endohedral metallofullerenols for MRI contrast agents *J. Am. Chem. Soc.* **125** 4391

[9] Lu X, Akasaka T and Nagase S 2013 Carbide cluster metallofullerenes: structure, properties, and possible origin *Acc. Chem. Res.* **46** 1627

[10] Yang S F, Kalbac M, Popov A and Dunsch L 2006 A facile route to the non-IPR fullerene $Sc_3N@C_{68}$: Synthesis, spectroscopic characterization, and density functional theory computations (IPR=isolated pentagon rule) *Chem. Eur. J.* **12** 7856

[11] Beavers C M, Zuo T M, Duchamp J C *et al.* 2006 $Tb_3N@C_{84}$: an improbable, egg-shaped endohedral fullerene that violates the isolated pentagon rule *J. Am. Chem. Soc.* **128** 11352

[12] Popov A A, Chen C, Yang S *et al.* 2010 Spin-flow vibrational spectroscopy of molecules with flexible spin density: electrochemistry, ESR, cluster and spin dynamics, and bonding in $TiSc_2N@C_{80}$ *ACS Nano* **4** 4857

[13] Shi Z Q, Wu X, Wang C R *et al.* 2006. Isolation and characterization of $Sc_2C_2@C_{68}$: a metal-carbide endofullerene with a non-IPR carbon cage *Angew. Chem. Int. Ed.* **45** 2107

[14] Tan Y, Xie S, Huang R and Zheng L 2009 The stabilization of fused-pentagon fullerene molecules *Nat. Chem.* **1** 450

[15] Slanina Z, Chen Z, Schleyer P V *et al.* 2006 $La_2@C_{72}$ and $Sc_2@C_{72}$: computational characterizations *J. Phys. Chem. A* **110** 2231

[16] Zywietz T K, Jiao H, Schleyer R and Meijere A 1998 Aromaticity and antiaromaticity in oligocyclic annelated five-membered ring systems *J. Org. Chem.* **63** 3417

[17] Kobayashi K, Nagase S, Yoshida M and Osawa E 1997 Endohedral metallofullerenes. are the isolated pentagon rule and fullerene structures always satisfied? *J. Am. Chem. Soc.* **119** 12693

[18] Beavers C M, Zuo T M, Duchamp J C *et al.* 2006 $Tb_3N@C_{84}$: an improbable, egg-shaped endohedral fullerene that violates the isolated pentagon rule *J. Am. Chem. Soc.* **128** 11352

[19] Zuo T, Walker K, Olmstead M M *et al.* 2008 New egg-shaped fullerenes: non-isolated pentagon structures of $Tm_3N@C_s(51\ 365)$-C_{84} and $Gd_3N@C_s(51\ 365)$-C_{84} *Chem. Commun.* **1067**

7

Oxide and Sulfide Metallofullerenes

Incorporating non-metal elements, together with metal oxides as dopants of the carbon source in the arc-burning process of generating endohedral metallofullerenes (EMFs), results in the formation of some novel EMF materials. Recently, two new families of EMFs have appeared, complementing the big class of metallofullerenes, which are nowadays divided into the following: (i) classic EMFs; (ii) metallic carbide EMFs; (iii) metallic nitride EMFs; (iv) metallic oxide EMFs; and (v) metallic sulfide EMFs. The latter two families are rather small, however, given the steady research growth in the area and the demand for production of new and novel forms of EMFs, it is expected that not only a plethora of metallic oxide and metallic sulfide EMFs but also bulk quantities of those EMFs will soon be available [1].

7.1 OXIDE METALLOFULLERENES

The very first EMF encapsulating a metal oxide was reported in 2008. Until that time, apart from traditional EMFs encapsulating only metal atoms inside various carbon cages, it was only possible to form EMFs containing metal nitrides [2], and metal carbides [3] inside the hollow carbon sphere. The key parameter that allowed the preparation of metal oxide EMFs is the presence of air in the fullerene reaction chamber. Thus, introducing an air

Endohedral Metallofullerenes: Fullerenes with Metal Inside, First Edition.
Hisanori Shinohara and Nikos Tagmatarchis.
© 2015 John Wiley & Sons, Ltd. Published 2015 by John Wiley & Sons, Ltd.

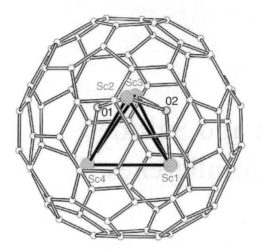

Figure 7.1 Structure of $Sc_4(\mu_3\text{-}O)_2@I_h\text{-}C_{80}$. Reproduced with permission from Ref. [5]. Copyright 2009 American Chemical Society

flow in the arc-burning apparatus as well as incorporating copper nitrate along with scandium oxide in the graphite rods, resulted in the preparation of $Sc_4(\mu_3\text{-}O)_2@I_h\text{-}C_{80}$, as obtained after recycling HPLC (high-performance liquid chromatography) and verified by MALDI-TOF mass spectrometry [4]. The structure of $Sc_4(\mu_3\text{-}O)_2@I_h\text{-}C_{80}$ was crystallographically predicted and it was found that the $I_h\text{-}C_{80}$ cage encapsulates a distorted tetrahedral set of Sc atoms with bridging O atoms asymmetrically positioned on two of the triangular faces of the Sc_4 tetrahedron (Figure 7.1).

Since the stabilization of the $I_h\text{-}C_{80}$ cage, forming a close shell, requires a charge transfer of six electrons, the encapsulated $Sc_4(\mu_3\text{-}O)_2$ cluster should possess a +6 charge in the form of $(Sc^{3+})_2(Sc^{2+})_2(O^{2-})_2$. The latter is indicative that Sc atoms are present in two different valence states in the encapsulated cluster unit. Thus, the Sc atoms interacting with two O atoms are in the +3 valence state, while the other two Sc atoms interacting each other and with one O atom, are in the +2 state. In addition, $Sc_4(\mu_3\text{-}O)_2@I_h\text{-}C_{80}$ shows two sharp peaks at a ratio of 1:3 in the ^{13}C NMR spectrum, thus indicating that in the NMR time scale, the $Sc_4(\mu_3\text{-}O)_2$ cluster freely rotates inside the $I_h\text{-}C_{80}$ cage. Furthermore, in the ^{45}Sc NMR spectrum two peaks are also evident, thus proving that not only the $Sc_4(\mu_3\text{-}O)_2$ cluster is rather rigid but also that O atoms do not rearrange between different faces of the Sc_4 tetrahedron. As far as reduction/oxidation (redox) properties of $Sc_4(\mu_3\text{-}O)_2@I_h\text{-}C_{80}$ are concerned, the material exhibits two reversible reductions and two reversible oxidations with an electrochemical (EC) gap of 1.10 V, almost

Table 7.1 Redox potentials and EC gap of $Sc_4(\mu_3\text{-}O)_2@I_h\text{-}C_{80}$ compared with those of other TNT-based fullerenes as well as $La_2@C_{80}$ and $Sc_4C_2@C_{80}$

Materials	$E^3_{red.}$ (V)	$E^2_{red.}$ (V)	$E^1_{red.}$ (V)	$E^1_{ox.}$ (V)	$E^2_{ox.}$ (V)	EC gap (V)
$Sc_4(\mu_3\text{-}O)_2@C_{80}$	−2.35	−1.73	−1.10	0.00	0.79	1.10
$Sc_3N@C_{80}$	−2.37	−1.62	−1.26	0.59	1.09	1.85
$Sc_3NC@C_{80}$	—	−1.68	−1.05	0.60	—	1.65
$Dy_3N@C_{80}$	—	−1.86	−1.37	0.70	—	2.07
$Lu_3N@C_{80}$	−2.26	−1.80	−1.42	0.64	1.11	2.08
$TiSc_2N@C_{80}$	−2.21	−1.58	−0.94	0.16	—	1.10
$TiY_2N@C_{80}$	—	−1.79	−1.11	0.00	—	1.11
$CeLu_2N@C_{80}$	—	−1.92	−1.43	0.01	—	1.44
$La_2@C_{80}$	−2.13	−1.72	−0.31	0.56	0.95	0.87
$Sc_4C_2@C_{80}$	—	−1.65	−1.16	0.40	1.10	1.56

All potentials are in volts vs. Fc/Fc^+.

two times smaller than that of TNT (trimetallic nitride template)-based EMFs [6]. The redox potentials and the EC gap of $Sc_4(\mu_3\text{-}O)_2@I_h\text{-}C_{80}$ are collected in Table 7.1 and compared with those of other trimetallic nitride template (TNT)-based fullerenes as well as $La_2@C_{80}$ and $Sc_4C_2@C_{80}$.

It should be noted that along with $Sc_4(\mu_3\text{-}O)_2@I_h\text{-}C_{80}$, species such as $Sc_4O_3@C_{80}$ or $Sc_4O_2@C_{80}O$ were also found upon examination of the fullerene soot by MALDI-TOF mass spectrometry. However, based on the mixed-valence state of scandium in $Sc_4O_2@C_{80}$ and theoretical calculations, the more probable structure is $Sc_4(\mu_3\text{-}O)_3@C_{80}$ (Figure 7.2). In addition, the HOMO (highest occupied molecular orbital) in $Sc_4(\mu_3\text{-}O)_2@I_h\text{-}C_{80}$ is largely confined to the $Sc_4(\mu_3\text{-}O)_2$ cluster, while in $Sc_4(\mu_3\text{-}O)_3@I_h\text{-}C_{80}$ the HOMO is rather delocalized over the fullerene sphere [5]. The electron transfer from the encapsulated metal atoms to the carbon cage results in the positive charge of the former and therefore in strong Coulomb repulsion forces between those positively charge ions. However, those metal–metal repulsive interactions can be compensated by attractive interactions between the metal ions and the negatively charged non-metal atoms (i.e., oxygen) [8].

Although the yield of production of $Sc_4(\mu_3\text{-}O)_3@I_h\text{-}C_{80}$ was rather low, advances in the purification process especially by removing the abundantly produced empty cages from the soot, together with extensive recycling HPLC, allowed the isolation of sufficient amount of material. Crystallographic studies suggested the structure $Sc_4(\mu_3\text{-}O)_3@I_h\text{-}C_{80}$ in which the C_{3v} symmetrical $Sc_4(\mu_3\text{-}O)_3$ cluster encapsulated inside the $I_h\text{-}C_{80}$ is the largest known cluster trapped inside a fullerene cage [7,9].

Along with $Sc_4(\mu_3\text{-}O)_2@I_h\text{-}C_{80}$ and $Sc_4(\mu_3\text{-}O)_3@I_h\text{-}C_{80}$, another scandium oxide EMF was produced and identified as $Sc_2(\mu_2\text{-}O)@C_s\text{-}C_{82}$ [10].

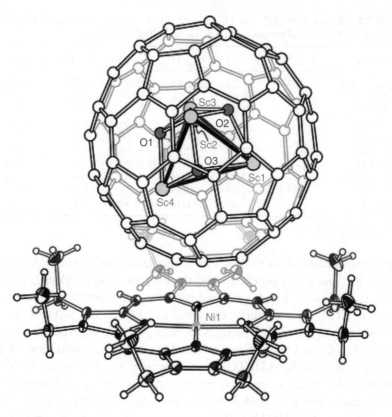

Figure 7.2   Structure of $Sc_4(\mu_3-O)_3@C_{80}$ together with the nickel porphyrin used for crystallization. Reproduced with permission from Ref. [7]. Copyright 2010 Royal Society of Chemistry

Although the HPLC retention time of $Sc_2(\mu_2-O)@C_s-C_{82}$ coincides with that of $Sc_3N@C_{78}$, thus hampering its isolation, aluminum(III) chloride treatment of the particular HPLC fraction as a novel and meaningful separation technique [11] overcomes this issue leading to the isolation of pure $Sc_2(\mu_2-O)@C_s-C_{82}$ as determined by mass spectrometry. X-ray crystallographic studies indicated the presence of a distorted Sc_2O cluster encapsulated near the center of a C_s-C_{82} cage, with the Sc ion positions arranged along the unique band of 10 contiguous hexagons present in the C_s-C_{82} cage (Figure 7.3). Density functional theory (DFT) calculations also suggested the stabilization of $Sc_2(\mu_2-O)@C_s-C_{82}$ by the charge transfer of four electrons from the encapsulated cluster to the carbon sphere. In Table 7.2 are listed the isolated oxide metallofullerenes and their molecular symmetry as revealed by crystallography.

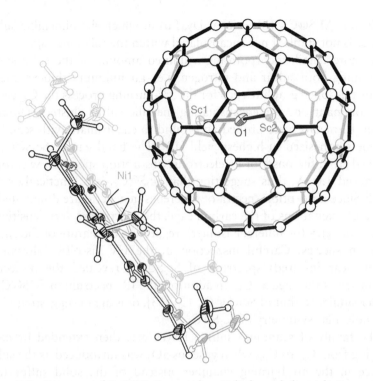

Figure 7.3 Structure of $Sc_2S@C_s$-C_{82} together with the nickel porphyrin used for crystallization. Reproduced with permission from Ref. [10]. Copyright 2010 American Chemical Society

Table 7.2 Oxide metallofullerenes isolated and crystallographically characterized

Oxide metallofullerene	Molecular symmetry	References
$Sc_4(\mu_3\text{-}O)_2@C_{80}$	I_h	[4–6]
$Sc_4(\mu_3\text{-}O)_3@C_{80}$	I_h	[5–7]
$Sc_2(\mu_2\text{-}O)@C_{82}$	C_s	[10]

7.2 SULFIDE METALLOFULLERENES

Initial attempts to produce sulfide EMFs using elemental sulfur combined with a metal were unsuccessful. On the other hand, TNT-based clusters encaged in fullerenes are abundantly produced in the presence of nitrogen, with the latter being responsible for producing the nitride anion [2]. With this in mind, recently, when guanidinium thiocyanate was introduced as the source of both nitrogen and sulfur in the metal-doped graphite powder, the first sulfide-based EMF was isolated. The structure of the new sulfide

EMF was $M_2S@C_{82}$ (M = Sc, Y, Dy, Lu) in which the bimetallic sulfide cluster is not stable if not encapsulated within the fullerene cage.

In more detail, incorporating a limited amount of the solid source containing both sulfur and nitrogen (i.e., guanidinium thiocyanate) in the metal-doped graphite rods for the arc-burning process the C_{82} cage is the dominant one for encapsulating metallic sulfur clusters. Different group III metals were probed and found to encapsulate with scandium sulfide synthesized in highest yield [12]. The band gap of $Sc_2S@C_{82}$ as derived from the onset of its electronic absorption spectrum was found to exceed 1.0 eV, thus suggesting that $Sc_2S@C_{82}$ is a kinetically stable EMF. Since absorption spectra of fullerenes in general are dominated by the $\pi-\pi^*$ excitations of the carbon cage, the spectra are very sensitive to cage symmetry (or fullerene isomer), regardless the nature of the encapsulation species. Careful inspection of the UV-Vis-NIR (ultraviolet-visible-near infrared) spectrum of $Sc_2S@C_{82}$ revealed the molecular symmetry of the cage as C_{3v}. In addition, the IR spectrum of $Sc_2S@C_{82}$ is rather similar to that of $Sc_2C_2@C_{3v}\text{-}C_{82}$ [13], thus again suggesting C_{3v} as the molecular symmetry of $Sc_2S@C_{82}$.

The family of scandium sulfide EMFs was then extended by cages ranging from C_{80} to C_{100} when gaseous SO_2 was introduced as the sulfur source in the arc-burning chamber, instead of the solid sulfur (and nitrogen) source described above. This process resulted in the isolation of a new isomer of $Sc_2S@C_{82}$ possessing C_s symmetry, in addition to the already known one of $Sc_2S@C_{3v}\text{-}C_{82}$ (Figure 7.4). In the new scandium sulfide EMF the Sc_2S cluster tends to transfer four electrons to the cage thus possessing an electronic structure of $(Sc^{3+})_2S^{2-}@(C_{82})^{4-}$. Furthermore,

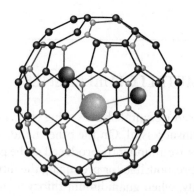

Figure 7.4   Structure of $Sc_2S@C_s\text{-}C_{82}$. Reproduced from Ref. [14], with permission of The Royal Society of Chemistry

Table 7.3 Redox potentials and EC gap for $Sc_2S@C_s-C_{82}$ and $Sc_2S@C_{3v}-C_{82}$

Materials	$E^4_{red.}$ (V)	$E^3_{red.}$ (V)	$E^2_{red.}$ (V)	$E^1_{red.}$ (V)	$E^1_{ox.}$ (V)	$E^2_{ox.}$ (V)	$E^3_{ox.}$ (V)	EC gap (V)
$Sc_2S@C_s-C_{82}$	—	−1.73	−1.12	−0.98	0.39	0.65	0.98	1.47
$Sc_2S@C_{3v}-C_{82}$	−2.49	−1.63	−1.19	−1.04	0.52	0.96	—	1.56

All potentials are in volts vs. Fc/Fc+.

contrary to the guanidinium thiocyanate method for the production of sulfide EMFs, the presence of SO_2 resulted in the production of $Sc_2S@$ C_{84}, $Sc_2S@C_{86}$, and other higher EMFs such as $Sc_2S@C_{2n}$ (n = 44–50) as identified by MALDI-TOF [14].

Electrochemistry studies revealed the redox properties of $Sc_2S@C_s-C_{82}$ and $Sc_2S@C_{3v}-C_{82}$. For the former, three reductions and three oxidations were identified. The cyclic voltammogram of the latter, however, was different, especially in terms of the fact that four reductions were found and that its second oxidation potential was dramatically shifted from 0.65 V for $Sc_2S@C_s-C_{82}$ to 0.96 V for $Sc_2S@C_{3v}-C_{82}$. All redox potentials for $Sc_2S@C_s-C_{82}$ and $Sc_2S@C_{3v}-C_{82}$ together with the EC gap are collected and presented in Table 7.3.

Furthermore, single-crystal X-ray studies on $Sc_2S@C_s-C_{82}$ and $Sc_2S@C_{3v}-C_{82}$ revealed that both contain fully ordered fullerene cages [15]. Interestingly, empty fullerenes were also produced in large quantities in the soot as identified by mass spectrometry, thus suggesting that the presence of SO_2 in the arc-burning process does not affect the abundant formation of the empty cages. The latter observation is different from the case when ammonia was used for the production of TNT-based EMFs in which case the production of empty fullerenes was highly suppressed [2].

Interestingly, C_{72} although possessing an isolated pentagon rule (IPR) structure, has never been isolated as an IPR isomer, probably because that particular IPR isomer with D_{6d} symmetry is the second most stable one after the non-IPR isomer with C_{2v} symmetry [16,17]. On the other hand, non-IPR structures of $La@C_2-C_{72}$ [18], $La_2@D_2-C_{72}$ [19], and $C_{2v}-C_{72}Cl_4$ [20,21] have been isolated and characterized. Therefore, the recent production of $Sc_2S@C_{72}$ and its isolation via extensive recycling HPLC was greeted with great excitement.

The UV-Vis-NIR spectrum of $Sc_2S@C_{72}$ showed characteristic absorptions in the visible and NIR regions, which however were substantially different from the absorptions of $La@C_2-C_{72}$, $La_2@D_2-C_{72}$, and $C_{2v}-C_{72}Cl_4$, thus suggesting that the symmetry of the cage in the $Sc_2S@C_{72}$ should be different. Indeed, crystallographic studies helped to identify the structure of $Sc_2S@C_{72}$ as non-IPR C_s (Figure 7.5). Further crystallographic analysis showed that the Sc_2S unit was tilted, while Sc atoms came closer to one end of the pentalene bond present where two pentagon rings of the fullerene cage abut.

Combination of computations with the crystallographic results revealed that the unique geometry of the scandium sulfide cluster together with the formal charge transfer of four electrons from the cluster to the cage play an important role in the stabilization of this new non-IPR C_s-C_{72} cage. Electrochemistry studies on $Sc_2S@C_s$-C_{72} revealed major differences in the redox properties, in terms of reversibility and anodic shifts, compared with those of other sulfide EMFs, thus highlighting the effect of cage structure and encaged cluster [22].

Another new bimetallic sulfide cluster fullerene $Sc_2S@C_{70}$ in which the Sc_2S moiety is encapsulated inside a non-IPR C_2-C_{70} cage (Figure 7.6) was very recently reported. Although the production yield of $Sc_2S@C_2$-C_{70} is even smaller than that of $Sc_2S@C_s$-C_{72} it was enough to allow a complete characterization [23]. Thus, after recycling HPLC a fraction corresponding to pure $Sc_2S@C_{70}$ was isolated as confirmed by MALDI-TOF and spectroscopically characterized. The electronic absorption spectrum of $Sc_2S@C_{70}$ is rather different not only from that of empty C_{70} but also from the non-IPR $Sc_3N@C_{2v}$-C_{70} [24], thus suggesting a different cage symmetry, which with the aid of theoretical calculations was predicted to be C_2. It is interesting to note that the C_2-C_{70} cage is structurally closely related to that of C_s-C_{72}, which is the one that hosts another Sc_2S

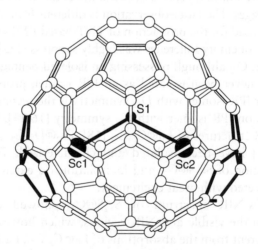

Figure 7.5 Structure of $Sc_2S@C_s$-C_{72}. Reproduced with permission from Ref. [22]. Copyright 2012 American Chemical Society

Figure 7.6 Structure of $Sc_2S@C_2$-C_{70}. Reproduced from Ref. [23], with permission of The Royal Society of Chemistry

Table 7.4 Redox potentials and EC gap for $Sc_2S@C_s$-C_{72} and $Sc_2S@C_2$-C_{70}

Materials	$E^4_{red.}$ (V)	$E^3_{red.}$ (V)	$E^2_{red.}$ (V)	$E^1_{red.}$ (V)	$E^1_{ox.}$ (V)	$E^2_{ox.}$ (V)	EC gap (V)
$Sc_2S@C_s$-C_{72}	—	−2.24	−1.53	−1.14	0.64	1.21	1.78
$Sc_2S@C_2$-C_{70}	−2.45	−1.99	−1.87	−1.44	0.14	0.65	1.57

All potentials are in volts vs. Fc/Fc⁺.

cluster. In fact, C_s-C_{72} can be obtained by simply adding a C_2 unit to a hexagon of C_2-C_{70}, thus suggesting that interconversion between $Sc_2S@C_s$-C_{72} and $Sc_2S@C_2$-C_{70} can occur by extrusion/addition of a C_2 unit without further atomic rearrangements.

The cyclic voltammogram of $Sc_2S@C_2$-C_{70} shows similarities and differences with that of other sulfide EMFs. Briefly, a reversible oxidation followed by an irreversible one is evident, a behavior which is very similar to the oxidation of $Sc_2S@C_s$-C_{72}, however, different from those of $Sc_2S@C_s$-C_{82} and $Sc_2S@C_{3v}$-C_{82}, which exhibit two reversible oxidative steps. On the other hand, while $Sc_2S@C_s$-C_{72} shows only reversible reductions, the reductive behavior of $Sc_2S@C_2$-C_{70} shows similarities with that of the two isomers of $Sc_2S@C_{82}$, which exhibit irreversible reductive processes. The redox potentials for $Sc_2S@C_s$-C_{72} and $Sc_2S@C_2$-C_{70} are collected and presented in Table 7.4. The EC gap of $Sc_2S@C_2$-C_{70} (1.57 V) is smaller compared with that of $Sc_2S@C_s$-C_{72} (1.78 V), but is slightly larger compared with that of $Sc_2S@C_s$-C_{82} (1.47 V) and $Sc_2S@C_{3v}$-C_{82} (1.56 V). Finally, likewise for other sulfide-based EMFs, there is a formal electron transfer of four electrons from Sc_2S to the C_2-C_{70} cage.

REFERENCES

[1] Rivera-Nazario D M, Pinzon J R, Stevenson S and Echegoyen L A 2013 Buckyball maracas: exploring the inside and outside properties of endohedral fullerenes *J. Phys. Org. Chem.* **26** 194

[2] Stevenson S, Rice G, Glass T *et al.* 1999 Small band-gap endohedral metallofullerenes in high yield and purity *Nature* **401** 55

[3] Iiduka Y, Wakahara T, Nakahodo T *et al.* 2005 Structural determination of metallofullerene Sc_3C_{82} revisited: a surprising finding *J. Am. Chem. Soc.* **127** 12500

[4] Stevenson S, Mackey M A, Stuart J P *et al.* 2008 A distorted tetrahedral metal oxide cluster inside an icosahedral carbon cage. Synthesis, isolation and structural characterization of $Sc_4(\mu_3\text{-}O)_2@I_h\text{-}C_{80}$ *J. Am. Chem. Soc.* **130** 11844

[5] Popov A A, Chen N, Pinzon J R *et al.* 2012 Redox-active scandium oxide cluster inside a fullerene cage: spectroscopic, voltammetric, electron spin resonance spectroelectrochemical, and extended density functional theory study of $Sc_4O_2@C_{80}$ and its ion radicals *J. Am. Chem. Soc.* **134** 19607

[6] Valencia R, Rodriguez-Fortea A, Stevenson S *et al.* 2009 Electronic structures of scandium oxide endohedral metallofullerenes, $Sc_4(\mu_3\text{-}O)n@I_h\text{-}C_{80}$ (n = 2, 3) *Inorg. Chem.* **48** 5957

[7] Popov A A and Dunsch L 2009 Bonding in endohedral metallofullerenes as studied by quantum theory of atoms in molecules *Chem. A Eur. J.* **15** 9707

[8] Popov A A, Avdoshenko S M, Pedas A M and Dunsch L 2010 Bonding between strongly repulsive metal oxides: an oxymoron made real in a confined space of endohedral metallofullerenes *Chem. Commun.* **46** 8031

[9] Mercado B Q, Olmstead M M, Beavers C M *et al.* 2010 A seven atom cluster in a carbon cage, the crystallographically determined structure of $Sc_4(\mu_3\text{-}O)_3@I_h\text{-}C_{80}$ *Chem. Commun.* **46** 279

[10] Mercado B Q, Stuart M A, Mackey M A *et al.* 2010 $Sc_2(\mu_2\text{-}O)$ trapped in a fullerene cage: the isolation and structural characterization of $Sc_2(\mu_2\text{-}O)@C_s(6)\text{-}C_{82}$ and the relevance of the thermal and entropic effects in fullerene isomer selection *J. Am. Chem. Soc.* **132** 12098

[11] Stevenson S, Mackey M A, Pickens J E *et al.* 2009 Selective complexation and reactivity of metallic nitride and oxometallic fullerenes with Lewis acids and use as an effective purification method *Inorg. Chem.* **48** 11685

[12] Dunsch L, Yang S, Zhang L *et al.* 2010 Metal sulfide in a C_{82} fullerene cage: a new form of endohedral clusterfullerene *J. Am. Chem. Soc.* **132** 5413

[13] Krause M, Kuzmany H, Dennis T J S *et al.* 1999 Diatomic metal encapsulates in fullerene cages: a Raman and infrared analysis of C_{84} and Sc_2C_{84} with D_{2d} symmetry *J. Chem. Phys.* **111** 7976

[14] Chen N, Chaur M N, Moore C *et al.* 2010 Synthesis of a new endohedral fullerene family, $Sc_2S@C_{2n}$ (n = 40–50) by the introduction of SO_2 *Chem. Commun.* **46** 4818

[15] Mercado B Q, Chen N, Rodriguez-Fortea A *et al.* 2011 The shape of the $Sc_2(\mu_2\text{-}S)$ unit trapped in C_{82}: crystallographic, computational, and electrochemical studies of the isomers $Sc_2(\mu_2\text{-}S)@C_s(6)\text{-}C_{82}$ and $Sc_2(\mu_2\text{-}S)@C_{3v}(8)\text{-}C_{82}$ *J. Am. Chem. Soc.* **133** 6752

[16] Kobayashi K, Nagase S, Yoshida M and Osawa E 1997 Endohedral metallofullerenes. Are the isolated pentagon rule and fullerene structures always satisfied? *J. Am. Chem. Soc.* **119** 12693

[17] Slanina Z, Ishimura K, Kobayashi K and Nagase S 2004 C_{72} isomers: the IPR-satisfying cage is disfavored by both energy and entropy *Chem. Phys. Lett.* **384** 114

[18] Wakahara T, Nikawa H, Kikuchi T *et al.* 2006 La@C_{72} having a non-IPR carbon cage *J. Am. Chem. Soc.* **128** 14228

[19] Lu X, Nikawa H, Nakahodo T *et al.* 2008 Chemical understanding of a non-IPR metallofullerene: stabilization of encaged metals on fused-pentagon bonds in La$_2$@C_{72} *J. Am. Chem. Soc.* **130** 9129

[20] Ziegler K, Mueller A, Amsharov K Y and Jansen M 2010 Disclosure of the elusive C_{2v}-C_{72} carbon cage *J. Am. Chem. Soc.* **132** 17099

[21] Tan Y Z, Zhou T, Bao J *et al.* 2010 $C_{72}Cl_4$: a pristine fullerene with favorable pentagon-adjacent structure *J. Am. Chem. Soc.* **132** 17102

[22] Chen N, Beavers C M, Mulet-Gas M *et al.* 2012 Sc$_2$S@C_s(10528)-C_{72}: a dimetallic sulfide endohedral fullerene with a non isolated pentagon rule cage *J. Am. Chem. Soc.* **134** 7851

[23] Chen N, Mulet-Gas M, Li Y Y *et al.* 2013 Sc$_2$S@C_2(7892)-C_{70}: a metallic sulfide cluster inside a non-IPR C_{70} cage *Chem. Sci.* **4** 180

[24] Yang S F, Popov A A and Dunsch L 2007 Violating the isolated pentagon rule (IPR): the endohedral non-IPR C_{70} cage of Sc$_3$N@C_{70} *Angew. Chem. Int. Ed.* **46** 1256

8

Non-metal Endohedral Fullerenes

The preparation of endohedral fullerenes incorporating non-metal atoms or molecules has mainly relied on different synthetic approaches from that based on the co-evaporation of carbon and metal atoms (i.e., arc-discharge method for endohedral metallofullerenes). Thus, ion implantation and glow discharge are the methods that apply equally for entrapping a N or P atom in a C_{60} or C_{70} cage. With respect to the production of noble gases inside fullerenes, initial studies dealt with collision experiments, though with meaningless yield. However, the breakthrough was achieved with the molecular surgery approach, which allowed the incorporation of other guests inside fullerenes. According to that method, encapsulation of a guest such as H_2, H_2O, CO, CH_4, and NH_3 is possible. The molecular surgery methodology is made up of steps, based on rational organic chemistry, for opening a hole in the fullerene sphere, incorporating the guest, and then closing the hole with retention of the guest inside the fullerene. All these novel, non-metal endohedral fullerenes are reviewed and analyzed below.

8.1 NITROGEN-CONTAINING N@C_{60}

The first paramagnetic species encapsulated into fullerenes was produced by nitrogen ion implantation in solid C_{60}. Note, although atomic nitrogen is an open shell and thus highly reactive, it is stabilized

Endohedral Metallofullerenes: Fullerenes with Metal Inside, First Edition.
Hisanori Shinohara and Nikos Tagmatarchis.
© 2015 John Wiley & Sons, Ltd. Published 2015 by John Wiley & Sons, Ltd.

when inside fullerenes without forming any covalent bond with the cage. The paramagnetic complex derived, $N@C_{60}$, found soluble in toluene and stable at ambient conditions, though reported later that degradation occurs with time when exposed to air, light, or heat. In any case, the electron paramagnetic resonance (EPR) spectrum of $N@C_{60}$ is characteristic in terms that a triplet line, due to the electron spin of 3/2 in the ground state of atomic nitrogen, is observed at room temperature [1]. Furthermore, the high I_h symmetry of C_{60} remains unchanged when nitrogen is encapsulated as concluded by EPR studies due to the absence of spin relaxation processes. The latter together with the absence of any charge-transfer interactions between N and C_{60}, suggested that nitrogen resides at the center of the cage in $N@C_{60}$ [2].

Although the N atom is highly reactive, it remains inert when inside fullerene, thus, without forming any covalent bond with the fullerene cage in $N@C_{60}$, while at the same time being isolated from the external environment. The shielding of encapsulated nitrogen from the outside environment is also demonstrated by the extremely sharp EPR lines and the long electron-spin lifetimes, with a phase coherence time $T_2 = 240$ μs at 170 K and a spin-lattice relaxation time T_1 of up to 4.5 min. Thus, it is understood that C_{60} plays the role of a trap to protect and stabilize an extremely reactive species such as atomic nitrogen [3–5]. Fairly sharp EPR lines are observed for $N@C_{60}$ even in the solid state, thus, confirming the spherical symmetry of the system, while the fine-structure and quadrupole interactions are zero and the g-factor and the hyperfine interaction are isotropic [6].

Furthermore, advanced EPR measurements on $N@C_{60}$ revealed the ability of the quartet spin system to sense small local fields at the site of the atom. Such EPR studies showed the non-vanishing zero-field splitting (ZFS) in the low temperature phase of crystalline C_{60} sensed by the quartet electron-spin state of nitrogen [7,8]. However, when $N@C_{60}$ is heated at temperatures as low as 260 °C, its characteristic EPR signal starts to decrease, thus suggesting instability toward heating [9]. This result was also supported by theoretical calculations, which predicted that the escape of nitrogen proceeds via bond formation with the cage. According to the calculations, the following mechanism for the liberation of nitrogen is proposed: initially, nitrogen moves off-center, forming a covalent bond with two carbons of the cage and then swings through the bond to the external environment of the cage. Overall, this is a favorable process since theoretical calculations showed that energy for that escape path of nitrogen is lower than that involving one through a six- or five-membered ring of C_{60} [3].

An alternative method for producing $N@C_{60}$ is by using radio-frequency (rf) plasma. Briefly, according to the experimental procedure applied, operation of the rf plasma reactor at 13.56 MHz and 50 W for 10 minutes is enough to produce $N@C_{60}$, which is initially separated from insoluble matter by CS_2 extraction and column chromatography [10,11].

Aligning $N@C_{60}$ is closely related to the principles of applying endohedral fullerenes as quantum bit carriers in the construction of a quantum computer. This is due to long spin lifetimes of $N@C_{60}$, which makes it a useful embodiment of a qubit for quantum computing. At this point it should be mentioned that the high sensitivity of EPR and the thermal electron spin polarization result in an improved signal-to-noise ratio, which is a prerequisite for efficient quantum computation [12]. In this context, the magnetic dipolar interaction may serve as mediator between the encapsulated paramagnetic species thus providing the necessary entanglement of the paramagnetic states. This is achieved upon embedding $N@C_{60}$ in a liquid matrix, where all spherical fullerene moieties can be aligned thus allowing control of the interaction angle. Regardless of the absence of any particular axis in $N@C_{60}$ for alignment in a liquid crystal, a splitting of the EPR lines is observed thus suggesting certain alignment of $N@C_{60}$ when dissolved at room temperature in the liquid crystal 4-methoxybenzylidene-4'-n-butylaniline.

This EPR splitting is justified in terms of a minute deviation from the spherical shape of the spin-density distribution of the encapsulated nitrogen in $N@C_{60}$. A slight deformation of the electron shell of C_{60} is induced upon interaction with the liquid crystal, thus affecting the electron-density distribution of the encapsulated N atom. On the other hand, when it comes to the non-spherical but rather ellipsoid in shape $N@C_{70}$, the alignment is a natural consequence [13,14].

In order to perform a thorough study of $N@C_{60}$ and determine its properties in full, the acquisition of purified material is of great importance. This will also allow more reactions to be performed, and thus to synthesize new hybrid materials carrying an active spin and also enhance its immediate applications. Initial attempts to enrich the sample of $N@C_{60}$ to a high concentration were carried out by sequential and/or recycling high-performance liquid chromatography (HPLC) separations [15]. Interestingly, the latter process allowed the separation of dinitrogen encapsulated species such as $N_2@C_{60}$ (which is EPR-inactive), however at very low yield [16]. Eventually, it became possible to separate a purified sample of $N@C_{60}$ and $N@C_{70}$ based on a two-stage HPLC methodology.

According to that HPLC procedure, 10 mL of 200 times enhanced material with 2% purity with respect to C_{60} was obtained in the first step

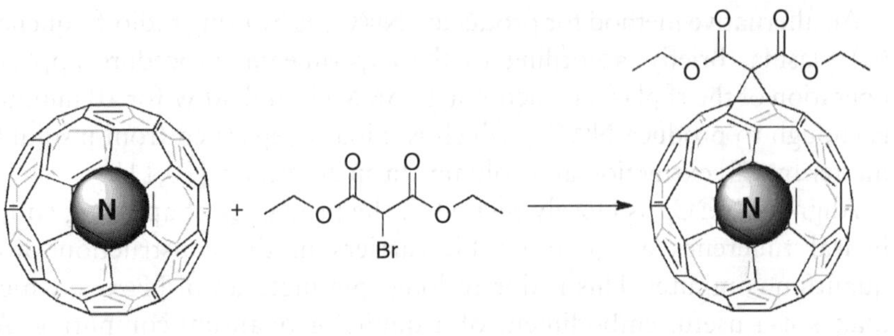

Figure 8.1 Bingel cyclopropanation reaction of N@C60–C60 with diethyl bromomalonate

which was further subjected to recycling HPLC in the second step to eventually isolate pure N@C$_{60}$ and N@C$_{70}$. Having purified N@C$_{60}$ and N@C$_{70}$ their corresponding electronic absorption spectra were recorded. Although the absorption spectrum of N@C$_{70}$ is identical to that of intact C$_{70}$, in the absorption spectrum of N@C$_{60}$, the characteristic absorption of intact C$_{60}$ in the visible region around 440–640 nm is missing. The latter is responsible for N@C$_{60}$ appearing yellowish-brown in solution [17].

The very clear hyperfine splitting in the EPR of N@C$_{60}$ makes it an ideal probe for monitoring reactions of C$_{60}$ via observing changes in the EPR spectrum. The first chemical reaction of N@C$_{60}$ performed was the Bingel cyclopropanation reaction (Figure 8.1). The reaction of N@C$_{60}$ with diethyl bromomalonate and a base proceeds similarly to that of intact C$_{60}$. The EPR spectrum of the functionalized N@C$_{60}$ material remained unchanged, showing the same EPR triplet as intact N@C$_{60}$, thus suggesting that the encapsulated nitrogen does not interfere with the chemical reaction occurring on the outer of the fullerene cage [18]. Similar results were obtained for Bingel multiple functionalized adducts of N@C$_{60}$, that is, without the participation of the encapsulated nitrogen in the chemical reaction of the fullerene [19].

On another functionalization reaction, the dimer N@C$_{60}$–C$_{60}$ was formed (Figure 8.2), purified by HPLC and characterized by UV-Vis (ultraviolet-visible) and IR spectroscopy. Examination of the EPR spectrum of N@C$_{60}$–C$_{60}$ showed that the nitrogen retains its atomic configuration as no changes is observed in the *g*-factor and the hyperfine coupling constant [20].

In another dimerization reaction, a photo-switchable dimer of N@C$_{60}$ incorporating the azobenzene moiety as linker, but with the encapsulated nitrogen located in only one of the two fullerene spheres, was prepared. The synthesis was based on the 1,3-dipolar cycloaddition reaction of azomethine

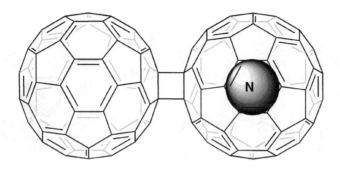

Figure 8.2 Chemical structure of N@C_{60}—C_{60} dimer

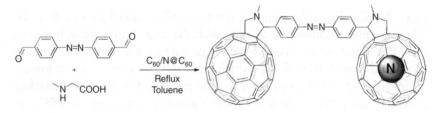

Figure 8.3 1,3-Dipolar cycloaddition reaction of azomethine ylides, in-situ generated by the thermal reaction of N-methylglycine and 4,4′-azobenzaldehyde, onto a mixture of C_{60}/N@C_{60}, yielding N@C_{60}–azo–C_{60} dimer

ylides, in-situ generated by the thermal reaction of N-methylglycine and 4,4′-azobenzaldehyde (Figure 8.3). Photoisomerization of the N@ C_{60}–azo–C_{60} can be induced by UV and Vis light irradiation. Based on EPR measurements, the presence of ZFS in N@C_{60}–azo–C_{60} was found [21].

Recently, another nitrogen-containing fullerene dimer was synthesized via the cyclopropanation route (Figure 8.4). The new material was purified by recycling HPLC and its EPR spectra were obtained and compared with that of intact N@C_{60}. It was found that the hyperfine constant and g-factor remained unchanged in the dimer thus indicating the presence of encapsulated nitrogen in the center of the fullerene cage. However, half of the spin signal was lost [22].

The photochemical functionalization of N@C_{60} with disilirane was also successfully performed. Photoirradiation of a 1:1.6 mixture of N@C_{60} and C_{60} gave after HPLC purification the bis-silylated adduct which was fully characterized by NMR, UV-Vis, and LD-TOF (laser-desorption time-of-flight) mass spectrometry. The hyperfine coupling constant and g-factor of the mono-adduct were found not to differ from those values recorded for

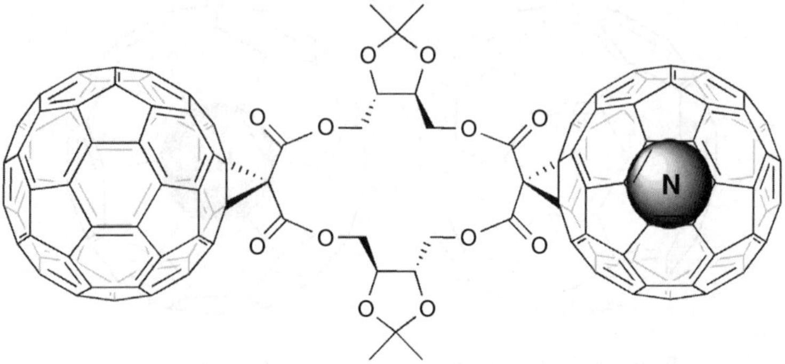

Figure 8.4　N@C_{60}-containing dimer

intact N@C_{60} thus indicating that nitrogen still occupied the center of the fullerene sphere without interacting with the cage. Importantly, however, the EPR signal of the bis-silylated mono-adduct was lower in intensity compared with that of N@C_{60} which is justified in terms of lower photochemical reactivity than the empty fullerene C_{60}. The latter is a marked difference since N@C_{60} shows similar thermal reactivity with N@C_{60} as described above for the Bingel cyclopropanation reactions.

Since the photochemical bis-silylation to C_{60} proceeds via the triplet excited state of C_{60}, the lower photochemical reactivity of N@C_{60} could be rationalized by considering a shorter lifetime of N@$^3C_{60}^*$ compared with that of $^3C_{60}^*$, which is also suggested by theoretical calculations [23]. In order to experimentally prove the role of N@$^3C_{60}^*$ in the lower photoreactivity of N@C_{60} with disilirane, laser flash photolysis studies of N@C_{60} were performed and the conclusion was that the non-radiative decay from an excited metastable species at the ground state of N@C_{60} is enhanced by the encapsulated N atom [24].

Pyrrolidine adducts of N@C_{60} were synthesized by applying the 1,3-dipolar cycloaddition reaction of in-situ generated azomethine ylides. Thus, as expected, the pyrrolidine modified N@C_{60} showed similar EPR characteristics, namely hyperfine coupling constant of 5.68 G and g-factor of 2.003, to those of unmodified N@C_{60} [25]. However, when a pyrrolidine adduct of N@C_{60} was independently synthesized, contrary not only to the previous studies of pyrrolidine derivatives [25], but also to studies based on cyclopropanated analogs of N@C_{60} [19], it showed an anisotropy in the ^{14}N hyperfine interaction. The presence of ZFS and hyperfine anisotropy was further confirmed by the observation of biexponential decays in both T_1 and T_2 measurements with pulsed EPR [26].

The photochemical stability of pyrrolidine adducts of $N@C_{60}$ were studied and found to be reasonably stable in solution, in the absence of oxygen, while stored in the dark. However, they start to degrade by losing nitrogen, as observed by the disappearance of the EPR signal, when exposed to ambient light. In this context, a mechanism was proposed for the spin loss in $N@C_{60}$, according to which the encapsulated N atom reacts with the sp^3 C atoms of the pyrrolidine functionalized fullerene, with the simultaneous breakage of the corresponding carbon–carbon bond, and then swings out of the cage [27]. Moreover, this mechanism may well explain the lower photoreactivity observed in the reaction of $N@C_{60}$ with disilirane [23]. In other words, it may be that the derivatized bis-silylated $N@C_{60}$ lost the nitrogen spin activity due to exposure to light in the course of the photoreaction. However, it should be noted that cyclopropanated adducts of $N@C_{60}$ are in general more stable than the pyrrolidine ones, since the nitrogen loss as described in the above-mentioned mechanism, is unfavored by the intermediate narrow four-membered ring formed and/or further blocked by the presence of the sterically demanding organic addends.

The epoxide $N@C_{60}O$ was also prepared and studied by EPR spectroscopy. It was found that at room temperature, the hyperfine constant and g-factor of $N@C_{60}O$ were the same as those corresponding to intact $N@C_{60}$, thus suggesting that the incorporation of the epoxide unit in the fullerene skeleton does not significantly interact with the encapsulated N atom. However, at 77 K, ZFS was observed and satellites in the triplet EPR line evolved [28].

Recently, dyads of endohedral nitrogen-containing fullerenes and porphyrins have been synthesized. The covalent linkage of the porphyrins to $N@C_{60}$ was carried out by 1,3-dipolar cycloaddition reaction, thus introducing pyrrolidine addend onto the fullerene cage (Figure 8.5). When a free-base porphyrin was used, the EPR spectrum of the dyad was similar to that of intact $N@C_{60}$, while ZFS features were observed at low temperature. On the contrary, when copper metallation of the porphyrin occurred, the EPR spectrum of the corresponding dyad was similar to that of copper porphyrin, while the characteristic triplet due to $N@C_{60}$ was absent, even at low temperature. Since demetallation resulted on recovering the EPR signal of $N@C_{60}$, it was suggested due to $N@C_{60}$ binding to copper porphyrin in the dyad, strong spin–spin interactions between the unpaired electrons of the nitrogen and the copper occur, thus resulting in quenching the EPR signal of $N@C_{60}$ [29].

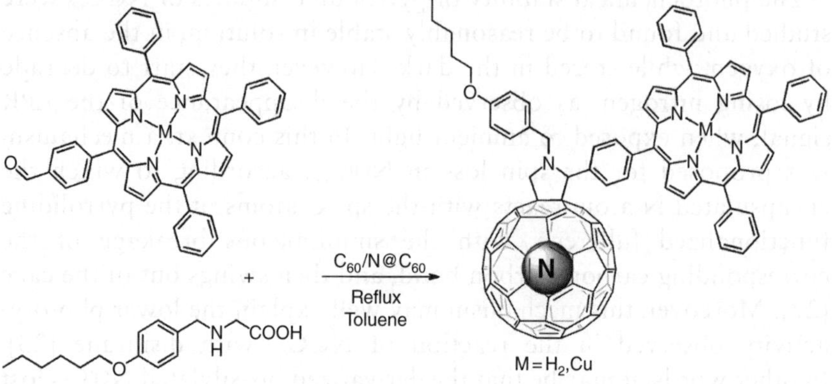

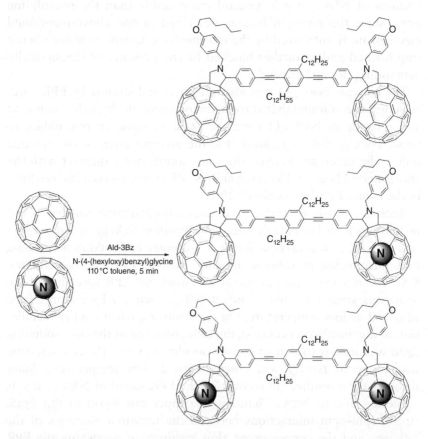

Figure 8.5 1,3-Dipolar cycloaddition reaction of azomethine ylides onto a N@C$_{60}$ yielding N@C$_{60}$-porphyrin dyad

Figure 8.6 1,3-Dipolar cycloaddition reaction of azomethine ylides in-situ generated by the thermal reaction of *N*-(4-(hexyloxy)benzyl)glycine and dibenzaldehyde-terminated oligophenylene-ethylene, onto N@C$_{60}$

The first two-spin-center dimer $N@C_{60}$–$N@C_{60}$ was recently made available by employing an oliphenylene-ethylene as a linking unit. After extensive HPLC purification, an estimated just below 5 µg of $N@C_{60}$ sample was allowed to thermally react with N-(4-(hexyloxy)benzyl)glycine and dibenzaldehyde-terminated oligophenylene-ethylene, affording the two-spin-center dimer $N@C_{60}$–$N@C_{60}$ as a mixture with the corresponding all-empty fullerene dimer and the half-filled dimer, according to Figure 8.6. Upon EPR examination, the derivatized $N@C_{60}$–$N@C_{60}$ showed ZFS and line broadening [30].

8.2 PHOSPHOROUS-CONTAINING $P@C_{60}$

The non-metal endohedral fullerene encapsulating phosphorous $P@C_{60}$ is produced by ion implantation. According to this method, C_{60} is continuously evaporated onto a substrate and at the same time bombarded with low-energy phosphorous ions. The material that is obtained, which is soluble in toluene, contains both intact C_{60} and $P@C_{60}$ with the ratio of empty-to-filled being 10^{-4}. The concentration of $P@C_{60}$ can be increased by applying recycling HPLC and up to now this is the method of choice to produce analytical samples of $P@C_{60}$. The presence of atomic phosphorous inside the C_{60} cage can be viewed by electron spin resonance (ESR). Although atomic phosphorous is extremely reactive, it is almost freely suspended, located at the center of the cage. Based on ESR studies, it is found that phosphorous in $P@C_{60}$ keeps its ground state configuration possessing a quartet spin. Importantly, the hyperfine interaction of phosphorous, as derived from the ESR line splitting, is found larger for $P@C_{60}$ than that of the free P atom [31]. The latter is indicative of a contraction in the electron cloud of the P atom, while the electronic structure of C_{60} remains virtually unchanged [9,32].

Analysis of the spin relaxation data for $P@C_{60}$ revealed that fluctuating ZFS interaction induced by collision-induced deformations of the carbon shell constitutes the dominant relaxation process. Due to the atom-like state of phosphorous when encapsulated inside C_{60}, the fullerene features a nearly isotropic high spin state. Thus, 2,4,6,-tris-(4-bromophenoxy)-1,3,5-triazine was used to create an ordered alignment of $P@C_{60}$ in a crystalline host matrix and it was found that an axial deformation of the fullerene guests leads to a sizeable ZFS of the quartet ground state of the encapsulated phosphorous. The latter opens the way for future quantum computing experiments by utilizing transition selective pulses in this multi-level system [33,34].

8.3   INERT GAS ENDOHEDRAL FULLERENES He@C_{60}, Ne@C_{60}, Ar@C_{60}, Kr@C_{60}, AND Xe@C_{60}

Initial studies for the incorporation of noble gases inside fullerene cages were dealing with collision experiments of fullerene cations in a mass spectrometer. According to those experiments, insertion of the He atom inside $C_{60}^{\bullet+}$ and $C_{70}^{\bullet+}$, in the course of the collision, became possible [35]. Although the availability of those materials was negligible, as they were only detected during mass spectrometry, theoretical calculations were used to estimate the energy of fullerenes containing noble gas atoms and the dynamics of motion inside the cage [36–38]. Later on, it was shown that He@C_{60} and Ne@C_{60} could be prepared during the arc-burning process generating fullerenes in the usual manner, however, under high temperature and a high-pressure noble gas atmosphere for long periods. For example, the incorporation of He inside C_{60} was almost 1 ppm, while heating of fullerenes in the presence of Ne also lead to its inclusion in similar yield [39]. In this process, the encapsulation of He, Ne, Ar, Kr, and Xe under pressure was proposed to occur by the reversible breaking of one or more bonds of the fullerene cage, thus opening an orifice large enough to allow the entrance of the particular atom.

As the size of the noble gases increases from He to Xe, it was no surprise that the Xe@C_{60} and Xe@C_{70} was produced at lower yields, if any at all, compared with the production of He, Ne, Ar, and Kr encapsulated inside C_{60} [40,41]. However, according to some extensive theoretical calculations, the incorporation of He through an orifice was re-examined and found to be less likely, since for the latter incorporation a barrier of more than 200 kcal/mol relative to non-interacting C_{60} and He in the ground state must be overcome. Thus, the formation of intermediates species in which He is inserted into a metastable fullerene isomer was theoretically suggested, however, experimental findings to justify this suggestion are not available [42].

Later, two other methods were developed for the production of noble gas atoms inside fullerenes. According to the first one, impinging beams of He$^+$ or Ne$^+$ generated in a vacuum chamber onto solid C_{60} surfaces results in incorporation of the noble gas inside the fullerene cage, with the ion's probability of penetrating the cage being a function of the ion energy [43]. On the other hand, the most recently developed method for producing noble gas encapsulated within fullerenes deals with an explosion technique. The latter method was chosen because explosions are high-energy processes releasing energy in a very short period of time.

Briefly, fullerenes together with an explosive reagent and a noble gas are enclosed in a high-strength steel reaction vessel. Importantly, a "movable plate" is placed within the apparatus to convert the explosion energy into kinetic energy of the gas molecules, which then bombard fullerenes. The amount of explosives can be adjusted so that the noble gas can attain sufficient kinetic energy to penetrate the fullerenes and be trapped in their empty space. Thus, He@C_{60} and He$_2$@C_{60} have been produced, enriched, and purified via extensive HPLC, while other noble gases can similarly be encapsulated [44].

Although the above described methods for producing noble gases inside fullerenes brought forth significant insight into the properties of those materials, they are limited in scope and still produce small amounts of encapsulated noble gas. Therefore, a chemical approach based on so-called "molecular surgery" attracted the interest of scientists in order to overcome those handicaps. According to this strategy, an orifice is created on the fullerene cage, through which the noble gas can enter as long as it is the right size compared with the opened cage.

The very first example of such an approach, in which a He atom was inserted through an open window into a fullerene, concerns bislactam fullerene derivative **1** (Figure 8.7). The opening in **1** is large enough to allow the incorporation of He at temperatures around 100–120 °C [45]. Incorporation of He was achieved even under low pressure (i.e., 3–4 atm, 100 °C, 8 h), though by applying higher He pressure it was possible

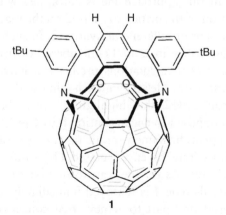

1

Figure 8.7 Structure of bislactam fullerene derivative **1** possessing an open window to allow the incorporation of the He atom

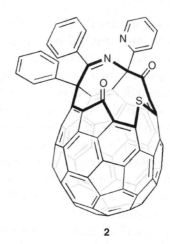

2

Figure 8.8 Structure of bislactam fullerene derivative **2** possessing a large open window. Since the orifice is large enough, injection, and ejection of He is possible

to increase the encapsulation yield. Importantly, He can be ejected from the cage by heating at 150 °C.

An even larger opening on C_{60} was made possible in material **2** (Figure 8.8) [46,47], which allowed not only the insertion of ^{3}He but also its ejection too. Importantly the encapsulation of He in **2** can be performed at nearly room temperature, reaching an incorporation fraction of 0.1% [48].

In order to minimize the rate of He ejection and thus to isolate it in a higher encapsulation ratio, a fullerene derivative with smaller opening was considered. In this approach He is entrapped within the cage by immediate reduction of the orifice size based on the reactivity of the two carbonyl units present in fullerene derivative **2**. In such a way, from the 13-membered ring orifice in **2** an 11-membered orifice is formed in **3** (Figure 8.9). Theoretical calculations predicted a barrier of 50.4 kcal/mol for He release from He@**3**, which is almost double that for the ejection of He from He@**2**, thus highlighting the stability of He when inserted into **3** as opposed to encapsulation into **2** [49].

In a similar approach, namely reducing the size of the orifice for reducing the release rate of encapsulated He, sulfoxide derivative **4** (Figure 8.10) was used. As for the transformation from **2** to **3**, the SO unit in open-cage fullerene **4** is removed immediately after insertion of He. A brief experimental part to achieve that concerns treatment of **4** under pressurized He gas (115 °C, 1230 atm, 60 min) to afford endohedral fullerene **5** (Figure 8.10) possessing a 12-membered ring orifice [50].

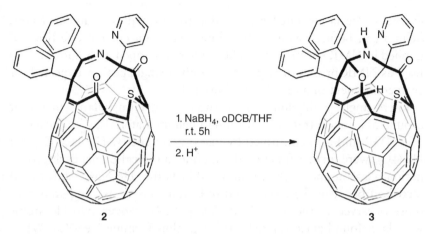

Figure 8.9 Open-caged fullerene derivatives – a 13-membered ring orifice exists in 2, which gives rise to an 11-membered orifice in 3

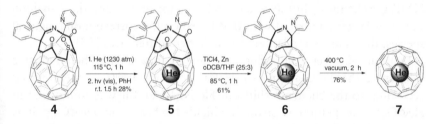

Figure 8.10 Chemical transformation of open-cage fullerene derivative 4–6 possessing a smaller orifice, thus preventing the ejection of the encapsulated He atom, before eventually closing the cage forming He@C_{60} 7

Further chemical transformations toward the closure of the opening in He@5 afforded initially the eight-membered ring He@6 and then He@ C_{60} material 7 (Figure 8.10), as verified by ^{3}He NMR spectroscopy. Importantly, during those closure reactions, no escape of He occurs as confirmed by mass spectrometry. The same methodology when applied to C_{70} results in the formation of He@C_{70}.

Importantly, since ^{3}He has a spin of ½ and is an excellent NMR nucleus, it can be used to probe the magnetic shielding environment inside fullerene cages. In this way, the chemical shift of ^{3}He@C_{60} was found at around –6 ppm, shielded (upfield shift by 6.0 ppm) as compared with free ^{3}He. Similarly, for ^{3}He@C_{70} the chemical shift of ^{3}He was found at –29 ppm, more than four times greater as compared with that based on C_{60}. Those significantly high chemical shifts of ^{3}He call for substantial aromatic ring currents inside C_{60} and C_{70}, though to a lesser extent in the latter. Furthermore, due to its small size and inertness, helium can be used as a probe of magnetic molecular properties and applied in NMR spectroscopy. Thus, fullerenes doped with ^{3}He could be used as NMR probes of reactivity as only a single peak is expected for each component of a reaction mixture [51].

Interestingly, a dihelium species was found encapsulated in C_{70} at around 5% concentration of the mono-helium materials, with the aid of ^{3}He NMR and also confirmed by mass spectrometry. The He@C_{70} and He$_2$@C_{70} were further functionalized under a typical Bingel cyclopropanation reaction and in the resulting adducts the ^{3}He NMR shift found increased between the mono- and di-helium species as compared with those observed in the intact materials [52]. Moreover, two He atoms were later found encapsulated inside C_{60}, thus forming He$_2$@C_{60}. When mixtures of He@C_{60} and He$_2$@C_{60}, and of He@C_{70} and He$_2$@C_{70} were reduced forming the corresponding hexa-anion species of He@C_{60}^{6-} and He$_2$@C_{60}^{6-}, and of He@C_{70}^{6-} and He$_2$@C_{70}^{6-}, respectively, in the ^{3}He NMR the chemical shift difference between the mono- and the di-helium signals is larger compared with that based on the corresponding neutral species, thus suggesting larger magnetic field differences [53].

In addition, when comparing the effect of encapsulated helium inside fullerenes, the hexa-anions of He@C_{60}^{6-} and He@C_{70}^{6-} were tested according to the chemical shift on ^{3}He NMR and it was found that in He@C_{60}^{6-} the helium is strongly shielded whereas in He@C_{70}^{6-} it is strongly deshielded, relative to ^{3}He gas, respectively. Thus, the chemical shift of ^{3}He@C_{60}^{6-} appears at -48.7 ppm, indicating that the hexa-anion is more aromatic than the parent fullerene (the latter is also understood in terms of the substantially increased electron delocalization) [54].

Beyond the previously described effect of fullerene anions on the ^{3}He chemical shift, the preparation of the fullerene cation $C_{60}Ar^+$ [55] allowed evaluation of the effect of the cation on the helium chemical shift. The latter was found to be downfield shifted, which is expected due to the fact that the fullerene cation possesses less aromatic character than its precursor species, which is in contrast with the upfield shift observed for the case of hexa-anion C_{60}^{6-} [56].

Aside from ^{3}He, the only other noble gas isotope with the NMR favorable spin of ½ is ^{129}Xe. As Xe is highly polarizable, it strongly interacts with other atoms and molecules, thus its chemical shift is strongly affected. In addition, the lower gyromagnetic ratio of ^{129}Xe makes it less sensitive compared with ^{3}He. Moreover, as pointed out earlier, Xe is the largest in size noble gas to insert in C_{60} and therefore a modified method was used to achieve that. Briefly, a finely powdered mixture of C_{60} was initially ground with potassium cyanide before applying the high-temperature and high-pressure process [57]. In the ^{13}C NMR spectrum of Xe@C_{60}, a single peak shifted downfield by almost 1 ppm compared with that of intact C_{60} is present, thus revealing weak electronic interactions between Xe and the fullerene cage. On the other hand, in ^{129}Xe NMR the chemical shift of xenon in Xe@C_{60} is at 179 ppm, however, based on theoretical calculations, ^{129}Xe is predicted to be deshielded in Xe@C_{60} by around 70 ppm [58]. This discrepancy is overcome by considering that in the earlier theoretical calculations, any electronic interactions between Xe and C_{60} have not been taken into account.

As mentioned earlier, the production yield for noble gas atoms inside fullerene cages is rather small. Considering that those species are also produced together with empty cages, it is essential and desirable to separate the doped fullerenes from the intact ones, as more concentrated samples of the labeled samples will be available which in turn will facilitate detailed investigation of their properties. To this end, initial purification attempts based on chromatography showed that partial separation of Kr@C_{60} from empty C_{60} is possible. However, such a separation and enrichment of Kr@C_{60} is highly affected by the stationary phase and the solvent used for the elution [59]. Nevertheless, the enriched Kr@C_{60} crystallized and its X-ray crystal structure revealed that the Kr atom is centered inside the fullerene cage without inferring changes in the size of C_{60} [60].

Moreover, enrichment of Ar@C_{60} synthesized by heating primarily C_{60} under a high pressure of Ar gas, was achieved by HPLC on a 5PYE column and with toluene as eluent. The enriched Ar@C_{60} material was further characterized by X-ray photoelectron spectroscopy and found to be sensitive to radiation damage and subsequent depletion [61]. Similarly, an enriched sample of Kr@C_{60} was prepared with the aid of multi-stage recycling HPLC and spectroscopically studied. It was found that in its ^{13}C NMR spectrum there is a single peak slightly shifted downfield as compared with that of intact C_{60}, thus suggesting not only the conservation of the icosahedral molecular symmetry of the cage but also the presence of weak electronic interactions between Kr and C_{60}. Those findings are further corroborated upon examination of the UV-Vis-NIR (ultraviolet-visible-near infrared) and IR spectra of Kr@C_{60} [62].

Functionalization reactions were performed on He@C_{60} without affecting or releasing the encaged noble gas atom. Thus, the 1,3-dipolar cycloaddition reaction of in-situ generated azomethine ylides took place on He@C_{60} and the reaction was easily followed by ^{3}He NMR spectroscopy [63]. Moreover, the addition of a 1,3-biradical generated by thermolysis of cyclopropa[b]naphthalene to ^{3}He@C_{60} was performed. When the ^{3}He NMR

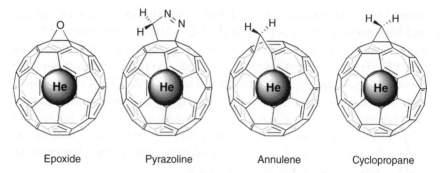

Epoxide   Pyrazoline   Annulene   Cyclopropane

Figure 8.11 Structures of He@C_{60} epoxide, pyrazoline, annulene, and cyclopropane derivatives

spectrum of the adduct was measured, an upfield shift of 3 ppm was found, thus suggesting that even though a double bond was removed from the fullerene cage, the adduct possesses a larger net ring current compared with that of the parent C_{60} [64]. Also, other derivatives of He@C_{60} and He@C_{70} were synthesized, namely the epoxide He@C_{60}O and the pyrazoline, annulene, and cyclopropane adducts of He@C_{60} (Figure 8.11) as well as the four isomers of cyclopropanated He@C_{70}. In all of the above-mentioned He-doped fullerene derivatives the ^{3}He NMR chemical shifts are nicely correlated with the aromatic character of the fullerenes, since the magnetic field inside the fullerene cage is affected by ring currents of the π-electrons which are altered themselves in the corresponding adducts [65].

Importantly, ^{3}He NMR spectroscopy has been applied to examine bis-addition to fullerenes containing an encapsulated He atom. Since functionalization of fullerenes results in changes of the magnetic field felt by the encapsulated ^{3}He, thus producing substantial changes in the corresponding ^{3}He chemical shifts, relative to those in intact fullerenes, ^{3}He NMR is a meaningful tool in studying functionalized fullerenes. In this way, different functionalization reactions were studied, namely (i) Bingel cyclopropanation reaction with diethyl malonate, (ii) 1,3-dipolar cycloaddition of in-situ generated azomethine ylides, and (iii) reduction by hydroboration.

As a general conclusion, bis-adducts with appended groups on opposite hemispheres of the fullerene cage tend to show ^{3}He NMR peaks downfield compared with bis-adducts with appended groups on the same hemisphere of the cage. In addition, there is large difference in the chemical shifts of the bis-adduct isomers, thus suggesting that the magnetic field felt by the ^{3}He atom due to the ring currents in the residual

π-electron system is extremely sensitive to the pattern of the added group on the fullerene skeleton [66]. Furthermore, up to hexakis adducts of the Bingel functionalized $^3He@C_{60}$ have been prepared and studied by ^3He NMR to determine the influence of the degree of functionalization and addition pattern on the ^3He chemical shift [67]. Later, the Diels-Alder cycloaddition reaction of dimethyl anthracene (DMA) to $^3He@C_{60}$ and $^3He@C_{70}$ was also studied, and when an excess of DMA was considered it tended to yield higher adducts. Although upon addition of DMA to C_{60} a single mono-adduct is formed, there are eight different regioisomers when bis-addition is considered, while a more complex mixture of tris- and other higher adducts can also be produced.

Furthermore, the situation with cycloaddition of DMA to C_{70} is more complicated, considering the lower symmetry of the cage compared with that of icosahedral C_{60}. Thus, applying that reaction to $He@C_{60}$ and then monitoring the ^3He NMR spectrum of the reaction mixture, one can understand the reaction pattern and estimate the number of adducts formed. This is easily achieved by simply assigning ^3He NMR peaks to the number of DMA addends. According to the latter, if the ratio of NMR intensities of two peaks is independent of DMA concentration, then the two peaks correspond to adducts with the same number of DMA molecules. On the other hand, if the ratio changes with DMA concentration then the peak that increases more rapidly corresponds to an adduct with more DMA molecules. Based on that, the ^3He NMR examination of the Diels-Alder cycloaddition of DMA to $He@C_{60}$ resulted in the formation of a mono-adduct, 6 bis-adducts, 11 tris-adducts, and 10 tetrakis-adducts, while other higher adducts were absent, even when higher concentration of DMA was used. Similarly, when examining the cycloaddition of DMA to $He@C_{70}$ a mono-adduct and three bis-adducts were identified with ^3He NMR spectroscopy [68].

Beyond C_{60} and C_{70}, higher fullerenes can be similarly doped and encapsulate He, thus forming materials such as $He@C_{76}$, $He@C_{78}$, and $He@C_{84}$. Since higher fullerenes exist on multiple isomers, ^3He NMR studies on $He@C_{76}$, $He@C_{78}$, and $He@C_{84}$ revealed the presence of a single and sharp peak for each isomer. However, it is not clear whether ^3He labeling of the isomers of the higher fullerenes occurs with the same efficiency for all isomers [69]. Moreover, the dimer C_{120} encapsulating ^3He in one of the C_{60} cages was synthesized. The preparation of the new material was followed by ^3He NMR and according to those measurements the peak at -8.8 ppm corresponds to $^3He@C_{120}$ [70]. Interestingly, the procedure of the solid-state synthesis for the dimerization of C_{60}, namely grinding with potassium cyanide, was adopted to enhance the incorporation of noble gases inside fullerenes. Although the mechanism that induces higher incorporation of noble gas inside C_{60} is not clear yet, it is believed that the addition of cyanide to the fullerene is a key step, most likely because it weakens the neighboring bonds which can then break more easily to yield the orifice to allow the noble gas atom to enter [71].

On the other hand, it is interesting to know if the encapsulated noble gas atoms can be released from inside fullerenes. Thus, a detailed study in which the rate of release of the noble gas from the fullerene cage by heating for different periods was carried out. It was found that the half-life of Ne@ C_{60} is at least several weeks upon heating at 630 °C, while the release of gas is also related to the presence or absence of entrapped solvent in the crystal lattice [72].

8.4 HYDROGEN-CONTAINING H_2@C_{60}

Molecules or molecular systems based on carbon that can spontaneously absorb/adsorb and also eject hydrogen are considered as novel materials potentially useful in hydrogen storage. In this context, the empty space of spherical fullerenes is rage enough to host H_2, provided that an orifice to allow the incorporation is initially made. To this end, the "molecular surgery" approach, which was first applied to incorporate noble gases inside fullerenes, was the technique that gave the first fullerene material to encapsulate molecular hydrogen [45]. Thus, H_2 was incorporated into bislactam fullerene derivative 1 (cf. Figure 8.7), however, it required more drastic conditions than that applied for inserting He. The H_2@1 material was prepared in 5% yield, purified by column chromatography and in the ^{1}H NMR spectrum a peak at −5.43 ppm confirmed the success of the preparation. Also, the mono-deuterated species HD was similarly encapsulated inside C_{60} and accordingly characterized.

In order to increase the encapsulation yield or better achieve quantitative incorporation of H_2 a fullerene possessing a bigger orifice was synthesized. Thus, fullerene derivative 8 (Figure 8.12) with a 13-membered ring orifice was synthesized [47]. Then, treatment of 8 in an autoclave with high pressure H_2 (800 atm) at 200 °C for 8 h, yielded H_2@8 in 100% yield, with the encapsulation yield being highly dependent on the H_2 pressure applied [46]. In the ^{1}H NMR spectrum of H_2@8, acquired in solution, the characteristic signal for the encapsulated H_2 appears at −7.25 ppm due to the strong shielding effect from the fullerene cage. Additionally, solid-state ^{1}H NMR measurements showed that a small anisotropy of the H_2 rotation inside C_{60} exists [73].

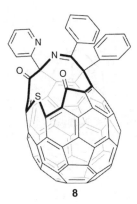

8

Figure 8.12 Open-cage fullerene derivative **8** with a 13-membered ring orifice

Furthermore, synchrotron X-rays diffraction studies revealed that H_2 is located at the center of the fullerene cage in $H_2@8$ [74]. When MALDI-TOF mass spectrometry of $H_2@8$, with dithranol as matrix, was conducted at a laser power around the ionization of **8**, the molecular ion peak of $H_2@8$ was observed. Interestingly, when the laser power in the MALDI-TOF measurement was enhanced, the molecular ion peak for $H_2@C_{60}$ was observed, thus suggesting that the closure of the orifice can occur while simultaneously molecular hydrogen remains entrapped inside the cage. Indeed soon after, the chemical transformation of $H_2@8$ to $H_2@C_{60}$ was accomplished according to the procedure shown in Figure 8.13 [75]. Briefly, this four-step reaction scheme, involves (i) oxidation of the sulfide unit in $H_2@8$ to the corresponding sulfoxide in $H_2@9$, (ii) photochemical removal of the sulfoxide moiety furnishing $H_2@10$, (iii) reduction of the two carbonyl groups, furnishing $H_2@11$ with a reduced orifice size (eight-membered ring), and (iv) annealing under vacuum at 340 °C for 2 h, to yield $H_2@C_{60}$ [76].

It should be noted that in all those reactions the presence of H_2 inside C_{60} was evident by its characteristic signal in the ¹H NMR spectrum. At the end, $H_2@C_{60}$ was purified and separated from empty C_{60} by recycling HPLC and isolated as $H_2@C_{60}$. Although the ¹³C NMR, UV-Vis, and IR spectra of $H_2@C_{60}$ are virtually the same as those of intact C_{60}, repulsive interactions between the encapsulated H_2 and the outer π-electron system of C_{60} in its highly reduced states are evident by electrochemistry.

In the ¹H NMR spectrum of $H_2@C_{60}$, the signal for encapsulated H_2 appears at −1.44 ppm, which is upfield shifted by 5.98 ppm relative to the signal of dissolved free H_2. The extent of the upfield shift for $H_2@C_{60}$ is comparable with that found for ³He@C_{60} (i.e., 6.36 ppm) in ³He NMR

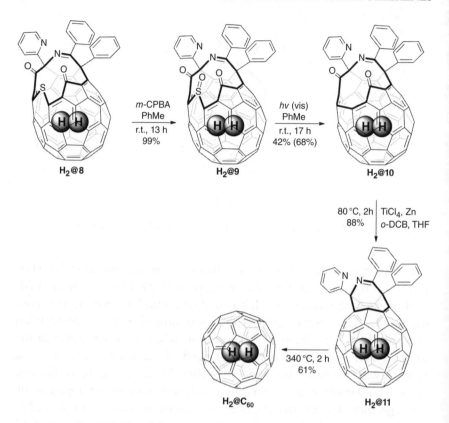

Figure 8.13   Chemical transformation of $H_2@8$ to $H_2@C_{60}$

relative to free ^3He [41,51]. The latter upfield shift demonstrates that the shielding effect of the ring currents of C_{60} is unaffected by the encapsulated paramagnetic species. Furthermore, in order to identify if an encapsulated H_2 inside C_{60} can communicate with the electronically excited walls of its fullerene, the nuclear spin-lattice relaxation of H_2 and $H_2@C_{60}$ as a function of solvent and temperature were studied. Solution NMR was used to study the relaxation of the encapsulated protons and found that the proton T_1 displays an unusual T_1-maximum vs. temperature, which was found to be similar to that of dissolved H_2 (i.e., not encapsulated in fullerene) thus suggesting a common physical mechanism [77].

Furthermore, the accessibility of the interior of C_{60} to external agents was studied by measuring the corresponding relaxivity to the spin-lattice relaxation rate for the protons of H_2 and $H_2@C_{60}$ in the presence of paramagnetic nitroxide radicals and found to be enhanced for $H_2@C_{60}$. Thus, it seems that the encapsulated dihydrogen in $H_2@C_{60}$ remains magnetically coupled with the exterior, and importantly it interacts strongly with paramagnetic species. This is rationalized in terms of a hyperfine coupling between the unpaired electron of the paramagnets and the proton modulated by the relative motions of the two species [78]. In addition, the photophysical properties of the triplet-state of $H_2@C_{60}$ were studied and compared with those of the triplet-state of C_{60} and it was found that (i) interactions of H_2 with the walls of triplet C_{60} are too weak and (ii) significant interactions between singlet oxygen ($^1O_2$) and the encapsulated H_2 in $H_2@C_{60}$ exist. The latter is rationalized in terms of a long-lived exciplex between $^1O_2$ and $H_2@C_{60}$, allowing sufficient time for contact and effective electronic–vibration interactions [79].

Another interesting property of $H_2@C_{60}$ concerns the study of chemical transformations of the allotropes of the encapsulated dihydrogen. Thus, it was possible to experimentally demonstrate the spin catalyzed conversion of $oH_2@C_{60}$ having a nuclear triplet state with parallel nuclear spins to $pH_2@C_{60}$ having a nuclear singlet state with antiparallel nuclear spins and vice versa, on a zeolite surface. This result may be useful for applications concerning the liquefication and storage of liquid hydrogen. However, the equilibrium between pH_2 and oH_2 takes very long periods, around months [80]. However, the presence of a so-called spin catalyst, in the form of an electronic or nuclear spin that can interact stronger with one of the nuclear spins of H_2 than the other can effectively induce a selective flip of one spin. Thus, the effect of spin conversion was examined on a derivatized fullerene possessing either a diamagnetic or a paramagnetic species. For the former case, it was found that the substitution of the cage does not play a significant role in that process.

However, in the case when C_{60} was functionalized with a nitroxide radical as a paramagnetic species the situation was different. Actually, the rate of interconversion of encapsulated H_2 is markedly increased, thus suggesting great potential in applying $H_2@C_{60}$ to magnetic resonance imaging and dynamic nuclear polarization (when the fullerene cage in $H_2@C_{60}$ is modified with paramagnetic species) [81]. However, the above method can only be effective at a temperature as low as 77 K, and therefore a better system is needed to switch the spin catalyst on and off at any temperature. The latter was achieved by considering the inherent electronic property of fullerenes: absorption of a photon by C_{60} or C_{70} generates a very short-lived excited singlet state that undergoes intersystem crossing to a triplet state. Then, the paramagnetism while fullerene exists in the triplet state can serve as a suitable means for converting the nuclear spins of dihydrogen in $H_2@C_{60}$. Thus, the paramagnetic nature of the triplet excited state of fullerene can spin-catalyze the interconversion from oH_2 to pH_2. However, this is the case only for $H_2@C_{70}$ and not for $H_2@C_{60}$ since C_{60} possesses shorter triplet lifetime compared with that for C_{70} [82].

With quantities of $H_2@C_{60}$ being available, the exohedral chemical modification of the carbon cage was next attempted. To this end, the regioselective penta-addition reaction of an organocopper material was successfully applied to modify $H_2@C_{60}$. Thus, treatment of $H_2@C_{60}$ with a phenylcopper compound (prepared in-situ from PhMgBr and CuBr·SMe$_2$) afforded material 12, which gave immediate access to materials 13 and 14 according to Figure 8.14 [83].

A series of $H_2@C_{60}$ mono-adducts bearing the nitroxide radical as a substituent was synthesized. Different methodologies for the functionalization of $H_2@C_{60}$ were applied, namely 1,3-dipolar cycloaddition of azomethine ylides and cyclopropanation reactions. In the different mono-adducts formed, the distance of the nitroxide radical from the encapsulated

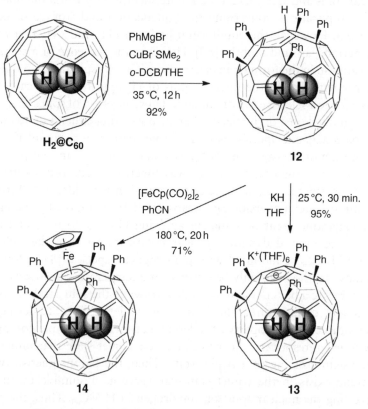

Figure 8.14 Exohedral chemical modifications of $H_2@C_{60}$

dihydrogen was varied and this allowed a distance-dependent spin nuclear relaxivity of H_2 to be performed [84]. In addition, bis-adducts of $H_2@C_{60}$ with nitroxide radicals were synthesized. The synthesis was based on the formation of pyrrolidine rings fused onto 6,6-double bonds of the fullerene cage via the 1,3-dipolar cycloaddition reaction of in-situ generated azomethine ylides. Five different bis-adduct isomers were isolated and their molecular structure was identified by UV-Vis examination as being *trans*-1, *trans*-2, *trans*-3, *trans*-4, and *e*-5. A striking observation during ^{1}H NMR measurements was that the chemical shifts for the bis-adducts were well-separated, thus making this technique suitable for assigning their structures and identifying the purity of the bis-adducts [85].

Encapsulation of H_2 is also possible in C_{70}. Actually, following standard conditions applied to generate the C_{60}-based open-cage 8 (cf. Figure 8.12), the corresponding C_{70}-based open-cage was synthesized and quantitatively filled with H_2 upon treatment with high-pressure H_2 gas in an autoclave. Upon ^{1}H NMR measurement, a sharp singlet was identified at −16.51 ppm, followed by a smaller signal at −15.22 ppm, with a relative ratio 97:3. Obviously, due to the larger internal space of C_{70}, as compared with C_{60}, it is possible to trap two H_2 molecules. The signal at −15.22 ppm is due to one H_2 molecule entrapped within the open-cage C_{70}, since it is subjected to the strong shielding effect of the fullerene cage. Therefore, the signal at −16.51 ppm is due to the two H_2 encapsulated molecules in the open-cage C_{70} [86]. Closure of the orifice in the open-cage C_{70} containing either one or two H_2 molecules results in the formation of $H_2@C_{70}$ and $(H_2)_2@C_{70}$, respectively. In the ^{1}H NMR, the chemical shifts of hydrogen for $H_2@C_{70}$ and $(H_2)_2@C_{70}$ appear at −23.80 and −23.97 ppm, respectively [87]. Furthermore, Diels-Alder derivatives of $H_2@C_{70}$ and $(H_2)_2@C_{70}$ with DMA were prepared and it was found that $(H_2)_2@C_{70}$ is less reactive. The chemical shifts of H_2 and $(H_2)_2$ in the modified C_{70} were found at −22.22 and −21.80 ppm, respectively.

8.5   WATER-CONTAINING $H_2O@C_{60}$

As described above, opening holes on the fullerene skeleton gives immediate access to the internal hollow space of the carbon nanomaterial and subsequently allows the encapsulation of atoms or small molecules. Therefore, widening the orifice window is an important step toward the encapsulation of bigger molecules. In this way, bowl-shaped fullerene 15 (Figure 8.15) possessing a 20-membered ring orifice was constructed and a water molecule was encapsulated within it [88].

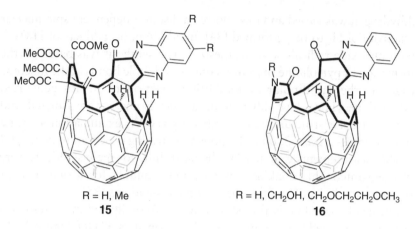

R = H, Me   R = H, CH_2OH, $CH_2OCH_2CH_2OCH_3$

15   **16**

Figure 8.15   Bowl-shaped fullerenes **15** and **16**, possessing 20- and 19-membered ring orifices, respectively

The presence of a singlet at −11.4 ppm in the ^1H NMR spectrum justified the presence of H_2O inside the open-cage **15** by taking into account the strong shielding effect of fullerene and the disappearance of the NMR signal upon treatment with D_2O. However, it was evident from mass spectrometry that water can escape from the open-cage. Therefore, a slightly smaller orifice was constructed, in order to more efficiently trap water, thus giving rise to the synthesis of open-cage fullerene **16** (Figure 8.15) incorporating a 19-membered ring. When water was encapsulated inside **16**, its proton signal in the NMR was found at −10 ppm, reflecting the difference in the shielding effect due to the different structure of the fullerene cage as compared with $H_2O@$**15** [89].

In other reports, cage-open fullerene derivatives with 18- and 19-membered ring orifice were prepared. The interesting feature of those open-cages is that during the chemical transformation for opening the cage a carbon atom was lost, thus, they are described as [59] fullerenones. In any case, water was found entrapped in both the 18- and 19-membered ring [59] fullerenones, however, at low yields [90–92]. The encapsulation of water inside open-cage [59] fullerenones was also theoretically studied. Based on density functional theory (DFT) calculations the equilibrium for encapsulating–releasing the water molecule was calculated [93].

The complete chemical transformation of opening an orifice in C_{60}, trapping a water molecule inside the empty space of the fullerene and finally closing the cage, giving rise to the first $H_2O@C_{60}$ was recently achieved (Figure 8.16) [94]. Direct evidence for the successful formation of $H_2O@C_{60}$ was acquired from single crystal X-ray diffraction studies.

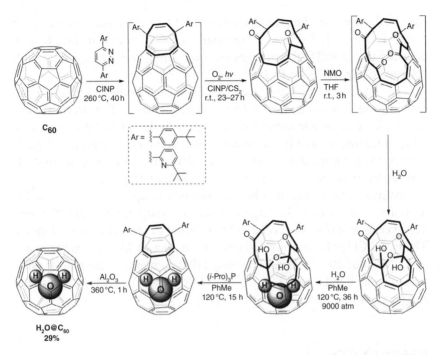

Figure 8.16 Complete chemical transformation of opening an orifice in C_{60}, trapping a water molecule inside the empty space of fullerene and finally closing the cage, giving rise to $H_2O@C_{60}$

Notably, it was found that the encapsulated water remains intact without leaking to the outer environment even when $H_2O@C_{60}$ was heated at 420 °C under vacuum. Furthermore, electronic absorption spectroscopy showed the absence of appreciable interactions between the encapsulated water and the C_{60} molecules in $H_2O@C_{60}$. Finally, in the ^1H NMR spectrum of $H_2O@C_{60}$ (deuterated o-dichlorobenzene), the water protons were identified at –4.81 ppm, reflecting a strong shielding effect from C_{60}. Notably, the entrapped water does not alter the fullerene structure, though it gives an appreciable dipole moment [95].

The open-cage fullerene **15** was also used to encapsulate one molecule of CO. The preparation of CO@**15** was achieved by heating $H_2O@$**15** and empty **15** dissolved in 1,1,2,2-tetrachloroethane under pressurized CO (9 MPa) [96]. Examination of the reaction mixture by ^1H NMR revealed that the characteristic signal at –11.4 ppm due to the protons of the encapsulated water disappeared and the presence of CO@**15** was confirmed by mass spectrometry. Furthermore, when labeled $^{13}CO@$**15** was prepared and examined by ^{13}C NMR a signal at 174.3 ppm was observed corresponding to encapsulated ^{13}CO (i.e., the resonance of "free" CO is at 184.6 ppm). Eventually, CO@15 found to lose CO, in contrast to H_2O@15, thus suggesting that water interacts stronger with the cage compared with CO.

Later, ammonia was introduced into open-cage fullerene 15, giving rise to endohedral fullerene NH_3@15 in moderate yield. The protons from encapsulated ammonia appear at -12.3 ppm in the ^1H NMR spectrum of NH_3@15, however, as a broad signal as compared with the protons signals derived from encapsulated water in H_2O@15. Importantly, gaseous NH_3 was found to escape over time from the orifice of 15 [97].

Finally, methane may also be incorporated in open-cage fullerene 15. Methane is up to now the largest molecule incorporated into fullerenes, however, its entrapment was initially predicted by theoretical calculations. The novel CH_4@15 was formed upon treatment of 15 with methane gas (19.2 MPa) at 200 °C. After recycling HPLC a methane content of 66% was calculated in CH_4@15. In the ^1H NMR spectrum of CH_4@15 a sharp singlet at -12.32 ppm was observed [98].

REFERENCES

[1] Almeida Murphy T, Pawlik Th,Weidinger A *et al.*1996 Observation of atomlike nitrogen in nitrogen-implanted solid C_{60} *Phys. Rev. Lett.* **77** 1075

[2] Knapp C, Dinse K-P, Pietzak B *et al.* 1997 Fourier transform, EPR study of N@C_{60} in solution *Chem. Phys. Lett.* **272** 433

[3] Mauser H, van Eikema Homes N J R, Clark T *et al.* 1997 Stabilization of atomic nitrogen inside C_{60} *Angew. Chem., Int. Ed. Engl.* **36** 2835

[4] Pietzak B, Waiblinger M, Almeida Murphy T *et al.* 1997 Buckminsterfullerene C_{60}: a chemical Faraday cage for atomic nitrogen *Chem. Phys. Lett.* **279** 259

[5] Morton J J L, Tyryshkin A M, Ardavan A *et al.* 2006 Electron spin relaxation of N@ C_{60} in CS_2 *J. Chem. Phys.* **124** 014508/1-5

[6] Weidinger A, Waiblinger M, Pietzak B and Almeida Murphy T 1998 Atomic nitrogen in C_{60}:N@C_{60} *Appl. Phys. A* **66** 287

[7] Weiden N, Kass H and Dinse K-P 1999 Pulse electron paramagnetic resonance (EPR) and electron-nuclear double resonance (ENDOR) investigation of N@C_{60} in polycrystalline C_{60} *J. Phys. Chem. B* **103** 9826

[8] Dinse K-P, Kass H and Weiden N 2000 EPR investigation of atoms in chemical traps *Carbon* **38** 1635

[9] Waiblinger M, Lips K, Harneit H *et al.* 2001 Thermal stability of the endohedral fullerenes N@C_{60}, N@C_{70}, and P@C_{60} *Phys. Rev. B* **64** 159901/1-5

[10] Huang H, Ata M and Ramm M 2002 ^{14}N@C_{60} formation in a nitrogen rf-plasma *Chem. Commun.* 2076

[11] Ata, M, Huang H and Akasaka T 2004 Nitrogen radio frequency plasma processing of fullerenes *J. Phys. Chem. B* **108** 4640

[12] Harneit W 2002 Fullerene-based electron-spin quantum computer *Phys. Rev. A* **65** 032322-1-6

[13] Mayer C, Herneit W Lips K and Weidinger A 2002 Alignment of the endohedral fullerenes N@C_{60} and N@C_{70} in a liquid-crystal matrix *Phys. Rev. A* **65** 061201-1-4

[14] Jakes P, Weiden N, Eichel R-A *et al.* 2002 Electron paramagnetic resonance investigation of endohedral fullerenes N@C_{60} and N@C_{70} in a liquid crystal *J. Magn. Reson.* **156** 303

[15] Jakes P, Dinse K-P, Meyer C *et al.* 2003 Purification and optical spectroscopy of N@C_{60} *Phys. Chem. Chem. Phys.* **5** 4080

[16] Suetsuma T, Dragoe N, Harneit W *et al.* 2002 Separation of N_2@C_{60} and N@C_{60} *Chem. Eur. J.* **8** 5080

[17] Kanai M, Porfyrakis K, Briggs G A D and Dennis T J S 2004 Purification by HPLC and the UV/Vis absorption spectra of the nitrogen-containing *incar*-fullerenes *i*N@ C_{60}, and *i*N@C_{70} *Chem. Commun.* 210

[18] Pietzak B, Waiblinger M, Almeida Murphy T *et al.* 1998 Properties of endohedral N@C_{60} *Carbon* **36** 613

[19] Dietel E, Hirsch A, Pietzak B *et al.* 1999 Atomic nitrogen encapsulated in fullerenes: effects of cage variations *J. Am. Chem. Soc.* **121** 2432

[20] Goedde B, Waiblinger M, Jakes P *et al.* 2001 Nitrogen doped C_{60} dimers (N@C_{60}–C_{60}) *Chem. Phys. Lett.* **334** 12

[21] Zhang J, Porfyrakis K, Morton J J L *et al.* 2008 Photoisomerization of a fullerene dimer *J. Phys. Chem. C* **112** 2802

[22] Hormann F, Hirsch A, Porfyrakis K and Briggs G A D 2011 Synthesis and magnetic properties of a nitrogen-containing fullerene dimer *Eur. J. Org. Chem.* 117

[23] Wakahara T, Matsunaga Y, Katayama A *et al.* 2003 A comparison of the photochemical reactivity of N@C_{60} and C_{60}: photolysis with disilirane *Chem. Commun.* 2940

[24] Nikawa H, Arakai Y, Slanina Z *et al.* 2009 The effect of atomic nitrogen on the C_{60} cage *Chem. Commun.* **46** 631

[25] Franco L, Ceola S, Corvaja C *et al.* 2006 Synthesis and magnetic properties of N@C_{60} derivatives *Chem. Phys. Lett.* **422** 100

[26] Zhang J, Morton J J L, Sambrook M R *et al.* 2006 The effects of a pyrrolidine functional group on the magnetic properties of N@C_{60} *Chem. Phys. Lett.* **432** 523

[27] Liu G, Khlobystov A N, Ardavan A *et al.* 2011 Photochemical stability of N@C_{60} and its pyrrolidine derivatives *Chem. Phys. Lett.* **508** 187

[28] Jones M A G, Britz D A, Morton J J L *et al.* 2006 Synthesis and reactivity of N@C_{60}O *Phys. Chem. Chem. Phys.* **8** 2083

[29] Liu G, Khlobystov A N, Charalambidis G *et al.* 2012 N@C_{60}-porphyrin: a dyad of two radical centres *J. Am. Chem. Soc.* **134** 1938

[30] Farrington B J, Jevric M, Rance G A *et al.* 2012 Chemistry at the nanoscale: synthesis of an N@C_{60}-N@C_{60} endohedral fullerene dimer *Angew. Chem. Int. Ed.* **51** 3587

[31] Knapp C, Weiden N, Kass K *et al.* 1998 Electron paramagnetic resonance study of atomic phosphorus encapsulated in [60]fullerene *Mol. Phys.* **95** 999

[32] Larsson J A, Gree J C, Harneit W and Weidinger A 2002 Phosphorous trapped within buckminsterfullerene *J. Chem. Phys.* **116** 7849

[33] Naydenov B, Spudat C, Harneit W *et al.* 2006 Ordered inclusion of endohedral fullerenes N@C_{60} and P@C_{60} in a crystalline matrix *Chem. Phys. Lett.* **424** 327

[34] Naydenov B, Mende J, Harneit W and Mehring M 2006 Entanglement in P@C_{60} encapsulated in a solid state matrix *Phys. Status Solidi B* **254** 2002

[35] Weiske T, Bohme D K, Hrusak J *et al.* 1991 Endohedral cluster compounds: inclusion of helium within $C_{60}^{\bullet+}$ and $C_{70}^{\bullet+}$ through collision experiments *Angew. Chem. Int. Ed.* **30** 884

[36] Ross M M and Callahan J H 1991 Formation and characterization of $C_{60}He^+$ *J. Phys. Chem.* **95** 5720

[37] Caldwell K A, Giblin D E and Gross M L 1992 High-energy collisions of fullerene radical cations with target gases: capture of the target gas and charge stripping of $C_{60}^{\bullet+}$, $C_{70}^{\bullet+}$, and $C_{84}^{\bullet+}$ *J. Am. Chem. Soc.* **114** 3743

[38] Caldwell K A, Giblin D E, Hsu C S *et al.* 1991 Endohedral complexes of fullerene radical cations *J. Am. Chem. Soc.* **113** 8519

[39] Saunders M, Jimenez-Vazquez H A, Cross R. J. and Poreda R J 1993 Stable compounds of helium and neon: He@C_{60} and Ne@C_{60} *Science* **259** 1428

[40] Saunders M, Jimenez-Vazquez H A, Cross R J *et al.* 1993 Incorporation of helium, argon, krypton, and xenon into fullerenes using high pressure *J. Am. Chem. Soc.* **116** 2193

[41] Saunders M, Cross R J, Jimenez-Vazquez H A *et al.* 1996 Noble gas atoms inside fullerenes *Science* **271** 1693

[42] Patchkovskii S and Thiel W 1996 How does helium get into buckminsterfullerene? *J. Am. Chem. Soc.* **118** 7164

[43] Shimshi R, Cross R J and Saunders M 1997 Beam implantation: a new method for preparing cage molecules containing atoms at high incorporation levels *J. Am. Chem. Soc.* **119** 1163

[44] Peng R-F, Chu S-J, Huang Y-M *et al.* 2009 Preparation of He@C_{60} and He_2@C_{60} by an explosive method *J. Mater. Chem.* **19** 3602

[45] Rubin Y, Jarrosson T, Wang G-W *et al.* 2001 Insertion of helium and molecular hydrogen through the orifice of an open fullerene *Angew. Chem. Int. Ed.* **40** 1543

[46] Murata Y, Murata M and Komatsu K 2003 100% encapsulation of a hydrogen molecule into an open-cage fullerene derivative and gas-phase generation of H_2@C_{60} *J. Am. Chem. Soc.* **125** 7152

[47] Murata Y, Murata M and Komatsu K 2003 Synthesis, structure, and properties of novel open-cage fullerenes having heteroatom(s) on the rim of the orifice *Chem. Eur. J.* **9** 1600

[48] Stanisky C M, Cross R J, Saunders M. *et al.* 2005 Helium entry and escape through a chemically opened window in a fullerene *J. Am. Chem. Soc.* **127** 299

[49] Chuang S-C, Murata Y, Murata M and Komatsu K 2007 The outside knows the difference inside: trapping helium by immediate reduction of the orifice size of an open-cage fullerene and the effect of encapsulated helium and hydrogen upon the NMR of a proton directly attached to the outside *Chem. Commun.* 1751

[50] Morinaka Y, Tanabe F, Murata M *et al.* 2010 Rational synthesis, enrichment, and ^{13}C NMR spectra of endohedral C_{60} and C_{70} encapsulating a helium atom *Chem. Commun.* **46** 4532

[51] Saunders M, Jimenez-Vazquez H A, Cross R J *et al.* 1994 Probing the interior of fullerenes by ^{3}He NMR spectroscopy of endohedral ^{3}He@C_{60} and ^{3}He@C_{70} *Nature* **367** 256

[52] Khong A, Jimenez-Vazquez H A, Saunders M *et al.* 1998 An NMR study of He_2 inside C_{70} *J. Am. Chem. Soc.* **120** 6380

[53] Sternfeld T, Hoffman R E, Saunders M *et al.* 2002 Two helium atoms inside fullerenes: probing the internal magnetic field in C_{60}^{6-} and C_{70}^{6-} *J. Am. Chem. Soc.* **124** 8786

[54] Shabtai E, Weitz A, Haddon R C *et al.* 1998 ^{3}He NMR of He@C_{60}^{6-} and He@C_{70}^{6-}. New records for the most shielded and the most deshielded ^{3}He inside a fullerene *J. Am. Chem. Soc.* **120** 6389

[55] Avent A G, Birkett P R, Kroto H W et al. 1998 Stable [60]fullerenecarbocations Chem. Commun. 2153

[56] Birkett P R, Buhl M, Khong A et al. 1999 The ^{3}He NMR spectra of a [60]fullerene cation and some arylated [60]fullerenes J. Chem. Soc., Perkin Trans. 2 2037

[57] Syamala M S, Cross R J and Saunders M 2004 ^{129}Xe NMR spectrum of xenon inside C_{60} J. Am. Chem. Soc. 124 6216

[58] Buhl M, Patchkovskii S and Thiel W 1997 Interaction energies and NMR chemical shifts of noble gases in C_{60} Chem. Phys. Lett. 275 14

[59] Saunders M, Khong A, Shimshi S et al. 1996 Chromatographic fractionation of fullerenes containing noble gas atoms Chem. Phys. Lett. 248 127

[60] Lee H M, Olmstead M M, Suetsuna T et al. 2002 Crystallographic characterization of Kr@C_{60} in (0.09Kr@C_{60}/0.91C_{60})•{Ni^{II}(OEP)}•2C_6H_6 Chem. Commun. 1352

[61] DiCamillo B A, Hettich R L,Guiochom G et al. 1996 Enrichment and characterization of a noble gas fullerene: Ar@C_{60} J. Phys. Chem. 100 9197

[62] Yamamoto K, Saunders M, Khong A et al. 1999 Isolation and properties of Kr@C_{60}, a stable van der Waals molecule J. Am. Chem. Soc. 121 1591

[63] Saunders M, Jimenez-Vazquez H A, Bangerter B W et al. 1994 ^{3}He NMR: a powerful new tool for following fullerene chemistry J. Am. Chem. Soc. 116 3621

[64] Saunders M, Jimenez-Vazquez H A, Cross R S et al. 1994 Reaction of cyclopropa[b] naphthalene with ^{3}He@C_{60} Tetrahedron Lett. 35 3869

[65] Smith III A B, Strongin R M, Brard L et al. 1994 Synthesis and ^{3}He NMR studies of C_{60} and C_{70} epoxide, cyclopropane and annulene derivatives containing endohedral helium J. Am. Chem. Soc. 116 10831

[66] Cross R J, Jimenez-Vazquez H A, Li Q et al. 1996 Differentiation of isomers resulting from bisaddition to C_{60} using ^{3}He NMR spectrometry J. Am. Chem. Soc. 118 11454

[67] Ruttimann M, Haldimann R F, Isaacs L et al. 1997 π-electron ring-current effects in multiple adducts of ^{3}He@C_{60} and ^{3}He@C_{70}: a ^{3}He NMR study Chem. Eur. J. 3 1071

[68] Wang G-W, Saunders M and Cross R J 2001 Reversible Diels-Alder addition to fullerenes: a study of equilibria using ^{3}He NMR spectroscopy J. Am. Chem. Soc. 123 256

[69] Saunders M, Jimenez-Vazquez H A, Cross R J et al. 1995 Analysis of isomers of the higher fullerenes by ^{3}He NMR spectroscopy J. Am. Chem. Soc. 117 9305

[70] Komatsu K, Wang G-W, Murata Y et al. 1998 Mechanochemical synthesis and characterization of the fullerene dimer C_{120} J. Org. Chem. 63 9358

[71] Cross R J, Khong A and Saunders M 2003 Using cyanide to put noble gases inside C_{60} J. Org. Chem. 68 8281

[72] Shimshi R, Khong A, Jimenez-Vazquez H A et al. 1996 Release of noble gas atoms from inside fullerenes Tetrahedron 52 5143

[73] Carravetta M, Murata Y, Murata M et al. 2004 Solid-state NMR spectroscopy of molecular hydrogen trapped inside an open-cage fullerene J. Am. Chem. Soc. 126 4092

[74] Sawa H, Wakabayashi Y, Murata Y et al. 2005 Floating single hydrogen molecule in an open-cage fullerene Angew. Chem. Int. Ed. 44 1981

[75] Komatsu K, Murata M and Murata Y 2005 Encapsulation of molecular hydrogen in fullerene C_{60} by organic synthesis Science 307 238

[76] Murata M, Murata Y and Komatsu K 2006 Synthesis and properties of endohedral C_{60} encapsulating molecular hydrogen J. Am. Chem. Soc. 128 8024

[77] Sartori E, Ruzzi M, Turro N J et al. 2006 Nuclear relaxation of H_2 and H_2@C_{60} in organic solvents J. Am. Chem. Soc. 128 14752

[78] Sartori E, Ruzzi M, Turro N J et al. 2008 Paramagnet enhanced nuclear relaxation of H_2 in organic solvents and in $H_2@C_{60}$ J. Am. Chem. Soc. 130 2221

[79] Lopez-Gejo J, Marti A A, Ruzzi M et al. 2007 Can H_2 inside C_{60} communicate with the outside world? J. Am. Chem. Soc. 129 14554

[80] Turro N J, Marti A A, Chen J Y-C et al. 2008 Demonstration of a chemical transformation inside a fullerene. The reversible conversion of the allotropes of $H_2@C_{60}$ J. Am. Chem. Soc. 130 10506.

[81] Li Y, Lei X, Jockusch S et al. 2010 A magnetic switch for spin-catalyzed interconversion of nuclear spin isomers J. Am. Chem. Soc. 132 4042

[82] Frunzi M, Jockusch S, Chen J Y-C et al. 2011 A photochemical on-off switch for tuning the equilibrium mixture of H_2 nuclear spin isomers as a function of temperature J. Am. Chem. Soc. 133 14232

[83] Matsuo Y, Isobe H, Tanaka T et al. 2005 Organic and organometallic derivatives of dihydrogen-encapsulated [60]fullerene J. Am. Chem. Soc. 127 17148

[84] Li Y, Lei X, Lawler R G et al. 2010 Distance-dependent paramagnet-enhanced nuclear spin relaxation of $H_2@C_{60}$ derivatives covalently linked to a nitroxide radical J. Phys. Chem. Lett. 1 2135

[85] Li Y, Lei X, Lawler R G et al. 2011 Synthesis and characterization of bispyrrolidine derivatives of $H_2@C_{60}$: differentiation of isomers using ^1H NMR spectroscopy of endohedral H_2 Chem. Commun. 47 2282

[86] Murata Y, Maeda S, Murata M and Komatsu K 2008 Encapsulation and dynamic behavior of two H_2 molecules in an open-cage C_{70} J. Am. Chem. Soc. 130 6702

[87] Murata M, Maeda S, Morinaka Y et al. 2008 Synthesis and reaction of fullerene C_{70} encapsulating two molecules of H_2 J. Am. Chem. Soc. 130 15800

[88] Iwamatsu S, Uozaki T, Kobayashi K et al. 2004 A bowl-shaped fullerene encapsulates a water into the cage J. Am. Chem. Soc. 126 2668

[89] Iwamatsu S and Murata S 2004 $H_2O@$open-cage fullerene C_{60}: control of the encapsulation property and the first mass spectroscopic identification Tetrahedron Lett. 45 6391

[90] Xiao Z, Yao J, Yang D et al. 2007 Synthesis of [59]fullerenones through peroxide-mediated stepwise cleavage of fullerene skeleton bonds and X-ray structures of their water-encapsulated open-cage complexes J. Am. Chem. Soc. 129 16149

[91] Zhang Q, Jia Z, Liu S et al. 2009 Efficient cage-opening cascade process for the preparation of water-encapsulated [60]fullerene derivatives Org. Lett. 11 2772

[92] Zhang Q, Pankewitz T, Liu S et al. 2010 Switchable open-cage fullerene for water encapsulation Angew. Chem. Int. Ed. 49 9935

[93] Pankewitz T and Klopper W 2008 Theoretical investigation of equilibrium and transition state structures, binding energies and barrier heights of water-encapsulated open-cage [59]fullerenone complexes Chem. Phys. Lett. 465 48

[94] Kurotobi K and Murata Y 2011 A single molecule of water encapsulated in fullerene C_{60} Science 333 613

[95] Thilgen C 2012 A single water molecule trapped inside hydrophobic C_{60} Angew. Chem. Int. Ed. 51 587

[96] Iwamatsu, S, Stanisky C M, Cross R J et al. 2006 Carbon monoxide inside an open-cage fullerene Angew. Chem. Int. Ed. 45 5337

[97] Whitener Jr K E, Frunzi M, Iwamatsu S et al. 2008 Putting ammonia into a chemically opened fullerene J. Am. Chem. Soc. 130 13996

[98] Whitener Jr K E, Cross R J, Saunders M et al. 2009 Methane in an open-cage [60] fullerene J. Am. Chem. Soc. 131 6338

9

Scanning Tunneling Microscopy Studies of Metallofullerenes

9.1 STM STUDIES OF METALLOFULLERENES ON CLEAN SURFACES

Scanning tunneling microscopy (STM) has been a powerful technique for studying the structural and electronic properties of endohedral metallofullerenes. In fullerenes, the STM technique was first applied to study the morphology of C_{60} on Au(111) [1], highly oriented pyrolytic graphite [2], GaAs(110) [3], Si(100) [4], Si(111) [5], Cu(111) [6,7], and Au(110) [8] surfaces. In particular, ultra-high vacuum (UHV)-STM has been proven to be a crucial technique for the study of endohedral metallofullerenes, since metallofullerenes like empty fullerenes are easy to sublime onto clean surfaces under UHV conditions. The early studies on fullerenes and metallofullerenes are reviewed extensively by Sakurai *et al.* [9].

Figure 9.1 shows the first UHV-STM image of a small (141×141 Å^2) area of the Si(100)2 × 1 surface covered with $Sc_2C_2@C_{82}$ molecules with a coverage of approximately three mono-layers at room temperature [10,11]. Pure scandium fullerenes were evaporated from a tantalum boat heated to approximately 700 °C onto the clean Si(100)2 × 1 surface in

Endohedral Metallofullerenes: Fullerenes with Metal Inside, First Edition.
Hisanori Shinohara and Nikos Tagmatarchis.
© 2015 John Wiley & Sons, Ltd. Published 2015 by John Wiley & Sons, Ltd.

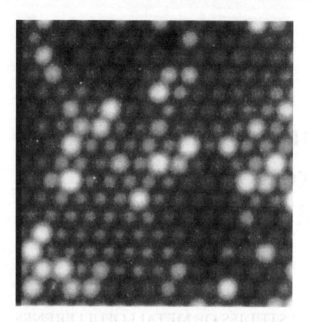

Figure 9.1   The first STM image of metallofullerenes: the third layer of $Sc_2C_2@C_{82}$ fullerenes on the Si(100)2 × 1 clean surface at a bias voltage of –3.0 V (tunneling current = 20 pA). The white contrasts correspond to the $Sc_2C_2@C_{82}$ fullerenes which are slightly ahead of the other close–packed fullerenes. Reproduced with permission from Ref. [10]. Copyright 1993 American Chemical Society

UHV conditions (5×10^{-11} Torr). The $Sc_2C_2@C_{82}$ molecules reside in the trough separated by the Si dimer rows and they are distributed randomly on the surface with a minimum separation of 11.7 Å. The STM image of the $Sc_2C_2@C_{82}$ molecules shows small deviation (within about 10%) from the perfectly circular shape. The first layer of $Sc_2C_2@C_{82}$ has provided only a short-range local ordering. When the first layer is completed, the second $Sc_2C_2@C_{82}$ layer begin to form and island formation is observed. The second layer is still somewhat irregular. However, the $Sc_2C_2@C_{82}$ molecule overlayers (the third layer and up) grown on the second layer are well ordered and perfectly close-packed, indicating that the overlayer film is basically formed by van der Waals interaction without interference from the Si substrate, similar to the case of the C_{60} [4] and C_{84} [5] depositions on the Si(100)2 × 1 surface.

The STM images present strong evidence that the two Sc atoms are indeed encapsulated by the C_{82} fullerene cage: the STM images of the $Sc_2C_2@C_{82}$ molecules show no characteristic bright (or dark) spots (which may correspond to the position of Sc atoms) on and around the carbon cage, and all images are essentially the same as those of hollow

C_{84} molecules [12,13]. The two Sc atoms are trapped securely inside the C_{82} cage, which is consistent with HRTEM (high-resolution transmission electron microscopy) [14], high-resolution ^{13}C NMR [15], and synchrotron X-ray diffraction [16–18] results.

In contrast to the STM observation on the Si(100)2 × 1 surface, Gimzewski [19] found that STM images of $Sc_2C_2@C_{82}$ on a Au(110) [8] surface show a characteristic internal structure. The appearance of the internal structure indicates that the interaction between $Sc_2C_2@C_{82}$ and the Si and Au surfaces differ from each other and that the electronic structures near the Fermi levels might also be different.

9.2 METALLOFULLERENES AS SUPERATOM

A typical large-scale (400 × 400 Å^2) STM image of a small amount [23 molecules (1000 Å^{-2})] of $Y@C_{82}$ on a Cu(111)1 × 1 surface at room temperature shows a preferential adsorption at the terrace edges (Figure 9.2) [20–22]. The $Y@C_{82}$ molecules are sublimated from the tantalum boat onto the Cu surface and impinge on the terrace of the surface with a kinetic energy corresponding to about 400 °C. The $Y@C_{82}$ molecules are mobile on the surface and segregate to the terrace edges. The impinging $Y@C_{82}$ molecules migrate to the edges following adsorption since the bonding to the substrate surface is relatively weak. The C_{60} adsorption on the Cu(111) 1 × 1 surface shows a similar mobile tendency [6]. This is in sharp contrast to the adsorption of fullerenes on Si(100) and Si(111) surfaces in which the fullerenes such as C_{60} [4], C_{70} [23], and C_{84} [13] do not freely migrate.

One of the most intriguing observations here is that the $Y@C_{82}$ fullerenes predominantly form clusters, $(Y@C_{82})_n$ (n = 2–6), and in particular dimers, $(Y@C_{82})_2$, on the Cu(111) surface even at the very initial stage of adsorption [20,21]. The distribution of the $(Y@C_{82})_n$ clusters has a maximum for dimers and more than 60% of the $Y@C_{82}$ molecules on the Cu(111) surface exist as dimers or larger clusters as shown in Figure 9.3 [22]. Previous STM results on higher fullerenes indicate that, like C_{60}, the higher fullerenes, such as C_{70} and C_{84}, also exist mostly as monomers in the initial stage of deposition on the Cu and Si surfaces [13,23]. These results indicate that the $Y@C_{82}$ metallofullerenes have a very special tendency to form dimers and larger clusters on the Cu surface. The dimerization energy of $Y@C_{82}$ on the Cu(111) surface was estimated to be about 0.18 eV [22]. In addition, large dipole moments of metallofullerenes may also play a crucial role in the dimerization, since the calculated and experimental dipole moments of $La@C_{82}$ are 3–4 D (debye) [24,25] and 4.4 ± 0.4 D [26], respectively.

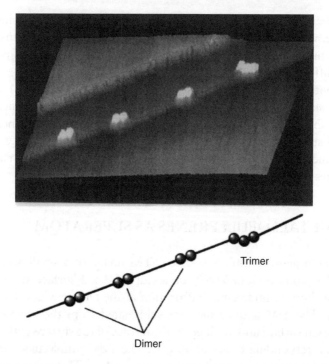

Figure 9.2 Typical large-scale (400 × 400 Å^2) STM image of (Y@C$_{82}$)$_2$ dimers adsorbed on a terrace edge of a Cu(111)1 × 1 clean surface (V_{sample} = −2.0 V). All Y@ C$_{82}$ molecules are adsorbed at mono-atomic step edges. Three dimers and one trimer are shown in this image. Reproduced with permission from Ref. [21]. Copyright 1995 American Chemical Society

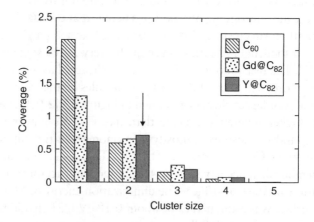

Figure 9.3 Cluster size distribution of C$_{60}$, Gd@C$_{82}$, and Y@C$_{82}$. Coverages are small: 4.1, 3.7, and 2.8% for C$_{60}$, Gd@C$_{82}$, and Y@C$_{82}$, respectively. Reproduced with permission from Ref. [22]. Copyright 1997 The American Physical Society

As discussed in Section 4.2, hyperfine splitting (hfs) analysis of the electron spin resonance (ESR) measurements of $Y@C_{82}$ indicates that the encaged Y atom donates three valence electrons to the C_{82} cage to form an endohedral metallofullerene of the type $Y^{3+}@C_{82}^{3-}$ [27,28]. An *ab initio* theoretical calculation [29,30] reveals that the charge on the encaged yttrium, that is, 3+, is little changed even when $Y@C_{82}$ ejects or accepts an additional electron. Namely, the $Y@C_{82}$ metallofullerene can be regarded as a positively charged core metal and a negatively charged carbon cage.

Such a molecule has a great similarity to the "superatom" concept proposed first by Watanabe and Inoshita [31,32] in a semiconductor heterostructure composed of a spherically symmetric positively charged core. Superatoms have also been discussed theoretically in relation to endohedral metallofullerenes by Rosén and Waestberg [33,34], Saito [35], and Nagase and Kobayashi [29,30].

The STM observation of the $Y@C_{82}$ dimers and clusters is direct experimental evidence that $Y@C_{82}$ molecules exhibit the superatom feature. The observed interfullerene distance is 11.2 Å, which is shorter than that of the simple $Y@C_{82}$–$Y@C_{82}$ van der Waals distance (11.4 Å), suggesting that the interfullerene interaction is not a simple dispersion type of weak interaction but a relatively strong interaction. A large dipole moment of $Y@C_{82}$ also plays an important role in the tight binding between $Y@C_{82}$ molecules, particularly in the solid state. In fact, a synchrotron X-ray diffraction study on a powder $Y@C_{82}$ sample [36] reveals the presence of such a charge-transfer-type interaction from the analysis of the total electron density distribution map of the $Y@C_{82}$ microcrystal.

In an $Y@C_{82}$–$Y@C_{82}$ fullerene interaction, the positively charged Y^{3+} core on one side of $Y@C_{82}$ attracts the negatively charged C_{82}^{3-} cage of the other $Y@C_{82}$ molecule. The $Y@C_{82}$–$Y@C_{82}$ "molecule" can be viewed just like a Li–Li molecule but with much weaker interaction. The superatom character of such a metallofullerene might in future lead to novel solid state properties.

9.3   STM/STS STUDIES ON METALLOFULLERENE LAYERS

Scanning tunneling spectroscopy (STS), together with STM, is an even more powerful technique to obtain electronic structures near Fermi levels of metallofullerenes [37–39]. Figure 9.4 shows an STS spectrum of $La@C_{82}$ films together with the corresponding ultraviolet photoelectron spectroscopy (UPS) spectrum as a reference [38]. The binding energies

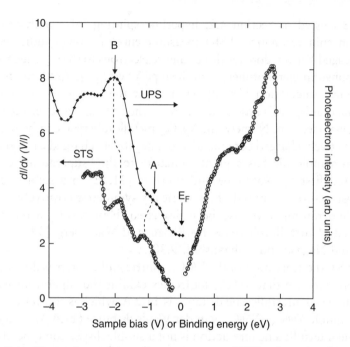

Figure 9.4 Comparison of tunneling spectrum and UPS valence band spectrum of La@C_{82} films. UPS was measured at 40 eV. Reproduced with permission from Ref. [38], Elsevier

for the UPS spectrum are referenced to the Fermi edge of a clean Cu(111) surface. The peaks in the STS spectrum match fairly well with those of the UPS spectrum in the occupied region (corresponding to the negative sample bias voltages) with a mutual peak shift of about 0.2 eV.

Hino and co-workers [40] observed spectral changes between La@C_{82} and C_{82} spectra and identified that peak A is due to the singly occupied molecular orbital (SOMO) level of La@C_{82}. The peak at 1.5 eV above the Fermi level can be assigned as the LUMO (lowest unoccupied molecular orbital) of La@C_{82} in the close-packed layer. Figure 9.4 strongly suggests that both the edges of the SOMO and LUMO peaks approach the Fermi level, leading to the La@C_{82} layer which possesses a very small energy band gap. The La@C_{82} layer is intrinsically semi-metallic.

Shinohara and co-workers [39] investigated mono- and multi-layer islands of La$_2$@C_{80} and La@C_{82} molecules grown on hydrogen-terminated Si(100)-2 × 1 surfaces by UHV-STM/STS. The observed La$_2$@C_{80} molecule has a spherical shape, consistent with the results from synchrotron X-ray measurements. The energy gap for the La$_2$@C_{80} multilayer islands is measured to be 1.3–1.5 eV, whereas that for La@C_{82} is 0.5 eV from STS measurements, indicating that the I_h cage of the $La_2@C_{80}$ molecule is highly stabilized by an electron transfer from the encaged La atoms. The $La_2@C_{80}$ layer has a large energy gap, and that of the $La@C_{82}$ layer is small. This can be explained by the difference of the number of electrons transferring to the fullerene cages:$(La_2)^{6+}@C_{80}^{6-}$ and $La^{3+}@C_{823}^-$.

9.4 STM/STS STUDIES ON A SINGLE METALLOFULLERENE MOLECULE

Imaging and characterizing an individual (single) metallofullerene molecule by STM/STS is an extremely interesting and important experiment, since this provides direct structural and electronic information on a single metallofullerene. Crommie and co-workers [41] achieved this for a single $Gd@C_{82}$ molecule on a Ag(001) clean surface by employing low temperature UHV-STM/STS at 80 K.

Figure 9.5 shows the STM topographic structure of two $Gd@C_{82}$ molecules residing next to a single C_{60} molecule on a Ag(001) surface. $Gd@C_{82}$ exhibits salient internal structure when imaged with a bias of 0.1 V (Figure 9.5a), but appears smooth and featureless when imaged at higher biases (Figure 9.5b, $V = 2.0$ V). This is in strong contrast to the internal structure of C_{60} which does not show any significant dependence on positive bias voltages.

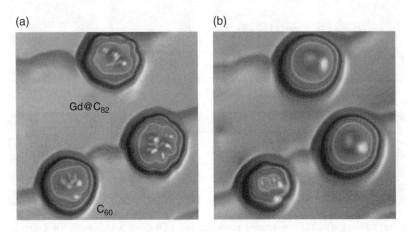

Figure 9.5 Constant current topographs (65×65 Å^2) of two $Gd@C_{82}$ molecules alongside a C_{60} reference molecule taken at (a) 0.1 V, 1 nA and (b) 2.0 V, 1 nA. Reproduced with permission from Ref. [41]. Copyright 2005 The American Physical Society

To probe the spatial and energetic distribution of the metallofullerene wave functions, they measured STS spectra at different positions of a single Gd@C$_{82}$ molecule and found six distinct molecular states. Based on density functional theory (DFT)-calculated spectra of Gd@C$_{82}$, Jiang *et al.* [42] reported that the internal structure in the STM image and corresponding STS spectra is strongly dependent on the position of the metal atom inside the cage although interaction with the Ag substrate also influenced the spectra. This is reasonable since the position of the encapsulated Gd atom is situated not at the center of the C$_{82}$ cage but close to the cage as revealed by X-ray diffraction [43].

Huang *et al.* [44] reported an STM/STS study of single Sc$_3$N@C$_{80}$ molecules adsorbed on a Cu(110)-(2×1)-O surface. At positive bias voltages of about 3.5–5.0 V, they identified a series of delocalized atom-like "superatom" molecular orbitals similar to those reported earlier for C$_{60}$ [45].

Figure 9.6a–d shows *dI/dV* images taken between 4.2 V and 4.8 V, in which the orbitals exhibit approximate mono- and di-polar symmetries. The Sc$_3$N@C$_{80}$ images with lower bias voltages in Figure 9.6e–h show strong and asymmetric LDOS (local density-of-states) localization within a molecule on account of the tunneling pathways mediated by the endohedral cluster.

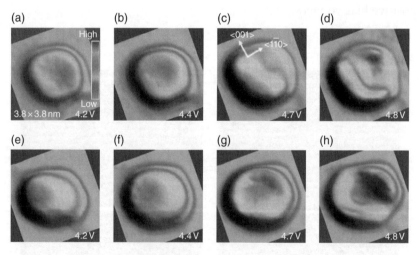

Figure 9.6 (a–d) *dI/dV* mapping showing the LDOS of SAMOs of an individual Sc$_3$N@C$_{80}$ molecule at different energies. (e–h) *dI/dV* mapping of the SAMOs of another Sc$_3$N@C$_{80}$ molecule showing different spatial distributions. The measurement bias voltages are indicated in the figures; the tunneling current $I_{setpoint}$ is 200 pA. Reproduced with permission from Ref. [44]. Copyright 2010 The American Physical Society

Unlike the C_{60} case [45], the encapsulated Sc_3N cluster in $Sc_3N@C_{80}$ distorts the almost spherical central potential of the carbon cage, providing an asymmetric spatial distribution to the superatom molecular orbitals (SAMOs) [44]. When $Sc_3N@C_{80}$ molecules form clusters such as dimers and trimers, however, the strong intermolecular hybridization results in highly symmetric hybridized SAMOs with clear bonding and antibonding characteristics. They reported that the electronic-structure calculations on $Sc_3N@C_{80}$ and its aggregates confirm the existence of SAMOs and reproduce their hybridization as observed in the experiment.

Huang $et\ al.$ [44] found pronounced changes in the dI/dV spectra as the $Sc_3N@C_{80}$ aggregation increases from monomers to dimers and trimers, suggesting that the intermolecular hybridization is significantly stronger for $Sc_3N@C_{80}$ compared with C_{60}. They concluded that the stronger hybridization of $Sc_3N@C_{80}$ aggregates may stem from the presence of the electronegative N atom in the center [44].

REFERENCES

[1] Wilson R J, Meijer G, Bethune D S $et\ al.$ 1990 Imaging C_{60} clusters on a surface using a scanning tunnelling microscope $Nature$ 348 621

[2] Wragg J L, Chamberlain J E, White H W $et\ al.$ 1990 Scanning tunnelling microscopy of solid C_{60}/C_{70} $Nature$ 348 623

[3] Li Y Z, Patrin J P, Chander M $et\ al.$ 1991 Ordered overlayers of C_{60} on GaAs(110) studied with scanning tunneling microscopy $Science$ 252 547

[4] Hashizume T, Wang X D, Nishina Y $et\ al.$ 1992 Field ion-scanning tunneling microscopy study of C_{60} on the Si(100) surface $Jpn.\ J.\ Appl.\ Phys.$ 31 L880

[5] Wang X D, Hashizume T, Shinohara H $et\ al.$ 1992 Scanning tunneling microscopy of Ca, on the Si(100)7x7 surface $Jpn.\ J.\ Appl.\ Phys.$ 31 L983

[6] Hashizume T, Motai K, Wang X D $et\ al.$ 1993 Intramolecular structures of C_{60} molecules adsorbed on the Cu(111)-(1 × 1) surface $Phys.\ Rev.\ Lett.$ 71 2959

[7] Cuberes M T, Schlittler R R and Gimzewski J K 1996 Room-temperature repositioning of individual C_{60} molecules at Cu steps: operation of a molecular counting device $Appl.\ Phys.\ Lett.$ 69 3016

[8] Joachim C, Gimzewski J K, Schlittler R R and Chavy C 1995 Electronic transparence of a single C_{60} molecule $Phys.\ Rev.\ Lett.$ 74 2102

[9] Sakurai T, Wang X D, Xue Q K $et\ al.$ 1996 Scanning tunneling microscopy study of fullerenes $Prog.\ Surf.\ Sci.$ 51 263

[10] Shinohara H, Hayashi N, Sato H $et\ al.$ 1993 Direct STM imaging of spherical endohedral discandium fullerenes $(Sc_2@C_{84})$ $J.\ Phys.\ Chem.$ 97 13438

[11] Wang X D, Xue Q, Hashizume T $et\ al.$ 1993 Scanning-tunneling-microscopy study of the solid-phase pure Sc_2C_{84} metallofullerene $Phys.\ Rev.\ B$ 48 15492

[12] Hashizume T, Wang X D, Nishina Y $et\ al.$ 1993 Field ion-scanning tunneling microscopy study of C_{84} on the Si(100) surface $Jpn.\ J.\ Appl.\ Phys.$ 32 L132

[13] Wang X D, Hashizume T, Shinohara H *et al*. 1993 Adsorption of C_{60} and C_{84} on the Si(100)2×1 surface studied by using the scanning tunneling microscope *Phys. Rev. B* **47** 15923

[14] Beyers R, Kiang C-H, Johnson R D *et al*. 1994 Preparation and structure of crystals of the metallofullerene $Sc_2@C_{84}$ *Nature* **370** 196

[15] Yamamoto E, Tansho M, Tomiyama T *et al*. 1996 ^{13}C-NMR study on the structure of isolated $Sc_2@C_{84}$ metallofullerene *J. Am. Chem. Soc.* **118** 2293

[16] Takata M, Nishibori E, Umeda B *et al* 1997 Structure of endohedral dimetallofullerene *Phys. Rev. Lett.* **78** 3330

[17] Iiduka Y, Wakahara T, Nakajima K *et al*. 2006 ^{13}C NMR spectroscopic study of scandium dimetallofullerene, $Sc_2@C_{84}$ *vs*. $Sc_2C_2@C_{82}$ *Chem. Commun.* **2057**

[18] Nishibori E, Terauchi I, Sakata M *et al*. 2006 High-resolution analysis of $(Sc_3C_2)@$ C_{80} metallofullerene by third generation synchrotron radiation X-ray powder diffraction *J. Phys. Chem. B* **110** 19215

[19] Gimzewski J K 1996 *The Chemical Physics of Fullerenes 10 (and 5) Years Later*, Vol. **316**, ed. W Andreoni (Dordrecht: Kluwer), pp. 117–136

[20] Shinohara H, Ohno M, Kishida M *et al*. 1995 *Fullerenes: Recent Advances in the Chemistry and Physics of Fullerenes and Related Materials*, Vol. **2**, eds K Kadish and R Ruoff (Pennington, NJ: Electrochemical Society), pp. 763–773

[21] Shinohara H, Inakuma M, Kishida M *et al*. 1995 An oriented cluster formation of endohedral $Y@C_{82}$ metallofullerenes on clean surfaces *J. Phys. Chem.* **99** 13769

[22] Hasegawa Y, Ling Y, Yamazaki S *et al*. 1997 STM study of one-dimensional cluster formation of fullerenes: dimerization of $Y@C_{82}$ *Phys. Rev. B* **56** 6470

[23] Wang X D, Yurov V Y, Hashizume T *et al*. 1994 Imaging of C_{70} intramolecular structures with scanning-tunneling-microscopy *Phys. Rev. B* **49** 14746

[24] Laasonen K, Andreoni W and Parrinello M 1992 Structural and electronic properties of $La@C_{82}$ *Science* **258** 1916

[25] Poirier D M, Knupfer M, Weaver J H *et al*. 1994 Electronic and geometric structure of $La@C_{82}$ and C_{82}: theory and experiment *Phys. Rev. B* **49** 17403

[26] Fucks D and Rietschel H 1996 *Fullerenes and Fullerene Nanostructures*, eds H Kuzmany, J Fink, M Mehring and S Roth (London: World Scientific), pp. 186–190

[27] Weaver J H, Chai Y, Kroll G H *et al*. 1992 XPS probes of carbon-caged metals *Chem. Phys. Lett.* **190** 460

[28] Shinohara H, Sato H, Saito Y *et al*. 1992 Mass spectroscopic and ESR characterization of soluble yttrium-containing metallofullerenes YC_{82} and Y_2C_{82} *J. Phys. Chem.* **96** 3571

[29] Nagase S and Kobayashi K 1994 Theoretical study of the lanthanide fullerene CeC_{82}. Comparison with ScC_{82}, YC_{82} and LaC_{82} *Chem. Phys. Lett.* **228** 106

[30] Nagase S and Kobayashi K 1994 The ionization energies and electron affinities of endohedral metallofullerenes MC_{82} (M = Sc, Y, La): density functional calculations *J. Chem. Soc., Chem. Commun.* 1837

[31] Watanabe H and Inoshita T 1986 Superatom: a novel concept in materials science *Optoelectron. Devices Technol.* **1** 33

[32] Inoshita T, Ohnishi S and Oshiyama A 1986 Electronic structure of the superatom: a quasiatomic system based on a semiconductor heterostructure *Phys. Rev. Lett.* **57** 2560

[33] Rosén A and Waestberg B 1988 First-principle calculations of the ionization potentials and electron affinities of the spheroidal molecules carbon (C_{60}) and lanthanum carbide (LaC_{60}) *J. Am. Chem. Soc.* **110** 8701

[34] Rosén A and Waestberg B 1989 Electronic structure of spheroidal metal containing carbon shells: study of the lanthanum-carbon and sixty-atom carbon (LaC_{60} and C_{60}) clusters and their ions within the local density approximation *Z. Phys. D: At. Mol. Clusters* **12** 387

[35] Saito S 1990 *Clusters and Cluster-Assembled Materials*, Vol. **206**, eds R S Averback, J Bernholc and D L Nelson (Pittsburgh, PA: Materials Research Society), pp. 115–120

[36] Takata M, Umeda B, Nishibori E *et al.* 1995 Confirmation by X-ray-diffraction of the endohedral nature of the metallofullerene $Y@C_{82}$ *Nature* **377** 46

[37] Klingeler R, Kann G, Wirth I *et al.* 2001 $La@C_{60}$: a metallic endohedral fullerene *J. Chem. Phys.* **115** 7215

[38] Ton-That C, Shard A G, Egger S *et al.* 2003 Structural and electronic properties of ordered $La@C_{82}$ films on Si(111) *Surf. Sci.* **522** L15

[39] Taninaka A, Shino K, Sugai T *et al.* 2003 Scanning tunneling microscopy/spectroscopy studies of lanthanum endohedral metallofullerenes *Nano Lett.* **3** 337

[40] Hino S, Takahashi H, Iwasaki K *et al.* 1993 Electronic structure of metallofullerene LaC_{82}: electron transfer from lanthanum to C_{82} *Phys. Rev. Lett.* **71** 4261

[41] Grobis M, Khoo K H, Yamachika R *et al.* 2005 Spatially dependent inelastic tunneling in a single metallofullerene *Phys. Rev. Lett.* **94** 136802

[42] Jiang J, Gao B, Hu Z P *et al.* 2010 Identification of metal-cage coupling in a single metallofullerene by inelastic electron tunneling spectroscopy *Appl. Phys. Lett.* **96** 253110

[43] Nishibori E, Iwata K, Sakata M *et al.* 2004 Anomalous endohedral structure of $Gd@ C_{82}$ metallofullerenes *Phys. Rev. B* **69** 113412

[44] Huang T, Zhao J, Feng M *et al.* 2010 Superatom orbitals of $Sc_3N@C_{80}$ and their intermolecular hybridization on Cu(110)-(2×1)-O surface *Phys. Rev. B* **81** 085434

[45] Feng M, Zhao J and Petek H 2008 Atom like, hollow-core-bound molecular orbitals of C_{60} *Science* **320** 359

10

Magnetic Properties of Metallofullerenes

10.1 MAGNETISM OF MONO-METALLOFULLERENES

Magnetic properties of powder samples of La@C_{82} and Gd@C_{82}, grown from toluene solution, have been studied by Funasaka *et al.* [1,2]. Magnetization data for a powder sample of Gd@C_{82} have been obtained employing a SQUID (superconducting quantum interference device) magnetometer. The results indicate that the data fall on a curve fitted to a Brillouin function consistent with J = 3.38 and g = 2: a signature of paramagnetic behavior. They also observed that the powder sample of La@C_{82} showed Curie–Weiss behavior at low temperature (< 40 K) (Figure 10.1). The observed effective magnetic moment per La@C_{82} was 0.38 μB, equivalent to 0.22 of an electron spin per fullerene molecule. This is quite different from the value of 1.0 which would be expected for S = 1/2 La@C_{82}. Similar magnetic properties for Gd@C_{82} were obtained by Dunsch *et al.* [3]: the susceptibility follows a Curie law and the magnetization can be described by a Brillouin function. They also reported that the two europium fullerenes, Eu@C_{74} and Eu@C_{82}, show the same M(H/T) magnetic behavior as that of Gd@C_{82} if all magnetization values are normalized to the saturation magnetization at high fields. Yang and co-workers [4] also found that isothermal magnetization curve of Gd@C_{82} follows the Brillouin function down to 8 K. The temperature-dependent magnetic susceptibility data of Gd@C_{82} obey the Curie–Weiss law above 40 K (Θ = 0.053 K).

Endohedral Metallofullerenes: Fullerenes with Metal Inside, First Edition.
Hisanori Shinohara and Nikos Tagmatarchis.
© 2015 John Wiley & Sons, Ltd. Published 2015 by John Wiley & Sons, Ltd.

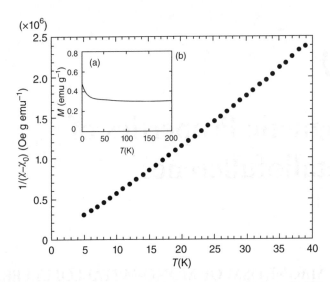

Figure 10.1 (a) Magnetization (emu g^{-1}) vs. temperature for a La@C$_{82}$ powder sample containing solvent molecules, measured at an applied field of 5 T. (b) Inverse susceptibility data as a function of temperature using the data in (a). Reproduced with permission from Ref. [1]. Copyright 1994 American Chemical Society

Powder samples of Y@C$_{82}$ exhibited localized-electron behavior both at temperatures above 200 K and temperatures below 90 K, but with different Curie–Weiss curves [5]. The Curie–Weiss curve at low temperature corresponded to 0.29(4) electrons per fullerene similar to that of La@C$_{82}$, with a small Curie constant of –2.7(8) K, and to 1.0(1) electrons at high temperature, with an extremely large Curie constant of 280(30) K. This single electron spin clearly agrees with the single unpaired electron which is expected, and as observed in solution electron spin resonance (ESR) experiments (see Section 4.2). The high-temperature susceptibility is weakly temperature dependent, and if it arises from a metal then the density of state (DOS) is 10(1) states molecule^{-1} eV^{-1} at 294 K.

The above magnetic measurements were, however, done on samples which contained solvent molecules, so magnetic behaviors of the powder samples of sublimed (thus solvent-free) pure metallofullerenes might be different. The magnetic properties on a solvent-free powder sample of La@C$_{82}$, which was prepared by sublimation, was reported by Watanuki *et al.* [6]. They performed ESR experiments on the sample between 7 K and 300 K. The results indicate that the paramagnetic component is almost temperature independent above 80 K, indicating Pauli paramagnetism of the conducting electrons. Based on the observed temperature dependence of the spin susceptibility, ESR linewidth, and the g-value, they concluded that solvent-free La@C_{82} is metallic between 80 K and 300 K. Magnetic measurements on solvent-free powder samples of metallofullerenes will be needed to further clarify the bulk magnetic properties of endohedral metallofullerenes.

The apparent discrepancy in the results between this ESR magnetic measurement (metallic) and the ultraviolet photoelectron spectroscopy (UPS) studies on La@C_{82} (insulator) [7,8] as described in Section 4.5 might be due to sample preparation, that is, sample purity together with the thickness of the sublimed samples. This is suggested because another UPS measurement by Eberhardt and co-workers [9] has shown DOSs at the Fermi level, suggesting a metallic behavior for La@C_{82} consistent with the ESR result.

Cerium endohedral metallofullerene (Ce@C_{82}) can be regarded as a π-f composite nanomagnet, where anisotropic f-electron spin is expected to couple with the rotational motion of the fullerene cage that has π-electron spin. Enoki and co-workers [10] found that in Ce@C_{82} crystals the crystal field effect in the metallofullerene cage is considerably reduced in contrast to that of ordinary rare-earth compounds. This is consistent with their study of a small electronic coupling between the f and π electrons, and the shallow potentials of the C_{82} fullerene cage surrounding the Ce ion. As a consequence, the crystal field effect is emphasized in the low-temperature range (below 100 K) (Figure 10.2). In the solid solvent-free film, Ce@C_{82} exhibited a Curie–Weiss behavior above 190 K with a Curie constant of 1.42 cm^3 mol^{-1} K ($\mu_{\text{eff}} = 3.36\mu_B$), where the Weiss temperature was observed to be only $\Theta = -1$ K.

Briggs, Shinohara, and co-workers [11] studied magnetic properties of solvent-free microcrystals of Sc@C_{82}. They found that the microcrystal is a paramagnet and the magnetic susceptibility decreases below 150 K with evidence of antiferromagnetic-like interactions by slow cooling. X-ray crystal analysis shows the presence of a phase transition at 150 K, which can be attributed to an orientational ordering transition of the fullerene molecules (Figure 10.3). In general, solvent-free metallofullerenes show paramagnetic behavior in magnetization curves.

At low temperatures, the slow cooling susceptibility decreases gradually from 4 to 150 K and follows the Curie–Weiss law between 60 K and 110 K (Figure 10.3). Fitting the SQUID susceptibility by a Curie–Weiss curve provides the Curie constant, $C = 0.96(0)$ emu K mol^{-1} Oe^{-1} and Curie temperature, $\theta = -107$ K, and the estimated spin density from the Curie constant is 0.25(6). The susceptibility then increases sharply between 150 K and 175 K. This antiferromagnetic-like behavior is occurring

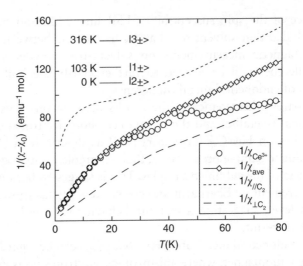

Figure 10.2 Reciprocal magnetic susceptibility of the contribution of the Ce^{3+} f electron experimentally obtained from $\chi_{Ce@C82}-\chi_{La@C82}$ (open circles) and the crystal field analyses. The solid line with diamonds is the average of the calculated susceptibility for the Ce^{3+} f electron using the crystal field parameters. The crystal-field level scheme is also shown. Reproduced with permission from Ref. [10]. Copyright 2003 American Chemical Society

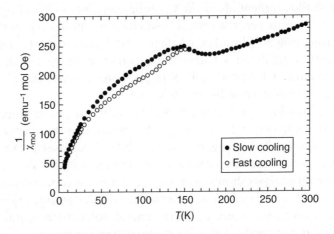

Figure 10.3 Temperature dependence of the inverse magnetic susceptibility of solvent-free $Sc@C_{82}$ micro-crystals for slow and fast cooling. An antiferromagnetic-like transition is seen around 150 K. Reproduced with permission from Ref. [11]. Copyright 2007 John Wiley & Sons

concomitant with a structural transition. A similar magnetic behavior was also reported for $La@C_{82}(CS_2)_{1.5}$ crystals [12].

At temperatures above 175 K, $Sc@C_{82}$ molecules are freely rotating in the crystal. By slowly cooling the crystal, the rotational motion of the

$Sc@C_{82}$ molecules gradually decreases. Below 175 K, the molecules are constricted by an antiferromagnetic-like interaction exerted among the $Sc@C_{82}$ molecules, and the rotational movement is almost frozen at 150 K. The origin of the antiferromagnetic-like interaction is most probably due to the electron spins residing on the C_{82} fullerene cage since the formal charge state of $Sc@C_{82}$ in a $Sc@C_{82}$-toluene (solvent) crystal has been reported as $Sc^{2.2+}@C_{82}^{2.2-}$ based on the charge density obtained from the maximum entropy method (MEM)/Rietveld X-ray analysis [13]. This suggests that the effective electron spin on the fullerene cage is 0.2. At temperatures below 150 K, the antiferromagnetic-like spin coupling increases substantially (Figure 10.3), which explains the observed increase of the susceptibility from 150 to 175 K.

At temperatures above 200 K, the susceptibility again follows the Curie–Weiss law. Fitting the observed susceptibility with a Curie–Weiss curve provides $C = 2.1(0)$ emu K mol^{-1} Oe^{-1} and $q = -309$ K, and the estimated spin density from the Curie constant is 0.56(0). The difference of the spin densities between the high (200–300 K) and low temperature (60–110 K) regions amounts to 0.21. This value corresponds to the observed effective spin on the C_{82} cage (i.e., 0.2), suggesting that the electron spins on the adjacent C_{82} cage tend to couple in an antiferromagnetic-like manner when the molecular rotation freezes below 150 K.

10.2 SXAS AND SXMCD STUDIES OF METALLOFULLERENES

Element-specific magnetism of metallofullerenes can be studied by synchrotron X-ray magnetic circular dichroism (SXMCD). Indeed, synchrotron X-ray absorption spectroscopy (SXAS) and SXMCD are ideal experimental techniques to investigate the electronic and magnetic properties of endohedral metallofullerenes since these high-energy spectroscopies are suitable for dealing with small quantities of material such as endohedral metallofullerenes [14].

This is a crucial advantage considering the time-consuming process of the extraction and purification of metallofullerenes from raw soot and solvent extracts using multi-step high-performance liquid chromatography (Section 2.3) and the consequent scarcity of the highly pure (>99%) metallofullerene materials required for electronic and magnetic studies. Furthermore, with SXAS and SXMCD techniques, it is possible to determine element-specific properties in a composite system such as metallofullerenes. With the advent of high-brilliance polarized soft X-ray synchrotron sources, SXMCD has been developed into a unique probe of the ground-state spin S and orbital L magnetic moments through the use of sum rules [15,16].

The first SXAS and SXMCD experiments on metallofullerenes were reported by De Nadai et al. [14] using the European Synchrotron Radiation Facility (ESRF) in Grenoble. Figure 10.4 shows the isotropic $M_{4,5}$ edges of Gd@C_{82}, Dy@C_{82}, Ho@C_{82}, and Er$_2$@C_{90} recorded at 6 K at normal incidence with respect to the sample surface (open circles). The spectra exhibit a multiplet structure due to electric-dipole transitions from the $3d$ core level to the unoccupied $4f$ levels. In all cases, the agreement between experimental and theoretical spectra is very good. This confirms the trivalent nature of these metallofullerenes studied in agreement with previous works.

The intermolecular magnetic coupling between the entrapped metal centers has been found to be paramagnetic and no magnetic anisotropy has been observed in agreement with earlier SQUID studies. A sum rule analysis of the SXMCD spectra showed that the orbital momentum of metal atoms was quenched by about 50% in comparison with the free M^{3+} ions. Furthermore, the reduction of $\langle L_z \rangle$ and $\langle S_z \rangle$ within the Gd, Dy, Ho, and Er series has been found similar to that of μ_{eff} observed by SQUID earlier [17] but to a greater extent, especially in the Gd@C_{82} and Dy@C_{82} cases where an anomalous branching ratio was observed.

Kitaura et al. [18] reported SXMCD spectroscopy of the Gd and Dy M_5 edges on metallofullerenes (M@C_{82}, M = Gd and Dy), and the

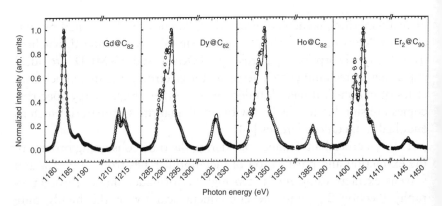

Figure 10.4 $M_{4,5}$ edges SXAS of Gd@C_{82}, Dy@C_{82}, Ho@C_{82}, and Er$_2$@C_{90} measured at normal incidence with respect to the sample surface (open circles). The spectra are normalized to a constant edge jump. The atomic calculations for the respective trivalent 3+ free ion are shown as solid lines. Reproduced with permission from Ref. [14]. Copyright 2004 The American Physical Society

corresponding nanopeapods [(M@C_{82})@SWNT, where SWNT represents single-wall carbon nanotube] in a temperature range between 10 K and 40 K.

Figure 10.5 shows the observed and calculated (normalized) XAS and the corresponding SXMCD spectra of the Gd and Dy M_5 edges on Gd@C_{82} and Dy@C_{82} and their corresponding nanopeapods at 10 K and 1.9 T.

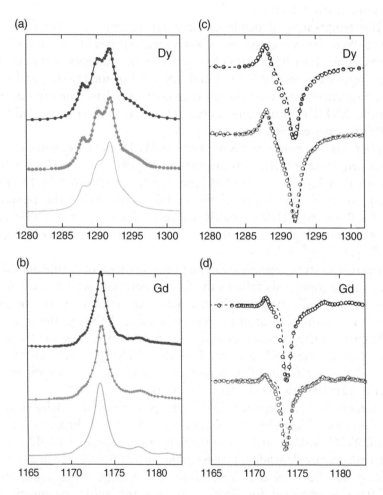

Figure 10.5 Theoretical and observed isotropic XAS and SXMCD spectra of Dy@C_{82}, (Dy@C_{82})@SWNT, Gd@C_{82}, and (Gd@C_{82})@SWNT. (a,b) Theoretical XAS spectra (solid line), and observed XAS spectra of metallofullerenes (upper solid circle) and nanopeapods (lower solid circle). (c,d) Theoretical SXMCD spectra (dotted line), and observed SXMCD spectra of metallofullerenes (upper open circle) and nanopeapods (lower open circle). Reproduced with permission from Ref. [18]. Copyright 2007 The American Physical Society

In spite of the small amount of samples used in these measurements, the observed SXAS and SXMCD spectra exhibit sufficient signal-to-noise ratios. The spectra show a multiplet structure due to electric dipole transitions from the $3d$ core level to the unoccupied $4f$ levels. The difference in SXAS and SXMCD spectral features between metallofullerenes and the corresponding nanopeapods is very small, which shows that the electronic states, such as charge state, of the encapsulated Gd and Dy ions are nearly the same.

The temperature dependence of magnetization was fitted by the Curie–Weiss law which provides Weiss temperatures of -4.6 ± 1.5 K for Dy@C_{82} and 3.5 ± 0.9 K for Gd@C_{82}. Magnetic moments obtained from the Curie–Weiss law (9.5 ± 0.4 and 6.8 ± 0.5 μ_B for Dy@C_{82} and Gd@C_{82}, respectively) are much larger than computed based on the sum rule in other SXMCD measurements but are similar to those estimated from SQUID measurements.

They also performed SXAS and SXMCD measurements on the so-called metallofullerene nanopeapods [19] (Figure 10.6) which encapsulate Dy@C_{82} and Gd@C_{82} molecules inside SWNTs. Interestingly, magnetic moments of the metal atoms inside the peapods, 11.4 ± 0.4 μ_B for (Dy@C_{82})@SWNT and 7.2 ± 0.3 μ_B for (Gd@C_{82})@ SWNT, were higher than those in bulk metallofullerenes (Figure 10.7) [18].

The temperature dependence of magnetic moments of the metallofullerenes and nanopeapods follows the Curie–Weiss law with a small Weiss temperature, indicating that the magnetic interaction between encapsulated rare-earth metal atoms is relatively weak. Although the observed differences in Curie constants and Weiss temperatures between Gd@C_{82} and (Gd@C_{82})@SWNT are small, those of Dy@C_{82} and (Dy@C_{82})@ SWNT are significant. This observation is consistently explained by charge transfer-induced crystal field effects.

Okimoto et al. [20] reported XMCD studies of erbium di-metallofullerenes such as Er@C_{82}-$C_{2v}(9)$, Er$_2$@C_{82}-$C_{2v}(9)$, Er$_2$C$_2$@C_{82}-$C_{2v}(9)$, Er$_2$C$_2$@C_{82}-$C_s(6)$, and ErYC$_2$@C_{82}-$C_s(6)$ at 1.9 T in the temperature range of 10–40 K [20]. The temperature-dependent magnetization at 1.9 T measured in a cooling process was fitted by the Curie–Weiss law. A large Weiss temperature of -18.8 K obtained for Er@C_{82} suggests the antiferromagnetic-like intermolecular interactions caused by the open-shell configuration of the carbon cage. In contrast, in erbium di-metallofullerenes with closed-shell cage configurations Weiss temperatures were reduced to -3 to -7 K.

(a)

(b)

(c)

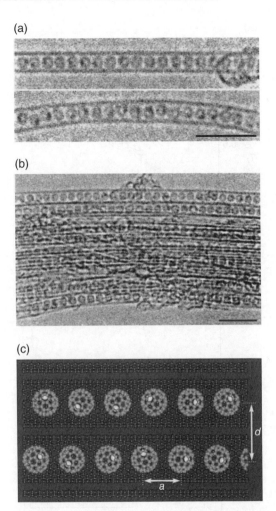

Figure 10.6 (a,b) High-resolution transmission electron microscopy (HRTEM) images of the isolated and bundled (Gd@C$_{82}$)@SWNT, respectively. Dark spots seen on most of the fullerene cages correspond to the encapsulated Gd atoms that are oriented randomly with respect to the tube axis (bar = 5 nm). (c) A schematic representation of the metallofullerene containing SWNTs: (Gd@C82)$_n$@SWNTs. Reproduced with permission from Ref. [19]. Copyright 2000 The American Physical Society

Effective magnetic moments determined from the fit were in the range of 8.4–8.5 ± 0.3 μ_B for all erbium di-metallofullerenes with C$_{82}$-C$_{2v}$(9) carbon cage and 7.8 ± 0.3 μ_B for ErC$_2$@C$_{82}$ and ErYC$_2$@C$_{82}$ with C$_s$(6) carbon cage.

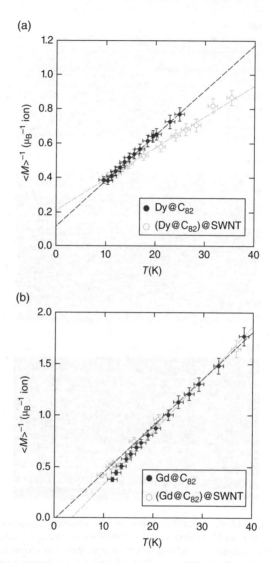

Figure 10.7 Plots of reciprocal magnetization in external magnetic field $B = 1.9$ T. (a) Plot of Dy@C$_{82}$ and (Dy@C$_{82}$)@SWNT and (b) plot of Gd@C$_{82}$ and (Gd@C$_{82}$)@ SWNT. The dotted line represents fitting results by the Curie–Weiss law. Reproduced with permission from Ref. [18]. Copyright 2007 The American Physical Society

REFERENCES

[1] Funasaka H, Sakurai K, Oda Y *et al.* 1994 Magnetic properties of Gd@C82 metallofullerene *Chem. Phys. Lett.* **232** 273

[2] Funasaka H, Sugiyama K, Yamamoto K and Takahashi T 1995 Magnetic properties of rare-earth metallofullerenes *J. Phys. Chem.* **99** 1826

[3] Dunsch L, Eckert D, Froehner J et al. 1998 *Fullerenes: Recent Advances in the Chemistry and Physics of Fullerenes and Related Materials*, Vol. 6, eds K Kadish and R Ruoff (Pennington, NJ: Electrochemical Society), pp. 955–966

[4] Huang H J, Yang S H and Zhang X X 1999 Magnetic behavior of pure endohedral metallofullerene Ho@C$_{82}$: a comparison with Gd@C$_{82}$ *J. Phys. Chem. B* **103** 528

[5] Allen, K. 1998 Intercalation chemistry of alkali metal fullerides PhD thesis, University of Oxford.

[6] Watanuki T, Suematsu H, Nakao Y et al. 1997 ESR of La@C$_{82}$ crystals The 13th Fullerene Symposium (Nagano), Abstract, p. 83

[7] Hino S, Takahashi H, Iwasaki K et al. 1993 Electronic structure of metallofullerene LaC$_{82}$: Electron transfer from lanthanum to C$_{82}$ *Phys. Rev. Lett.* **71** 4261

[8] Poirier D M, Knupfer M, Weaver J H et al. 1994 Electronic and geometric structure of La@xaC$_{82}$ and C$_{82}$: Theory and experiment *Phys. Rev. B* **49** 17403

[9] Kessler B, Bringer A, Cramm S et al. 1997 Evidence for incomplete charge transfer and la-derived states in the valence bands of endohedrally doped La@C$_{82}$ *Phys. Rev. Lett.* **79** 2289

[10] Inakuma M, Kato H, Taninaka A et al. 2003 Magnetic anisotropy of cerium endohedral metallofullerenes *J. Phys. Chem. B* **107** 6965

[11] Ito Y, Fujita W, Okazaki T et al. 2007 Magnetic properties and crystal structure of solvent-free Sc@C$_{82}$ metallofullerene microcrystals *Chem. Phys. Chem.* **8** 1019

[12] Nuttall C J, Inada Y, Nagai K and Iwasa Y 2000 Lanthanide metallofullerene crystals *Phys. Rev. B* **62** 8592

[13] Nishibori E, Takata M, Sakata M et al. 1998 Determination of the cage structure of Sc@C$_{82}$ by synchrotron powder diffraction *Chem. Phys. Lett.* **298** 79

[14] De Nadai C, Mirone A, Dhesi S S et al. 2004 Local magnetism in rare-earth metals encapsulated in fullerenes *Phys. Rev. B* **69** 184421

[15] Thole B T, Carra P, Sette F and van der Laan G 1992 X-ray circular dichroism as a probe of orbital magnetization. *Phys. Rev. Lett.* **68** 1943

[16] Carra P, Thole B T, Altarelli M and Wang X, 1993 X-ray circular dichroism and local magnetic fields *Phys. Rev. Lett.* **70** 694

[17] Huang H J, Yang S H and Zhang X X 2000 Magnetic properties of heavy rare-earth metallofullerenes M@C$_{82}$ (M = Gd, Tb, Dy, Ho, and Er). *J. Phys. Chem. B* **104** 1473

[18] Kitaura R, Okimoto H and Shinohara H 2007 Magnetism of the endohedral metallofullerenes M@C$_{82}$ (M=Gd, Dy) and the corresponding nanoscale peapods: Synchrotron soft x-ray magnetic circular dichroism and density-functional theory calculations *Phys. Rev. B* **76** 172409

[19] Hirahara K, Suenaga K, Bandow S et al. 2000 One-dimensional metallofullerene crystal generated inside single-walled carbon nanotubes *Phys. Rev. Lett.* **85** 5384

[20] Okimoto H, Kitaura R, Nakamura T et al 2008 Element-specific magnetic properties of di-erbium Er$_2$@C$_{82}$ and Er$_2$C$_2$@C$_{82}$ metallofullerenes: a synchrotron soft X-ray magnetic circular dichroism study *J. Phys. Chem. C* 2008 **112** 6103

11

Organic Chemistry of Metallofullerenes

The last couple of decades has seen significant effort put into the preparation of bulk amounts of endohedral metallofullerenes (EMFs) allowing the development of their chemical functionalization. The reactivity of EMFs can sometimes be significantly different from that of the empty cages and therefore it is fundamentally important to survey known functionalization methodologies, while also develop new ones. Then, essential information in terms of reactivity and reaction mechanisms, closely associated with the nature and number of the encapsulated species, may result which could enhance research in this hot field. Additionally, with the chemical functionalization of EMFs it is possible to introduce functional organic moieties onto the carbon cage and thus yield novel hybrid materials with interesting properties with potential applications in diverse fields. As expected, the organic chemistry of EMFs follows the well-established fullerene chemistry, though only a fraction of the known reactions for empty cages have been successfully applied to EMFs, with certain details yet to be revealed and understood [1,2].

11.1 CYCLOADDITION REACTIONS

11.1.1 Disilylation

The very first example of functionalization of EMFs was the reaction of $La@C_{82}$ with 1,1,2,2,-tetrakis(2,4,6-trimethylphenyl)-1,2-disilirane [3], yielding metallofullerene derivative **1**, according to Figure 11.1.

Endohedral Metallofullerenes: Fullerenes with Metal Inside, First Edition.
Hisanori Shinohara and Nikos Tagmatarchis.
© 2015 John Wiley & Sons, Ltd. Published 2015 by John Wiley & Sons, Ltd.

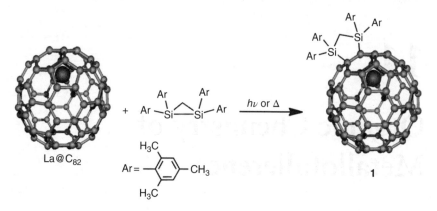

Figure 11.1 Disilylation reaction of La@C_{82} with 1,1,2,2,-tetrakis(2,4,6 -trimethylphenyl)-1,2-disilirane

This is a photochemical reaction, in which the reaction mixture was photoirradiated with a tungsten-halogen lamp (cut-off < 400 nm) in a degassed sealed tube. In addition, La@C_{82} shows higher reactivity toward disilirane as compared with C_{60} (or C_{70}), which can be rationalized in terms of its stronger electron donor and acceptor properties. Furthermore, La@C_{82} thermally also reacts with 1,1,2,2-tetrakis(2,4,6-trimethylphenyl)-1,2-disilirane, contrary to the case of empty fullerenes which do not react thermally.

The formation of the corresponding mono-adduct was justified by laser-desorption time-of-flight (TOF) mass spectrometry, in which the peak corresponding to the molecular ion of **1**, as well as of La@C_{82} as the major fragment, was evident, while peaks associated with higher adducts were simply missing. The electron spin resonance (ESR) spectrum of adduct **1** showed a set of octets, with hyperfine coupling constants relatively higher than the corresponding one for intact La@C_{82}, not only thus proving the formation of a regioisomeric mixture, but also suggesting that the La atom is less positively charged in **1** as compared with intact La@C_{82}.

In a similar fashion, Gd@C_{82} was also modified either upon photoirradiation or thermally, with disilirane, thus highlighting the applicability of the disilylation reaction [4]. Therefore, the next step was to test the reactivity of di-metallofullerenes La$_2$@C_{80} and Sc$_2$@C_{80} toward 1,1,2,2-tetrakis(2,4,6-trimethylphenyl)-1,2-disilirane. Again, both di-metallofullerenes M$_2$@C_{80} (M = La, Sc) were successfully functionalized applying either photochemical or thermal reaction conditions [5]. In addition, it was found that the two La atoms in disilirane functionalized

$La_2@C_{80}$ hop two-dimensionally along the equator of the C_{80} cage, thus suggesting the potentiality to induce unique electronic and magnetic fields in the direction perpendicular to the plane in such modified metallofullerenes.

In order to evaluate and compare the reactivity of $Sc_3N@C_{80}$ belonging to the trimetal nitride template (TNT) class of fullerenes, with that of $M_2@C_{80}$, the same disilylation reaction was carried out. Although both $Sc_3N@C_{80}$ and $La_2@C_{80}$ possess the same electronic structure described as $(Sc_3N)^{6+}@C_{80}^{6-}$, it was found that $Sc_3N@C_{80}$ only photochemically reacts with 1,1,2,2-tetrakis(2,4,6-trimethylphenyl)-1,2-disilirane, but not thermally. Then, by simply comparing the redox potentials of the two materials, it was easily understood that although $La_2@C_{80}$ and $Sc_3N@C_{80}$ show similar oxidation potentials, the first reduction potential of $Sc_3N@C_{80}$ is much more negative than that of $La_2@C_{80}$, thus suggesting that $Sc_3N@C_{80}$ is much less reactive toward nucleophiles such as 1,1,2,2-tetrakis(2,4,6-trimethylphenyl)-1,2-disilirane, which is in line with the thermal inactivity of $Sc_3N@C_{80}$ [6].

Structural characterization of the mono-adduct of $Sc_3N@C_{80}$ with 1,1,2,2-tetrakis(2,4,6-trimethylphenyl)-1,2-disilirane, based on NMR measurements, revealed the formation of a 3:2 isomeric mixture, which could not be resolved even with recycling high-performance liquid chromatography (HPLC). Following careful analysis of the ^1H NMR spectrum, it was concluded that the major isomer derived from the addition of disilirane on the bond junction between five- and six-membered rings of $Sc_3N@C_{80}$ resulted in the formation of a closed structure (1,2-adduct), while the minor isomer is the 1,4-adduct. Importantly, the former was found to thermally isomerize to the latter, thus meaning that the 1,2-adduct is thermodynamically less stable than the 1,4-adduct, though the former is more favorable kinetically [7]. Along the same lines, the bis-silylated adducts of $Lu_3N@C_{80}$ were synthesized and spectroscopic analyses revealed that the most stable isomers were the ones formed upon 1,4-addition [8]. Moreover, electrochemical and theoretical studies demonstrated that the highest occupied molecular orbital-lowest unoccupied molecular orbital (HOMO-LUMO) gap is smaller in the bis-silylated derivatives as compared with that of intact $Lu_3N@C_{80}$.

An interesting observation during the functionalization of di-metallofullerenes with disilirane was that the circular motion of the encapsulated metal atoms could be controlled by the exohedral addition. Based on ^{13}C NMR measurements, it was found that in $Ce_2@C_{80}$ the two Ce atoms freely circulate inside the I_h-C_{80} cage. When $Ce_2@C_{80}$ reacted with 1,1,2,2-tetrakis(2,4,6-trimethylphenyl)-1,2-disilirane, the silylated

$Ce_2@C_{80}$-based mono-adduct was isolated via HPLC. However, its ^1H NMR spectrum revealed two sets of signals, thus suggesting the presence of two conformers, which can interconvert one to the other by conformational changes of the disilirane moiety.

Thus, and in combination with X-ray analysis, it was revealed that the free random motion of the two Ce atoms in the $Ce_2@C_{80}$ is fixed at certain positions by the exohedral addition of disilirane – this is completely in contrast to the two-dimensional horizontal hopping observed in functionalized $La_2@C_{80}$ [9]. Notably, the attachment of the silicon substituent on $Ce_2@C_{80}$ regulates the position of the encapsulated metal atoms under the equator inside the fullerene cage. Furthermore, addition of disilirane can tune the electronic properties of metallofullerenes as the silylated cages become electron-rich as a result of the electron donation from the disilirane moiety [10].

Regardless of the great interest developed for $La_2@C_{80}$ and $Ce_2@C_{80}$ because of the three-dimensional random motion of the two encapsulated metal atoms, the dynamic behavior of the metals in different metallofullerene cages remained partially explored. In this context, the bis-silylation of $Ce_2@C_{78}$ was carried out under similar reaction conditions applied for the functionalization of $Ce_2@C_{80}$ [10]. Based on X-ray crystallographic analysis it was found that the corresponding mono-adduct results from the regioselective 1,4-addition of disilirane to $Ce_2@C_{78}$. Interestingly, the two Ce atoms in $Ce_2@C_{78}$ do not move facing the hexagonal rings at the equator of functionalized $Ce_2@C_{78}$, being more tightly localized on the C_3 axis of the C_{78} cage as compared with the case of the intact metallofullerene [11].

Notably, although the addition of disilirane was shown to proceed well with a variety of mono- and di-metallofullerenes as well as of $Sc_3N@C_{80}$, the situation with the reduced and oxidized forms of $M@C_{82}(M = Y, La, Ce)$ was totally different. Thus, reduced $[M@C_{82}]^-$ did not react at all with disilirane, either photochemically or thermally, while the oxidized form $[M@C_{82}]^+$ behaved similar to $M@C_{82}$, namely giving the corresponding mono-adducts. The former behavior was rationalized in terms of the redox properties and HOMO/LUMO levels of $[M@C_{82}]^-$, thus confirming that the reactivity of $M@C_{82}$ can be tuned by ionization [12].

11.1.2   Azomethine Ylides

Azomethine ylides are organic 1,3-dipoles possessing a carbanion next to an immonium ion and can be readily produced upon decarboxylation of the immonium salts derived from the condensation of α-amino acids

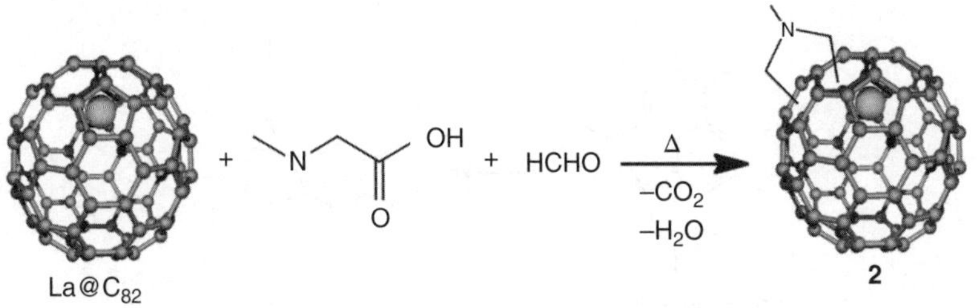

Figure 11.2 1,3-Dipolar cycloaddition reaction of in-situ generated azomethine ylides to La@C$_{82}$

with aldehydes or ketones. When they are added to fullerenes, a fullero-pyrrolidine mono-adduct is formed in which a pyrrolidine ring is fused to the junction between two six-membered rings of the fullerene [13,14]. Notably, functionalized aldehydes lead to the formation of 2-substituted fulleropyrrolidines, whereas reaction with N-substituted glycines leads to N-substituted fulleropyrrolidines.

The very first reaction of azomethine ylides with EMFs goes back to 2004. In that frame, azomethine ylide generated in situ from the thermal reaction of methyl glycine and formaldehyde adds to Gd@C$_{82}$ furnishing the corresponding pyrrolidine mono-adduct which was characterized by MALDI-TOF mass spectrometry and electronic absorption spectroscopy.

Thermal reaction of methyl glycine and formaldehyde with the major isomer of La@C$_{82}$ gave a mixture of products comprising of mono- as well as bis-adducts. Careful recycling HPLC of the reaction mixture resulted in the separation and isolation of two mono-adducts, as proved by mass spectrometry. Based on the mass spectrum, it became clear that the corresponding azomethine ylide was added to La@C$_{82}$ forming a fused pyrrolidine ring onto the carbon cage (Figure 11.2). From the UV-Vis-NIR (ultraviolet-visible-near infrared) spectra recorded, it was found that the characteristic absorptions of metallofulleropyrrolidine adduct 2 were slightly blue-shifted compared with the ones of intact metallofullerene, thus suggesting that the electronic structure of functionalized La@C$_{82}$ was slightly, if at all, altered compared with that of the parent La@C$_{82}$ material. However, at that time, information regarding the symmetry of the metallofulleropyrrolidine adduct and, importantly, of the reaction site to which the pyrrolidine ring was fused had yet to be obtained [15].

Having shown that the 1,3-dipolar cycloaddition of azomethine ylides is an efficient reaction for the functionalization of EMFs, it was then applied to modify TNT fullerenes. In this context, pyrrolidine derivatives of Sc$_3$N@C$_{80}$ and Er$_3$N@C$_{80}$ were prepared. In both Sc$_3$N@C$_{80}$ and

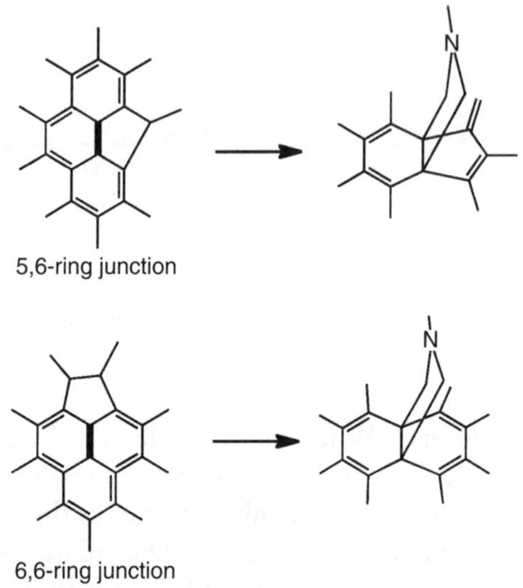

5,6-ring junction

6,6-ring junction

Figure 11.3 1,3-Dipolar cycloaddition reaction of azomethine ylides forming pyrrolidine rings fused to 5,6- and 6,6-ring junctions

$Er_3N@C_{80}$ materials, the cage was I_h symmetrical, meaning that there are two equal double bonds for reaction with the azomethine ylides, the 5,6- and the 6,6-ring junctions, namely corannulene- and pyrene-type, respectively. Based on ^{13}C NMR studies, it was then concluded that in the pyrrolidine adduct of $Sc_3N@C_{80}$ the 5,6-ring double bond was reacted and not the 6,6-ring, as schematically shown in Figure 11.3. The UV-Vis absorption spectra of the TNT fullerenes were similar to those of their mono-adducts, thus suggesting that the pyrrolidine derivatives retain the main aromatic cage features of the parent metallofullerenes [16,17].

On the other hand, addition of N-tritylazomethine ylide to I_h symmetrical $Sc_3N@C_{80}$ resulted in the formation of both 5,6- and 6,6-pyrrolidine mono-adducts **3** and **4**, respectively (Figure 11.4), which were fully characterized by NMR spectroscopy and X-ray crystallography. Moreover, the 6,6-isomer was ascribed as the kinetic product, while the 5,6-isomer was ascribed as the more stable thermodynamic product. The latter finding was rationalized in terms of NMR studies and the evaluation of the resonances obtained. In this context, it was realized that the 5,6-ring junction possesses a horizontal plane of symmetry, leading to a tritylpyrrolidine derivative with two equivalent methylene carbons each bearing nonequivalent germinal protons in the pyrrolidine ring, while due to the absence of a horizontal symmetric plane for the 6,6-ring unction, the corresponding pyrrolidine adduct possesses two

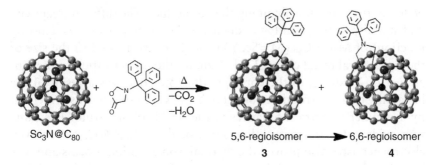

Sc$_3$N@C$_{80}$ 5,6-regioisomer ⟶ 6,6-regioisomer

3   **4**

Figure 11.4 1,3-Dipolar cycloaddition reaction of N-tritylazomethine ylide to Sc$_3$N@C$_{80}$ yielding 5,6- and 6,6-regioisomers **3** and **4**

nonequivalent methylene carbons, while the two sets of their protons are similarly nonequivalent. Furthermore, considering the nonplanar nature of the pyrrolidine ring, the 6,6-isomer was expected to exist as a pair of enantiomers that could then interconvert through inversion of the pyrrolidine ring [18].

However, when the D_{5h} symmetrical isomer of Sc$_3$N@C$_{80}$ was subjected to the 1,3-dipolar cycloaddition reaction with the N-tritylazomethine ylides, it showed significantly higher reactivity as compared with the corresponding I_h isomer. Similarly with the case of I_h-Sc$_3$N@C$_{80}$, two pyrrolidine mono-adducts for D_{5h}-Sc$_3$N@C$_{80}$ were formed, corresponding to the 5,6- and 6,6-regioisomers [19]. However, an independent theoretical study showed that actually only the 5,6-isomer of D_{5h}-Sc$_3$N@C$_{80}$ should be formed, ruling out the formation of the pyracylene adduct on the grounds of reaction energies and calculated NMR data [20].

Interestingly, when Y$_3$N@C$_{80}$ was subjected to the same azomethine ylides functionalization reaction applied for Sc$_3$N@C$_{80}$, the product obtained was unexpectedly different. Although a pyrrolidine ring was fused onto the carbon cage of Y$_3$N@C$_{80}$, the cycloaddition did not occur at the 5,6-double bond as in the case of Sc$_3$N@C$_{80}$, [17] but rather regioselectively at the 6,6-ring junction. This observation clearly indicated that the reactivity of TNT metallofullerenes toward exohedral 1,3-dipolar cycloaddition of azomethine ylides is profoundly affected and effectively controlled by the nature of the encapsulated metal cluster [21]. Isomerization of the 6,6-regioisomer of the Y$_3$N@C$_{80}$ derivative gives the corresponding 5,6-pyrrolidine mono-adduct through thermal rearrangement.

This result indicates that the 5,6-adduct is the most energetically stable one and the transition state to obtain the 6,6-regioisomer should be lower than that for obtaining the 5,6-adduct. The different reactivity and stability of the resulting adducts (i.e., 5,6-ring junction vs. 6,6-ring junction) of $M_3N@C_{80}$ ($M = Sc, Y$) should be also related to the size of the endohedral metal cluster [22]. Actually, the isomerization of the 6,6-adduct of $Y_3N@C_{80}$ to the corresponding 5,6- one was a result of the instability of the pyrrolidine adduct of $Y_3N@C_{80}$ in solution, as evidenced by a number of extra resonances in the ^1H NMR spectrum. The same was evident for the pyrrolidine functionalized $Er_3N@C_{80}$. Moreover, electrochemistry also proved the thermal isomerization process and was introduced as a meaningful probe to identify the 5,6- and 6,6-regioisomers of TNT fullerenes [23].

Taking into account theoretical studies, in which the planar encapsulated Sc_3N cluster causes a local outward pyramidalization of the C—C cage bonds close to the Sc nuclei, it was suggested that the corannulene 5,6-double bonds next to Sc have the highest strain of all the bonds in the I_h-C_{80} cage, and therefore they are the most reactive toward the 1,3-dipolar cycloaddition reaction [24]. It is also interesting to note that crystallographic studies indicated that the Y_3N cluster freely moves inside the cage, however, without interacting with the pyrrolidine addend [25].

Since it was claimed that the size of the encapsulated metal cluster in TNT fullerenes affects the formation and stability of certain regioisomers toward the 1,3-dipolar cycloaddition reaction of azomethine ylides, it was not surprising that the metallofullerene incorporating the largest lanthanide metal cluster, namely $Gd_3N@C_{80}$, was chosen as a suitable candidate to prove that theory. In addition, the mixed TNT metallofullerene $Sc_xGd_{3-x}N@C_{80}$ ($x = 1 - 3$), was also subjected to the same reaction in a comparative study. Thus, it was found that the 6,6-adduct of $Gd_3N@C_{80}$ is the major product, in sharp contrast to the case of $Sc_xGd_{3-x}N@C_{80}$ ($x = 1 - 3$), in which the 5,6-adducts are the major ones [26].

Recently, a comprehensive study was performed which shed light on the stability of pyrrolidine adducts and especially the 5,6- vs. 6,6-adduct formation. Thus, the 1,3-dipolar cycloaddition reaction of azomethine ylides with $M_3N@C_{80}$ ($M = Sc, Lu, Y, Gd$) gave rise to two mono-adducts, identified as expected as 5,6- and 6,6-adducts. Careful screening of the reaction products revealed that initially the 6,6-adduct is formed followed by the 5,6-adduct by complete or partial thermal isomerization through appropriate rearrangement. Importantly, the latter rearrangement and isomerization is dependent on the size of the encapsulated metal cluster. Moreover, it was shown that for the larger yttrium and gadolinium nitride clusters, the conversion is reversible [27].

Having shown that the size of metal cluster in TNT fullerenes governs the exohedral chemical reactivity toward the 1,3-dipolar cycloaddition reaction of azomethine ylides, enhancing the formation of the 6,6-regioisomer as the size becomes bigger, it was then interesting to reveal if the addition pattern is affected by the electronic properties of the TNT fullerenes. In this context, a series of $Sc_{3-x}Y_xN@C_{80}$ $(x=0-3)$ was synthesized and a comparative study on their electronic properties and reactivity was performed. Based on the similarities the electronic absorption spectra of $Sc_{3-x}Y_xN@C_{80}$ $(x=0-3)$ show, it is evident that the metal cluster donates six electrons to the I_h-C_{80} cage, namely the electronic structure is $(Sc_{3-x}Y_xN)^{6+}@(C_{80})^{6-}$ $(x=0-3)$. When $Sc_{3-x}Y_xN@C_{80}$ $(x=0-3)$ was reacted with N-ethylglycine and formaldehyde, it was shown that $Sc_3N@C_{80}$ and $Sc_2YN@C_{80}$ yield only the 5,6-regioisomer adduct, but in $ScY_2N@C_{80}$ the reactivity changes, and the 6,6-regioisomer appears as a minor fraction, while $Y_3N@C_{80}$ exclusively yields the 6,6-adduct [28].

Beyond the $M_3N@C_{80}$ series of TNT fullerenes that can be abundantly isolated in macroscopic quantities, significant efforts allowed the isolation of pure $Sc_3N@C_{78}$ possessing D_{3h} molecular symmetry in bulk amounts. X-ray crystallography revealed that the Sc atoms are localized over the three pyracylene patches [29,30]. Moreover, in $Sc_3N@C_{78}$ there are strong interactions between the metal cluster Sc_3N and the fullerene cage, namely the Sc_3N is constrained to the horizontal D_{3h} plane of the C_{78} cage most likely due to its ellipsoidal scheme [31,32]. The latter represents a significant property of $Sc_3N@C_{78}$ thus allowing the exploration of the regiocontrol of exohedral adduct docking on the metallofullerene sphere.

Along these lines, $Sc_3N@C_{78}$ was functionalized following the established protocol of 1,3-dipolar cycloaddition with the N-tritylazomethine ylide, furnishing two mono-adducts. The latter were separated and isolated in pure forms with the aid of recycling HPLC. Considering that D_{3h} ellipsoidal C_{78} has 8 different carbon atoms and 13 sets of C—C bonds, careful NMR studies revealed that the fused pyrrolidine ring was located on two different 6,6-ring junction sites, which are offset from the horizontal plane defined by the encapsulated metal cluster [33].

Another aspect of the versatility of the 1,3-dipolar cycloaddition reaction of azomethine ylides is that it can undergo a retro-cycloaddition which leads to the alkene and the azomethine ylide. Thus, when N-ethylpyrrolidino-$M_3N@C_{80}$ $(M=Sc, Y)$ was thermally treated in the presence of maleic anhydride, the retro-cycloaddition reaction occurred

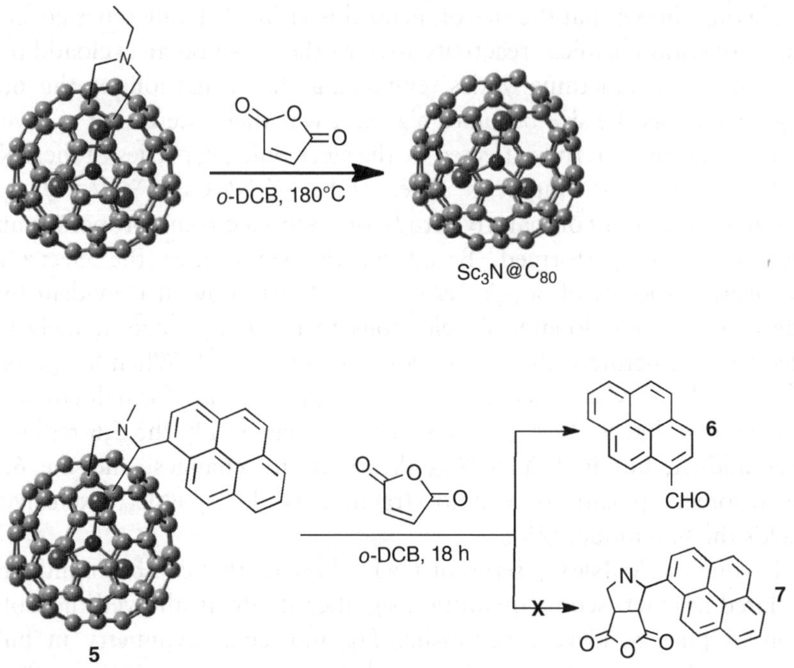

Figure 11.5 1,3-Dipolar retro-cycloaddition reactions of N-ethylpyrrolidino-$Sc_3N@C_{80}$ and pyrene-substituted pyrrolidino-$Sc_3N@C_{80}$ 5

quantitatively yielding $Sc_3N@C_{80}$ (Figure 11.5). Although the mechanism of the retro-cycloaddition has yet to be clarified, it is believed to proceed by the thermal formation of the azomethine ylide, which in turn is trapped by the maleic anhydride. Interestingly, when pyrrolidine-modified $Sc_3N@C_{80}$ material 5 possessing a pyrene unit as substituent at the α-C of the pyrrolidine ring was heated in the presence of maleic anhydride, the corresponding pyrene-substituted pyrrolidine adduct 6 was not formed. Instead, aldehyde 7 was isolated, thus suggesting that oxygen may play a role in the evolution of the 1,3-dipole [34], as shown in Figure 11.5.

In di-metallofullerene $La_2@C_{80}$ the two La atoms randomly rotate inside the C_{80} cage. In order to investigate if this random motion of the La atoms is affected by the 1,3-dipolar cycloaddition of azomethine ylides, $La_2@C_{80}$ was functionalized with 3-triphenylmethyl-5-oxazolidinone. As there are two different C—C bonds on the C_{80} cage to be attacked by the azomethine ylide, namely the coranulene-type 5,6- and the pyrene-type 6,6-, two isomers were produced. The major one was found to be the 6,6-adduct based on crystallographic X-rays analysis.

Interestingly, in the 6,6-regioisomer the two La atoms were found fixed, unlike the random circulation in the parent $La_2@C_{80}$, thus showing that the motion of encapsulated metals can be regulated by specific exohedral addition patterns [35].

Furthermore, in order to shed light on whether the position of the encapsulated metal atoms is controlled by the addition site of the addends, pyrrolidinodimetallofullerenes of $Ce_2@C_{80}$ were synthesized. The two regioisomers of 5,6- and 6,6-$Ce_2@C_{80}$ adducts, similar to the case of the $La_2@C_{80}$, were produced, however, in different abundancy, namely in a 1:1 ratio [35]. As the only difference between $Ce_2@C_{80}$ and $La_2@C_{80}$ is the number of f electrons of the encapsulated metal atom, the only explanation for this different ratio of formation might be due to stability grounds of the adducts. In any case, the conclusion is that the metal positions inside the fullerene cage can be controlled by means of the addition positions of the addends [36].

Evidently, exohedral functionalization of EMFs plays a significant role not only for improving solubility and processability of the modified metallofullerenes but also in controlling the position of encapsulated metal species and their corresponding properties. Since $Sc_3C_2@C_{80}$ is paramagnetic species, in order to record its ^{13}C NMR spectrum the metallofullerene had to be reduced to become diamagnetic [37]. However, the pyrrolidine mono-adduct of $Sc_3C_2@C_{80}$, synthesized by the 1,3-dipolar cycloaddition of azomethine ylides in-situ derived upon thermal condensation of N-methylglycine and ^{13}C-enriched formaldehyde, showed a singlet for the ^{13}C-labeled methylene carbon.

The latter finding indicates that the low spin density distribution of the metal carbide cluster, which is close to the addend, does not significantly affect the fused pyrrolidine moiety. In addition, density functional theory (DFT) calculations predicted that for the Sc_3C_2 cluster, one Sc atom is located at the bottom of the cage and far away from the fused pyrrolidine ring in the mono-adduct of $Sc_3C_2@C_{80}$ while the other two Sc atoms are positioned symmetrically and close to the pyrrolidine moiety. Also, based on ESR studies it was shown that although the Sc_3C_2 cluster freely rotates in the parent $Sc_3C_2@C_{80}$ metallofullerene, for the pyrrolidine mono-adduct the presence of the fused pyrrolidine ring on the C_{80} cage hinders that free rotation thus leading to inhomogeneous spin density distribution on the internal cluster [38].

In an attempt to elucidate the structure of the metal carbide EMF $Sc_2C_2@C_{82}$, which was initially assumed to be $Sc_2@C_{84}$, based on NMR studies and theoretical calculations [39], the regioselective 1,3-dipolar cycloaddition reaction with 3-triphenylmethyl-5-oxazolidinone was performed. With the aid of MALDI-TOF, the formation of the pyrrolidine mono-adduct of $Sc_2C_2@C_{82}$ was proved. Furthermore, X-ray crystallography revealed that the pyrrolidine moiety is fused to a 6,6-ring junction far from either of the two Sc atoms, which can be reasonably explained by considering the molecular orbital distribution and the bond strains of the metallofullerene. Additionally, it became evident that the addition of the pyrrolidine ring does not influence the orientation of the encapsulated metal carbide cluster, however, it significantly alters the electronic and electrochemical properties of $Sc_2C_2@C_{82}$ [40].

11.1.3   Bingel Cyclopropanation

Another widely used cycloaddition reaction for EMFs is the Bingel cyclopropanation reaction. This versatile modification involves the generation of carbon nucleophiles from α-halo esters and their subsequent regioselective addition to metallofullerenes. In general, the addition takes place exclusively on double bonds either between two six-membered rings or between a pentagon-hexagon junction of the fullerene skeleton, yielding methano-metallofullerenes.

Notably, in stark contrast to empty fullerenes, the Bingel addition to EMFs can lead to the formation of 6,6- open structures with metal atoms located adjacent to the site of the addition [41]. Obviously, the nature and size of the encapsulated metal play an important role in directing the addition sites to EMFs. Furthermore, as it was shown for the case of $Y_3N@C_{80}$, the Bingel cycloaddition reaction gave cycloadducts at the 5,6-junction as derived upon thermal isomerization of the less stable adducts at the closed 6,6-junction, Interestingly, computational studies revealed that the 6,6-closed adducts were less stable and tend to lead always to the open 6,6-adducts [41].

The novelty of this type of organic functionalization is explained in terms of the wide number of different methano-metallofullerene materials that can be reached due to the presence of ester moieties: either properly functionalized carbon nucleophiles can be constructed that ultimately would add to the metallofullerene, or additional chemical transformation to the added group of an already synthesized methano-metallofullerene could occur. Thus, theoretically, any chemical group carrying novel physical, electronic, magnetic, or mechanical properties that fulfill the appropriate requirements for the construction of novel metallofullerene-based materials can be properly designed and synthesized. In recent years, modifications of the original Bingel reaction

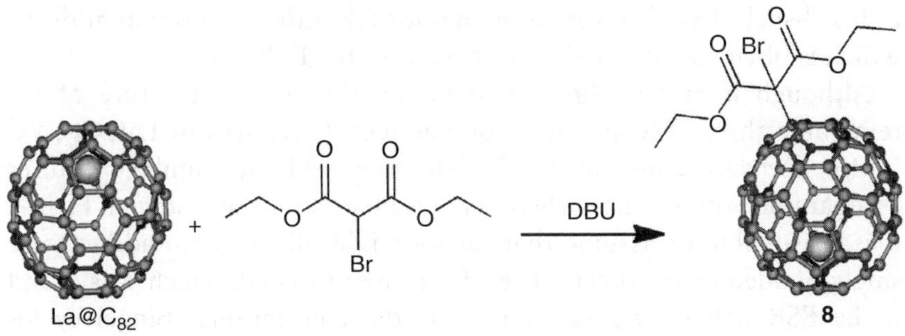

Figure 11.6 Bingel-type cycloaddition reaction on La@C$_{82}$ yielding the singly bonded La@C$_{82}$CBr(COOC$_2$H$_5$)$_2$ derivative **8**

have also appeared. In those modifications, novel strategies utilizing (i) carbanionic precursors to methano-metallofullerenes other than malonates and (ii) alternative pathways generating the reactive monohalomalonate intermediate in situ, have been developed.

The Bingel reaction conducted with La@C$_{82}$ and diethyl bromomalonate in the presence of 1,8-diazabicyclo[5,4,0]-undec-7-ene (DBU) yielded a mixture of mono- and multi-adducts [42]. The most abundantly produced mono-adduct was isolated with the aid of HPLC and found to be ESR-inactive in sharp contrast to the paramagnetic nature of the parent La@C$_{82}$. The structure of mono-adduct **8** (Figure 11.6), in which the bromomalonate moiety is attached to the C$_{82}$ cage by a single bond, contrary to the conventional Bingel cyclopropanation reaction yielding fused cyclopropane rings on empty fullerenes, was proved with ^{13}C NMR together with X-rays. Interestingly, the electronic absorption spectrum of **8** shows characteristic absorptions different from those intact La@C$_{82}$ displays, as the onset is blue-shifted. The latter suggests that the HOMO-LUMO band gap of **8** is larger than that of La@C$_{82}$, thus it can be more easily reduced or oxidized.

The presence of DBU as a base in the Bingel reaction could create problems, since charge-transfer phenomena with the metallofullerene might occur. Interestingly, the formation of the anion (La@C$_{82}$)$^-$ due to electron transfer from DBU is facilitated and therefore the amount of the base added in the Bingel functionalization reaction should be kept to a minimum. If this is the case, then five mono-adducts are formed and HPLC separated as the major products of the reaction. Unexpectedly, four out the five mono-adducts were found to be ESR silent and only one found to be paramagnetic and displaying the characteristic eight-line pattern in its ESR spectrum. X-ray crystallographic studies revealed a closed-shell electronic structure for one of the other four mono-adducts, which is obviously in good agreement with the ESR data.

Although it proved difficult to obtain the crystal structure of the remaining three ESR-inactive mono-adduct derivatives of La@C_{82}, ^{13}C NMR measurements revealed that they exhibit similar features comparable with the one where its crystal structure was solved. Hence, it is reasonable to assume that all four ESR silent mono-adducts are singly bonded regioisomers. The fifth isomer isolated, which was found to be ESR active, is suggested to be the conventional Bingel cyclopropanated adduct, however, possessing different physicochemical properties in accordance with its unique structure [43].

As stated earlier for the pyrrolidine adducts of $M_3N@C_{80}$, the icosahedral C_{80} cage has two different double bonds for addition reactions, namely at 5,6- and 6,6-ring junctions. When a series of $Gd_3N@C_{2n}$ $(n = 40, 42, 44)$ was allowed to react under typical Bingel conditions with bromodiethylmalonate and DBU, the most surprising finding was that $Gd_3N@C_{88}$ was unreactive, while the other two Gd_3N-based metallofullerenes gave products. Collectively those findings, and considering the pyramidialization angle effect, suggest that there is a decrease in reactivity of metallofullerenes as the size of the cage increases [44].

In fact, $Gd_3N@C_{80}$ was more reactive than $Gd_3N@C_{84}$ under the specified Bingel reaction conditions. The latter, however, yielded only one product and multi-adducts, while the former furnished two monoadducts and several multi-adducts. In addition, mono-addition also occurred for $Gd_3N@C_{82}$ when reacted with bromodiethylmalonate and DBU. Based on theoretical and computational calculations, the similar reactivity that both $Gd_3N@C_{82}$ and $Gd_3N@C_{84}$ display is rationalized in terms of large pyramidalization angles, leading to the more stable 5,6-adducts [45]. The redox behavior of the mono-methano-$Gd_3N@C_{80}$ material was found to be very similar to that of the corresponding derivatives of $Y_3N@C_{80}$ and $Er_3N@C_{80}$ [23]. Given that the observed electrochemical behavior for mono-methano-$Gd_3N@C_{80}$ is almost identical to that for all 6,6-adducts of $M_3N@C_{80}$, it is reasonable to assume that the current mono-methano-$Gd_3N@C_{80}$ is also a 6,6-adduct.

A newer member on the TNT metallofullerene family is $Lu_3N@C_{80}$, carrying high potentiality as an X-ray contrast agent [46]. Preparation of different Bingel mono-adducts 9–11 of $M_3N@C_{80}$ $(M = Lu, Sc)$ was possible when a catalytic amount of dimethylformamide (DMF) was added to the reaction mixture (Figure 11.7). ^{1}H NMR spectroscopy suggested that the reaction is regioselective affording the 6,6-adducts of the I_h cage. In addition, bis-adducts were also isolated via HPLC [47].

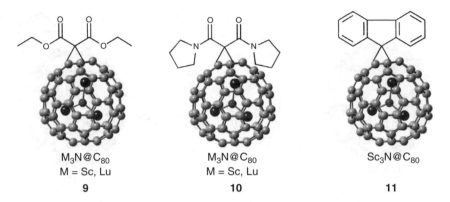

Figure 11.7   Bingel mono-adducts 9–11 of $M_3N@C_{80}$ (M = Lu, Sc)

11.1.4   [4+2] Diels-Alder Cycloaddition

In the classical [4+2] Diels-Alder cycloaddition reaction, 1,3-dienes add to metallofullerenes to form cyclohexene rings fused to the carbon sphere. Again, this type of cycloaddition offers the great advantage of controlling the degree as well as the site of addition. The several examples of such cycloadditions reported so far in the literature can be roughly divided into two main categories: (i) those in which the 1,3-dienes are cycloadded directly to the metallofullerene skeleton; and (ii) those in which the 1,3-dienes are generated in situ from the reactants.

Up to now the only example of a classical [4+2] Diels-Alder cycloaddition to metallofullerenes concerns the addition of cyclopentadiene to $La@C_{82}$. Interestingly, $La@C_{82}$ reacted only when an excess of cyclopentadiene was used (Figure 11.8), as shown by the formation of a new set of octet signals in the ESR spectrum of adduct 12 [48]. Considering the latter with the C_{2v} symmetry of $La@C_{82}$, the regioselectivity of the reaction with cyclopentadiene is highlighted. Mass spectrometry on 12 showed the molecular ion of the mono-adduct, followed by the loss of one cyclopentadiene unit. Furthermore, it should be also noted that 12 decomposes with time affording the original $La@C_{82}$ metallofullerene. The retro-reaction proceeds rather fast, with a half-life time of 1.8 h, thus competing with the forward reaction described in Figure 11.8.

The first example of an in-situ generated 1,3-diene and its Diels-Alder [4+2] cycloaddition to metallofullerenes involves the reaction of $Sc_3N@C_{80}$ TNT fullerene with 6,7-dimethoxyisochroman-3-one. According to Figure 11.9, the formation of monoadduct 13 was confirmed by MALDI-TOF mass spectrometry, while ^{13}C NMR showed that there is a plane of

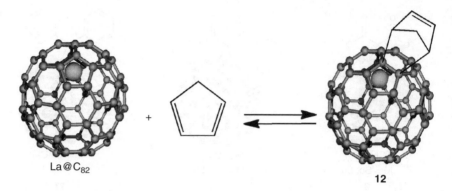

Figure 11.8 Diels-Alder cycloaddition reaction of cyclopentadiene to La@C_{82} furnishing adduct **12**

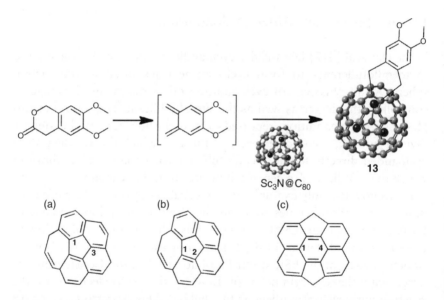

Figure 11.9 Diels-Alder cycloaddition of 6,7-dimethoxyisochroman-3-one to Sc_3 N@C_{80} and the three possible sites where the cycloaddition occurs, namely (a) the 1,3-position on a five-membered ring, (b) the 1,2-position on a five-membered ring, and (c) the 1,4-position on a six-membered ring

symmetry in **13**. The latter is consistent with the formation of three possible sites for cycloaddition to occur, namely (a) the 1,3-position on a five-membered ring, (b) the 1,2-position on a five-membered ring, and (c) the 1,4-position on a six-membered ring, with the more favorable one being at a 5,6-ring junction [49]. Further characterization of **13** with crystallographic analysis confirmed that the cycloaddition occurred at the 5,6-position, while the Sc_3N cluster is positioned away from the site of addition and causes the carbon atoms immediately neighboring the Sc atoms to protrude slightly from the surface of the fullerene cage [50]. In addition, DFT theoretical studies predicted the most reactive sites for the [4+2] Diels-Alder cycloaddition reaction onto $Sc_3N@C_{80}TNT$ metallofullerene, taking into account the double bond character and the pyramidalization angle of the carbon–carbon bonds [24].

11.1.5   [2+2] Cycloaddition

Cyclobutanefullerenes, presenting fused four-membered rings onto the fullerene sphere, are typically formed upon [2+2] cycloadditions. Basically these types of cycloadditions are less common, with the first reported example being the thermal [2+2] cycloaddition of benzyne to $Gd@C_{82}$. Benzyne is in-situ generated by the diazotization of anthranilic acid with isoamyl nitrite. Reaction of benzyne with $Gd@C_{82}$ resulted in the formation of two mono-adducts as suggested by MALDI-TOF mass spectrometry [51].

Surprisingly, when the same reaction was performed with $La@C_{82}$ metallofullerene, the molecular ion peak registered showed the presence of the $-NO_2$ group, thus suggesting that the metallofullerene derivative formed has the formula $La@C_{82}(C_6H_4)NO_2$. In fact, the presence of $-NO_2$ is responsible for the disappearance of the characteristic octet signal in the ESR spectrum of the mono-adduct. The benzyne group selectively adds to the 5,6-ring junction of $La@C_{82}$ to form a fused cyclobutenyl ring [52]. This is in sharp contrast with the benzyne [2+2] cycloaddition to C_{60}, where the addition occurs at the 6,6-double bond [53].

Benzyne generated in situ from anthranilic acid and isoamyl nitrite was also cycloadded to $Sc_3N@C_{80}$ TNT metallofullerene. Based on MALDI-TOF mass spectrometry measurements, two mono-adducts were identified and isolated via HPLC. With the aid of NMR, these two mono-adducts were registered as 5,6- and 6,6-adducts. In addition, crystallographic characterization verified the above conclusion [54].

Moreover, 4,5-diisopropoxybenzyne, generated in situ from the thermal reaction of 2-amino-4,5-diisopropoxybenzoic acid with isoamyl nitrite, cycloadded to $Sc_3N@C_{80}$. In this way, monoadducts 14 and 15, corresponding to the 5,6- and 6,6-ring junction cycloadditions, were formed, separated by HPLC and isolated in pure form as verified by MALDI-TOF mass spectrometry. Interestingly, when the reaction was performed in the presence of air, a third product **16** was isolated along with the above-mentioned mono-adducts, however, possessing a molecular ion peak with 16 atomic units higher. In fact, this third metallofullerene derivative becomes the major product of the cyclo-addition reaction when an excess of water is introduced in the reaction mixture.

Crystallography revealed the structure of that third adduct as being an open-cage derivative and found to possess a benzyne unit and an oxygen atom at 5,6-ring junctions in a single pentagon. Collectively the three different derivatives **14–16** formed, upon reaction of 2-amino-4,5-diisopropoxybenzoic acid with isoamyl nitrite and $Sc_3N@C_{80}$, are shown in Figure 11.10 [55]. The open-cage adduct **16** has a rather novel structure, in which the insertion of an oxygen atom at the carbon cage results in the simultaneous breakage of a C—C bond. Moreover, the C—C bond at the site where the addition of the benzyne unit occurs is also broken. Hence, this double rupture of C—C bonds results in the formation of an orifice involving a 13-membered ring in **16**.

11.1.6   Carbenes

Photolysis or thermolysis of 2-adamantane-2,3-[*3H*]-diazirine (AdN_2) yielded the corresponding adamantylidene carbene, which showed high reactivity for metallofullerenes. Particularly, the regiospecific reaction of $La@C_{82}$ with AdN_2 afforded single isomer **17**, according to Figure 11.11 [56]. Electronic absorption spectroscopy suggested that the metallofullerene derivative $La@C_{82}$-Ad **17** retained the novel and characteristic electronic structure of the parent $La@C_{82}$. In addition, cyclic voltammetry studies revealed that $La@C_{82}$-Ad **17** exhibits three one-electron reversible reductions and two oxidations, similarly to intact $La@C_{82}$. However, the redox potentials in **17** were found cathodically shifted as compared with $La@C_{82}$, thus indicating that the incorporation of the Ad moiety onto the cage of $La@C_{82}$ decreases its electron accepting character. Notably, X-ray single-crystal analysis and theoretical calculations of $La@C_{82}$-Ad **17** revealed that the La position remained practically unchanged by the Ad cycloaddition.

On the other hand, as already claimed by maximum entropy method (MEM)/Rietveld analysis, $Gd@C_{82}$ possesses an exceptional structure [57], in which the Gd metal is located near the C—C double bond on the opposite side of the C_{82}-C_{2v} cage along the C_2 axis. However, in disagreement with experimental results [58], the $Gd@C_{82}$-Ad

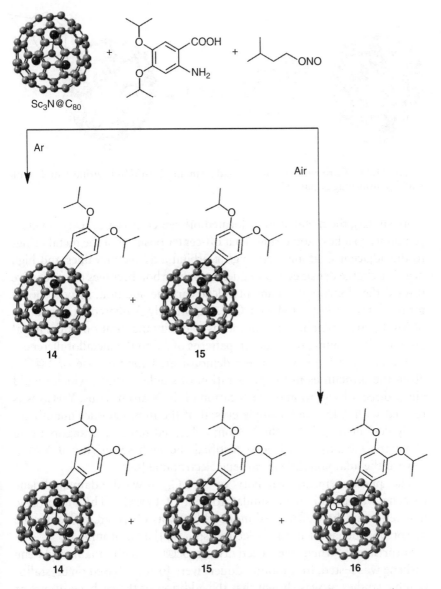

Figure 11.10 Thermal cycloaddition reaction of 2-amino-4,5-diisopropoxybenzoic acid to $Sc_3N@C_{80}$ affording different products **14–16** in the presence of Ar or air

cycloadduct was synthesized and its single crystal X-ray analysis was carried out. These results, together with theoretical calculations, found against the anomalous structure of $Gd@C_{82}$ predicted by the MEM/ Rietveld analysis [59].

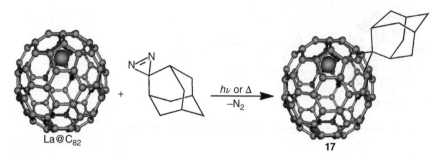

Figure 11.11   Carbene reaction of 2-adamantane-2,3-[*3H*]-diazirine (AdN$_2$ with La@C$_{82}$ furnishing adduct **17**

In M@C$_{82}$, the metal atom is located off-center of the C$_{82}$ cage, in close proximity to a hexagonal ring. That off-center position of the metal brings to the adjacent C atoms both high π-orbital axis vector values and high negative charge densities. Consequently, one carbon becomes most reactive toward the electrophile adamantylidene carbene addition, as was shown for the cases of M@C$_{82}$ (M = La, Gd). Importantly, Ad-based monoadducts of M@C$_{82}$ represent an efficient means to clarify the location of the metal atom by determining the addition patterns of Ad to the metallofullerene.

Actually, the latter was nicely demonstrated for the case of Y@C$_{82}$, given the difficulties in the preparation of single crystals, which would allow direct observation of the location of the Y atom. Thus, Y@C$_{82}$ was reacted with AdN$_2$ and a single crystal of the mono-adduct unambiguously demonstrated that the Y atom is located under a hexagonal ring along the C$_2$ axis, which also justified the high selectivity of Y@C$_{82}$ toward the adamantylidene carbene electrophile [60].

The photochemical reactivity of Ce@C$_{82}$ toward adamantylidene carbene was found to be similar to that of La@C$_{82}$. The anisotropic magnetism of the Ce@C$_{82}$-Ad monoadduct was observed, reflecting the anisotropic behavior of the *f*-electron spin on the Ce atom [61].

Surprisingly, when the reactivity of Sc@C$_{82}$ with adamantylidene carbene was tested, four mono-adducts were formed. Based on crystallographic studies, it was shown that the addition of the carbene occurs at a 6,6-junction very close to the encapsulated Sc atom [62]. The exceptional reactivity of Sc@C$_{82}$ is rationalized in terms of the small ionic radius of Sc^{3+} which in turn allows stronger metal–cage interactions.

Thermal and photochemical reactions of La$_2$@C$_{78}$ with AdN$_2$ were also carried out, however, different products were furnished depending on the nature of the reaction. Thus, thermal reaction yielded seven mono-adducts, due to the low symmetry of the carbon cage. On the contrary, the photochemical reaction, which proceeds faster than the thermal one, furnished not only a mixture of four mono-adducts (all of them have been produced thermally) but also some higher adducts. It was found that three of the mono-adducts possessed C_s molecular symmetry with the addition of the Ad unit occurring at both the 5,6- and 6,6-junctions around the pole and equator of $La_2@C_{78}$. Based on X-ray and theoretical studies, one of the mono-adducts has an open structure with the two La atoms located on the C_3 axis of $La_2@C_{78}$ with D_{3h} molecular symmetry [63].

In order to verify that the initially described $Sc_2@C_{84}$ (isomer III) [64] is a scandium carbide metallofullerene, namely $Sc_2C_2@C_{82}$ (III), it was functionalized with AdN_2 under irradiation. Then, X-ray single crystal analysis of the metallofullerene-Ad cycloadduct revealed a 5,6-addition pattern to a fullerene cage that originates from the C_{3v} isomer of C_{82} [65].

Along the same lines, another carbide cluster metallofullerene, the newly prepared $Sc_2C_2@C_{80}$, with C_{2v} molecular symmetry, was reacted with AdN_2 and furnished five mono-adduct isomers. Single crystal X-ray studies revealed that the Ad addition mostly took place at the cage region on top of the arch of the metal carbide cluster [66].

The photochemical reaction of $M_2@C_{80}$ $(M = La, Ce)$ with AdN_2 selectively afforded the corresponding adducts by adamantylidene carbene addition. The I_h symmetrical C_{80} possesses two different non-equivalent carbon atoms, and furthermore the Ad carbene addition can either break or not break the C—C bond, thus, suggesting the possibility of four addition patterns of the carbene to C_{80}. Based on extensive NMR studies as well as single crystal X-ray analysis, it was found that the adamantylidene carbene selectively adds at the 6,6-junction of the $M_2@C_{80}$ to yield the 6,6-open adduct $M_2@C_{80}$-Ad $(M = La, Ce)$. Especially for the $La_2@C_{80}$-Ad it was found that the two La atoms are collinear with the spiro carbon of the adduct with a long La—La length, unlike the three-dimensional random motion in $La_2@C_{80}$ [67].

An important aspect of metallofullerenes is the stabilization of the carbon cages violating the isolated pentagon rule (IPR) by the encapsulated metal species. On account of this, carbon atoms located at 5,5-junctions on non-IPR metallofullerenes are expected to show high reactivity as derived from the high surface curvature. In this way, D_2 symmetrical non-IPR $La_2@C_{72}$, possessing two pairs of fused pentagons, was found to readily react with AdN_2, thus, generating six isomers of $La_2@C_{72}$-Ad. Interestingly, and in accordance with theoretical calculations, those carbons at the 5,5-junctions of $La_2@C_{72}$ were less reactive than their neighboring ones due to their strong interactions with the two La atoms, a significant factor that causes enhanced stabilization. Since two pairs of fused pentagons exist in $La_2@C_{72}$, prolonged photo-irradiation resulted in the formation of a mixture of bis-adducts, in addition to the mono-adducts that formed instantaneously [68,69].

The electronic absorption spectra of the isolated isomers of $La_2@C_{72}$-$(Ad)_2$ bis-adducts are very similar to that of the parent $La_2@C_{72}$, showing prominent absorptions in the visible and the NIR regions, thus indicating that the electronic structure of $La_2@C_{72}$ remained unaltered after the covalent binding of two adamantylidene units onto the carbon cage. Moreover, the Vis-NIR spectra of the bis-adducts are similar to those of the corresponding mono-adducts, thus, suggesting that $La_2@C_{72}$-$(Ad)_2$ also possesses an open-cage structure.

In the previously described paradigm on the synthesis of bis-adduct $La_2@C_{72}$-$(Ad)_2$, the Ad addition occurred stepwise, meaning that the reaction site of the second Ad unit might be controlled by the first one. Meanwhile, $La_2@C_{80}$ possesses I_h symmetry, in which the two La atoms – each one donates three electrons to form the electronic structure $(La^{3+})_2@$$(C_{80})^{6-}$ – freely rotate in three dimensions inside the C_{80} cage [70–73]. Also, it was found that adamantylidene carbene selectively adds at the 6,6-junction of the $La_2@C_{80}$ in which the two La atoms in the 6,6-open $La_2@C_{80}$-Ad adduct are collinear with the spiro carbon of the adduct with a long La–La length [67]. Thus, in order to monitor the influence on the regioselectivity, the hetero-bis-adduct of $La_2@C_{80}$ by carbene addition was designed and synthesized. Derivatization was conducted stepwise using initially phenylchloro carbene, then followed by adamantylidene carbene.

It was found that the regioselectivity of the second carbene addition to the mono-adduct **18** took place at the 6,6-junction, along with bond cleavages on the C_{80} cage, to furnish the open-cage bis-adduct **19** (Figure 11.12). Notably, the La–La length in bis-adduct **19** was elongated compared with that of the mono-adduct **18**. The latter is rationalized by considering the expansion of the inner space of the cage caused by the bond cleavage allowing two La atoms to separate, thus further reducing the electrostatic repulsion between positively charged atoms [74].

11.2 RADICAL ADDITION REACTIONS

Generally, free radical reactions show low selectivity in the formation of mono-adducts due to high reactivity. The first C–C bond formation on EMFs was succeeded by the addition of excess diphenyldiazomethane

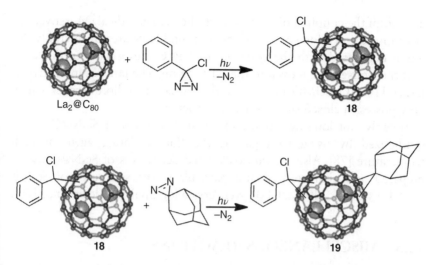

Figure 11.12 Mono- and bis-adduct formation of a carbene reaction on $La_2@C_{80}$

to a toluene solution of $La@C_{82}$. Mass spectrometry showed the formation of up to tris-adducts. On the other hand, due to the paramagnetic nature of $La@C_{82}$, ESR spectroscopy of the reaction mixture revealed the presence of several new octet signals, corresponding to mono-, bis-, and tris-adducts, with similar hyperfine coupling constant and g-values to those of intact $La@C_{82}$. The latter findings suggest that the electronic structure of $La@C_{82}$ is greatly retained regardless of the functionalization with diphenyldiazomethane [75].

The first methano-$Sc_3N@C_{80}$ derivative with free radicals generated from diethyl malonate catalyzed by manganese(III) acetate was prepared. The role of manganese acetate is to abstract the hydrogen from the malonate unit, and then the latter adds to the metallofullerene cage [76]. Two mono-adducts, together with bis-, tris-, and up to octa-adducts were obtained and isolated after HPLC separation. Although there are two reaction sites on the I_h-$Sc_3N@C_{80}$, namely the corannulene-type site, in which a double bond at a 5,6-ring junction is abutted by two hexagons, and the pyrene-site, in which a double bond at a 6,6-ring junction is abutted by a hexagon and a pentagon, the radical reaction is highly regioselective as NMR spectroscopy suggests that in both mono-adducts the addition occurs at the 6,6-ring junction.

Benzyl radicals, produced in situ by hydrogen abstraction from toluene in the presence of 3-triphenylmethyl-5-oxazolidinone, were added to $La@C_{82}$ to furnish two benzyl mono-adducts [77]. Although the parent $La@C_{82}$ is a paramagnetic species, the mono-adducts were ESR silent as a result of the coupling of La@C_{82} with the benzyl radical. Moreover, the electronic absorption spectra of the mono-adducts are considerably different compared with the spectrum of La@C_{82}, especially in the sense that their onsets are moved to the NIR region. The latter is indicative of larger HOMO-LUMO gap in the derivatized metallofullerenes which now possess a closed shell electronic structure.

Notably, not long ago, fluoroalkylated derivatives of Sc_3N@C_{80} were synthesized by using CF_3I gas as the fluoralkylating agent at high temperature [78]. Also, perfluoroalkylated derivatives of Sc_3N@C_{80} were prepared and characterized and particularly the structures of Sc_3N@$C_{80}(CF_3)_{10}$ and Sc_3N@$C_{80}(CF_3)_{12}$ were crystallographically determined [79].

11.3   MISCELLANEOUS REACTIONS

Apart from the previously described established methodologies for the functionalization of EMFs, some other reactions that do not immediately fall into those categories were also applied. For example, gas phase derivatization of EMFs M@C_{82} and M_2@C_{80} (M = Ce, Nd) was performed by ion/molecule reactions with the plasma from the self-chemical ionization of vinyl acetate. Under these conditions, the electrophile $C_2H_3O^+$ is formed in the gas phase, which then subsequently adds to the metallofullerene cage [80].

An interesting approach for introducing a cyclopropane ring onto metallofullerenes, besides the route using bromomalonates, is via the electrochemical generation of the Lu_3N@C_{80} dianion and its reaction with the electrophile $PhCHBr_2$. In this way, the methano-metallofullerene Lu_3N@C_{80}(CHPh) **20** was prepared, according to Figure 11.13 [81]. Based on MALDI-TOF mass spectrometry the presence of the mono-adduct was confirmed, however, due to the I_h symmetry of Lu_3N@C_{80} the formation of two regioisomers, the 5,6- and 6,6-,was expected.

Interestingly, cyclopropanated rings onto a metallofullerene cage can also be introduced by a copper(I)-catalyzed cycloaddition reaction. In this context, a copper(I)-catalyst mediated reaction of Tb@C_{82} with α-diazocarbonyl compounds was carried out to furnish not only mono-adduct **21** (Figure 11.14), but also some higher adducts when a large excess of the diazocarbonyl compound is used [82]. The electronic structure of those adducts was explored by X-ray photo-electron spectroscopy (XPS), UV-Vis-NIR, and photoluminescence spectroscopy. The XPS results suggested that the oxidation state of Tb in **21** is 3+, in accord with intact metallofullerene, thus revealing that

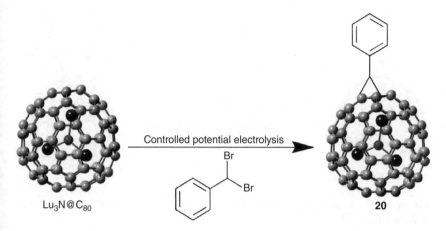

Figure 11.13 Electrochemical generation of $Lu_3N@C_{80}$ dianion and its reaction with the electrophile $PhCHBr_2$ forming methano-metallofullerene $Lu_3N@C_{80}(CHPh)$ adduct **20**

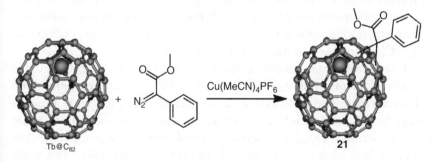

Figure 11.14 Copper(I) mediated reaction of $Tb@C_{82}$ with α-diazocarbonyl compound furnishing adduct **21**

exohedral functionalization has little effect on the valence of the encapsulated metal atom.

In a different approach, an EMF derivative bearing a single phosphorus substituent by a region- and chemo-selective reaction of $Dy@C_{82}$ with dimethyl acetylenedicarboxylate and triphenylphosphine was synthesized. The formation of the mono-adduct was identified by MALDI-TOF mass spectrometry, while its structure was determined by X-ray crystallography [83].

In the search to improve the electron-accepting character of metallofullerenes, functionalization of I_h symmetric $La_2@C_{80}$ with tetracyanoethylene oxide (TCNEO) at high temperature was performed. Since TCNEO generates carbonyl ylides via ring opening when the

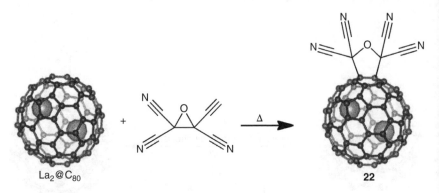

Figure 11.15   Functionalization of I_h-La$_2$@C$_{80}$ with tetracyanoethylene oxide furnishing adduct **22**

temperature is greater than 100 °C, adduct **22** (Figure 11.15) was obtained and isolated by HPLC as confirmed by mass spectrometry. The presence of four cyano units decorating La$_2$@C$_{80}$ enhances the electron-accepting properties of **22** as electrochemistry studies revealed. Moreover, X-ray crystallography suggests that the cycloaddition of TCNEO to La$_2$@C$_{80}$ takes place at the 5,6-ring junction [84].

A series of water-soluble EMF based on Gd and Pr encapsulated metals – metallofullerenols Pr@C$_{82}$O$_m$(OH)$_n$ – were prepared and characterized. The synthesis involves treatment either with aqueous sodium hydroxide and tetrabutylammonium hydroxide as a phase transfer catalyst [85,86] or nitric acid and subsequent hydrolysis, thus initially involving a NO$_2$ radical addition process [87]. The Gd@C$_{82}$(OH)$_x$, gadolinium fullerenols, were investigated as magnetic resonance imaging (MRI) contrast agents. The *in-vitro* water proton relaxivity of Gd@C$_{82}$(OH)$_x$ was found to be significantly higher than that of the commercially available MRI agent. This was also confirmed *in vivo* in the lung, liver, spleen, and kidney of mice [88]. At this point it should be emphasized that it is also possible to control the addition of hydroxyl groups added onto the metallofullerene and, in particular, the preparation of a dihydroxyl adduct of Gd@C$_{82}$ was reported. According to that report, when a 1,2-dichlorobenzene solution of the metallofullerene was treated with an aqueous hydrogen peroxide solution, in the absence of air, the Gd@C$_{82}$(OH)$_2$ adduct was obtained as confirmed by MALDI-TOF mass spectrometry [89].

A further systematic study was performed involving the investigation of lanthanide metallofullerenols, M@C$_{82}$(OH)$_x$ (M = La, Ce, Gd, Dy, Er) to obtain insight into the proton relaxation mechanism of these novel forms as MRI contrast agents. Overall, it was found that these metallofullere-nols possess longitudinal and transverse relaxivities for water protons significantly higher than those of the corresponding commercially available ones. The strong relaxivities of those lanthanoid-based metallofullerenols are rationalized in terms of dipole–dipole relaxation together with a substantial decrease in the overall molecular rotational motion [90].

In addition, when water-soluble holmium metallofullerenols were prepared and neutron irradiated, a different biodistribution than that of other fullerenes was observed, thus suggesting that certain derivatized metallofullerenes can be selectively tissue-targeted [91]. Moreover, $^{166}Ho@C_{82}(OH)_x$ as a radiotracer analog was intravenously administered to mice and found to be selectively localized in the liver but with slow clearance, as well as taken up by bones without clearance. Thus, the feasibility of using water-soluble metallofullerenols as radiotracers to elucidate their behavior in animals was shown [92].

The first water-soluble trimetallic nitride endohedral metallofullerols $Sc_3N@C_{80}(OH)_{10}O_{10}$ were synthesized from polyanionic radical inter-mediates. Briefly the procedure followed to obtain water-soluble adducts of $Sc_3N@C_{80}$ involves treatment of toluene solution of the metallofuller-ene with Na metal under an inert atmosphere and subsequent exposure of the reaction mixture to air and water. Initially, a precipitate from the toluene solution is formed containing polyanionic radical species, which is then oxidized by air in water to yield a golden-colored aqueous solution. These TNT-based metallofullerols were characterized by a variety of traditional spectroscopic methods including IR, XPS, and MALDI-TOF mass spectrometry [93]. Along the same lines, following similar experimental protocols, $Sc_xGd_{3-x}@C_{80}O_{12}(OH)_{26}$ $(x=1,2)$ were also synthesized. It is interesting to note that the production yield of $Sc_xGd_{3-x}@C_{80}$ $(x=1,2)$ is 10 and 50 times higher compared with that of $Gd@C_{82}$ and $Gd_3N@C_{80}$, thus highlighting their potential as novel MRI contrast agents [94].

In addition, water-soluble $Gd_3N@C_{80}$ and $Lu_3N@C_{80}$ functionalized with poly(ethylene glycol) and multihydroxyl groups were prepared. The new $M_3N@C_{80}(OH)_x\{C[COO(CH_2CH_2O)_{114}CH_3]_2\}$ $(M=Gd,Lu)$ were found to have T_1 and T_2 relaxivity values superior to those of the conventional gadolinium-containing MRI agent. The effectiveness was demonstrated by means of *in vitro* relaxivity and MRI studies, infusion experiments with agarose gel and *in vivo* rat brain studies. Notably, an inherent advantage of the water-soluble TNT-based adducts is that the water molecules are shielded by the fullerene cage from direct interaction with the metal atoms [95].

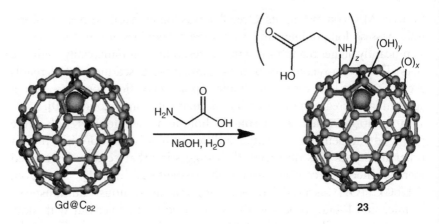

Figure 11.16 Nucleophilic addition of an alkaline solution of β-alanine to Gd@C_{82} furnishing water-soluble Gd@$C_{82}(O)_x(OH)_y(NHCH_2COOH)_z$ adduct **23**

In an attempt to utilize the highly insoluble and thus intractable from the fullerene soot Gd@C_{60} fraction, its chemical derivatization was performed. In such a way, new water-soluble and air-stable Gd@$C_{60}[C(COOH)_2]_{10}$ was prepared. The functionalization reaction involves multiple cycloadditions of bromomalonates, according to the well established methodology of Bingel cyclopropanation. Thus, Gd@$C_{60}[C(COOCH_2CH_3)_2]_{10}$ was formed which was then hydrolyzed to the corresponding acid Gd@$C_{60}[C(COOH)_2]_{10}$. The aqueous solubility of the latter allowed its evaluation as MRI contrast agent and relaxometry measurements revealed relaxivity values superior to those derived from the commercially available gadolinium chelate-based one [96].

In a different reaction strategy for the functionalization of EMFs, nucleophilic addition of glycine to esters to Gd@C_{82} was investigated. Based on MALDI-TOF mass spectrometry studies, the formation of multiple adducts possessing glycine esters or both glycine esters and hydroxyl groups was found [97]. In addition, nucleophilic addition of an alkaline solution of β-alanine to Gd@C_{82} was also performed and resulted in the production of water-soluble Gd@$C_{82}(O)_x(OH)_y(NHCH_2COOH)_z$ **23** (Figure 11.16).

The structure of the new adduct was confirmed by standard analytical techniques, including IR, XPS, and mass spectrometry [98]. Importantly, the free carboxylic acid moiety in **23** can be further derivatized, following classical peptide synthesis conditions. Thus, conjugation of the antibody of the green fluorescence protein (GFP) to **23** was achieved yielding a new gadolinium-based modified metallofullerene as a model for a tumor targeted imaging agent. In that model system, the activity of the antiGFP conjugate was conveniently recognized by GFP thus leading to the performance of more direct and facile *in vitro* experiments than those involving tumor antibodies. In addition, the 23-antiGFP conjugate was found to exhibit higher water proton relaxivity than that of **23**, thus suggesting that 23-antiGFP shows not only targeting potential but also exhibits high efficiency as a MRI contrast agent [99].

11.4 DONOR–ACCEPTOR DYADS

A fundamental aspect of EMFs is their applicability and performance in electron donor–acceptor systems [100]. Especially considering that (i) empty fullerenes, particularly C_{60} and its derivatives, exhibit low reorganization energy in charge-transfer phenomena, and high electron mobility and (ii) EMFs possess larger absorptive coefficients in the visible region and a low HOMO-LUMO energy gap, it is of no surprise that as soon as the chemical reactivity and functionalization methodologies of EMFs were established and developed, novel donor–acceptor dyads would be synthesized. The first covalent linkage of an electron donor to a TNT endohedral fullerene was achieved by performing the 1,3-dipolar cycloaddition reaction of azomethine ylide, as derived upon thermal condensation of ferrocene carboxylaldehyde and sarcosine, to $Sc_3N@C_{80}$ (Figure 11.17).

After column and HPLC purification, the mono-adduct $Sc_3N@C_{80}$-Fc pyrrolidine derivative **24** was isolated as proved by MALDI-TOF mass spectrometry [101]. Considering that (i) in the I_h symmetrical $Sc_3N@C_{80}$ two different C—C double bonds, namely at 6,6- and 5,6-ring junctions, are present, (ii) from the ^1H and ^{13}C NMR measurements two different pyrrolidine carbons exist, and (iii) no isomerization is observed after prolonged heating, adduct **24** should possess a fused pyrrolidine ring at a 5,6-ring junction. Structure **24** was further proved by electrochemistry measurements, as its cyclic voltammetry showed a reversible reduction which is a fingerprint for the mono-adducts 5,6-structures, whereas the 6,6- ones exhibit irreversible behavior. As far as electron-transfer interactions are concerned, photoexcitation of **24** resulted in a rapid decay of the so-formed singlet excited state. Moreover, significant changes were observed in the differential absorption spectrum of **24**, with the formation of new bands in the visible and NIR regions.

The new band appearing in the visible region is due to the one-electron oxidized ferrocenium ion. Importantly, the new NIR band evolved

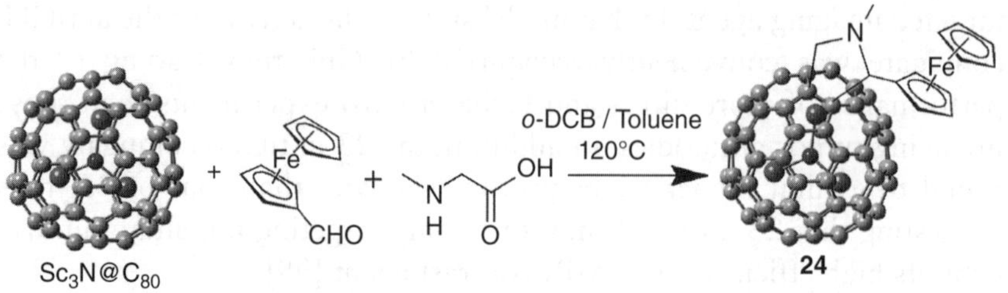

Figure 11.17   1,3-Dipolar cycloaddition reaction of azomethine ylides to $Sc_3N@C_{80}$ yielding $Sc_3N@C_{80}$-Fc dyad **24**

resembles the one observed for a radiolytically and spectroelectrochemically derived one-electron reduced $Sc_3N@C_{80}$ species. The radical ion pair $(Sc_3N@C_{80})^{\bullet -}$-$Fc^{\bullet +}$ **24** is formed, thus opening new avenues for applications of TNT-based fullerenes in photovoltaics.

Along the same lines, two isomeric 5,6-pyrrolidine donor–acceptor dyads of $Sc_3N@C_{80}$ with triphenylamine as the electron donor were prepared. Materials **25** and **26** were obtained upon reaction of $Sc_3N@C_{80}$ with (i) 4-(N-diphenylamino)-benzaldehyde and sarcosine and (ii) formaldehyde and N-(benzyl-4-diphenylaminophenyl)glycine, respectively (Figure 11.18). The electrochemical properties of **25** and **26** were evaluated by cyclic voltammetry and it was found that they both exhibit three reversible reductions – characteristic for the 5,6-adduct – and two irreversible oxidation processes for **25** and three for **26**. Insight into the charge-transfer interactions within dyads **25** and **26** arises from transient absorption measurements. Briefly, it was found that (i) when triphenylamine is the substituent of the N atom in the pyrrolidine ring, in dyad **25**, a significantly longer lived radical ion pair is yielded, as compared with dyad **26** in which the triphenylamine is the substituent of the α-carbon in the pyrrolidine ring and (ii) both dyads **25** and **26** have longer lived charged separated states than their corresponding C_{60} analogs [102].

En route to investigating charge-transfer interactions of TNT-based fullerenes with organic electron donors, a series of $Y_3N@C_{80}$ donor–acceptor dyads were synthesized via the 1,3-dipolar cycloaddition reaction of azomethine ylides or the Bingel cyclopropanation reaction, with two different electron donors, namely extended tetrathiafulvalene (exTTF) and phthalocyanine. Reaction of $Y_3N@C_{80}$ with a phthalocyanine aldehyde and sarcosine yielded TNT-based fulleropyrrolidine dyad **27** (Figure 11.19). On the other hand, reaction of $Y_3N@C_{80}$ with malonates bearing phthalocyanine and exTTF substituents, in the

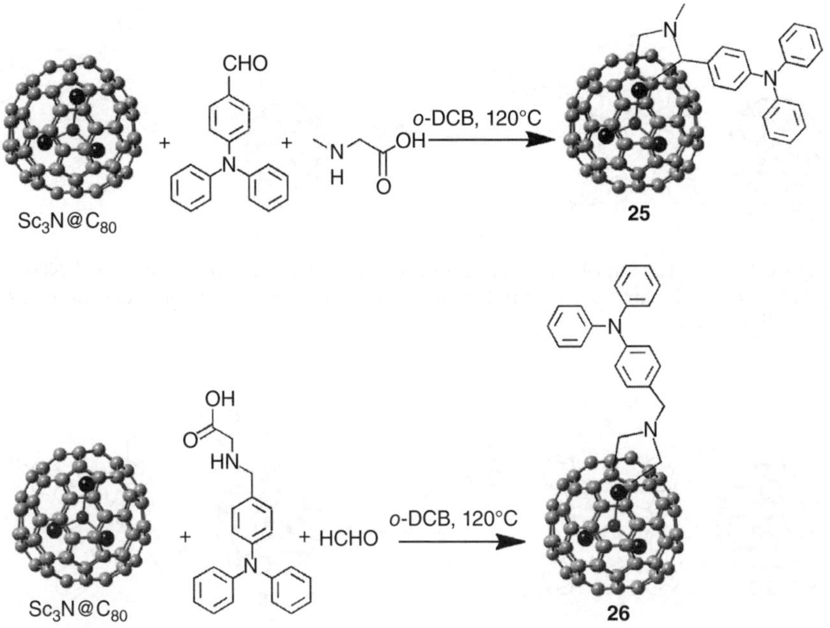

Figure 11.18 1,3-Dipolar cycloaddition reaction of azomethine ylides to $Sc_3N@C_{80}$ yielding $Sc_3N@C_{80}$-triphenylamine donor–acceptor dyads in which the triphenylamine unit is a substituent at the α-carbon (adduct **25**) and the nitrogen (adduct **26**) of the pyrrolidine ring

presence of DBU as a base, yielded dyads **28** and **29**, respectively (Figure 11.20) [103].

Mass spectrometry verified all **27–29** dyads were mono-adducts and NMR studies proved their molecular structures. In fact, in **27–29** the cycloaddition occurs at the 6,6-ring junction, which especially for **27** is characteristic of the encapsulated metal nitride cluster Y_3N, since in encapsulated Sc_3N the cycloaddition proceeds at the 5,6-ring junction. Interestingly, dyad **27** is rather unstable and undergoes a retro-cycloaddition reaction at room temperature. Photoinduced electron transfer interactions from the organic addends to the TNT fullerene were studied by means of electrochemistry, photoluminescence, and transient absorption spectroscopy and confirmed with the aid of theoretical calculations.

As $Y_3N@C_{80}$-exTTF dyad **29** proved to be quite unstable and the exTTF moiety easily oxidized to the corresponding anthraquinone, a new exTTF adduct based on $La_2@C_{80}$ metallofullerene was synthesized. Following the established strategy of 1,3-dipolar cycloaddition for the functionalization of metallofullerenes via the in-situ formation of

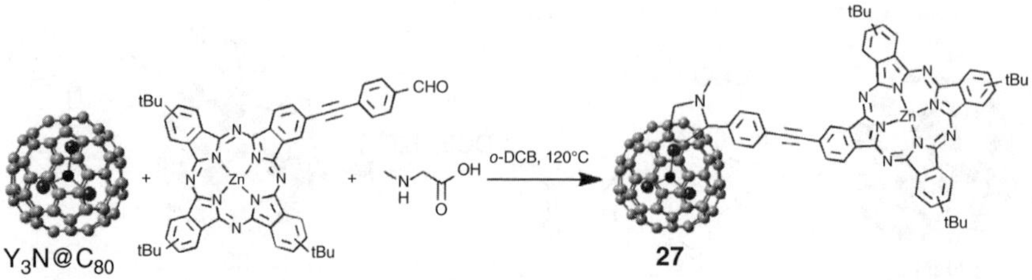

Figure 11.19 1,3-Dipolar cycloaddition reaction of azomethine ylides to $Y_3N@C_{80}$ yielding dyad **27** carrying zinc phthalocyanine as substituent at the α-carbon of the pyrrolidine ring

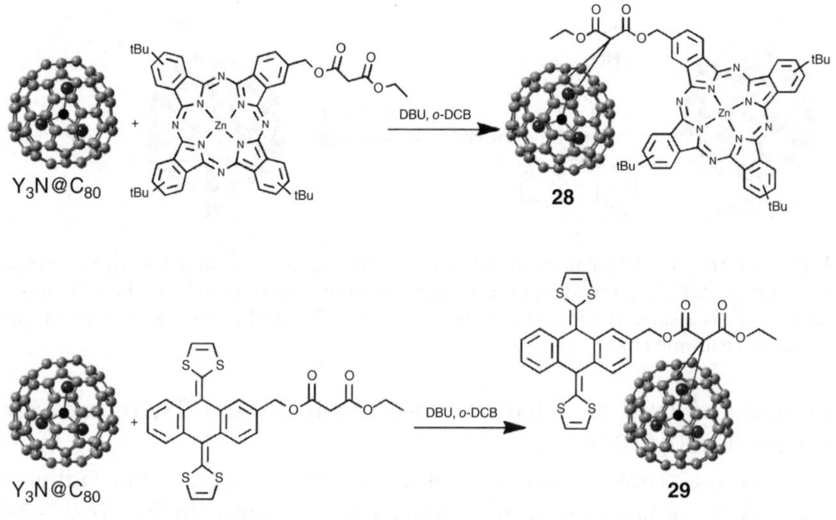

Figure 11.20 TNT-based donor–acceptor dyads $Y_3N@C_{80}$-Pc **28** and $Y_3N@C_{80}$-exTTF **29** as formed upon Bingel cyclopropanation

azomethine ylides, exTTF was incorporated as a substituent of the α-carbon in the fused pyrrolidine ring on $La_2@C_{80}$ [104]. The fused pyrrolidine moiety in $La_2@C_{80}$-exTTF adduct **30** (Figure 11.21) was cycloadded on the 5,6-double bond, in accordance with other similarly modified metallofullerenes. Combination of absorption spectroscopy and electrochemistry measurements showed that electronic interactions between exTTF and $La_2@C_{80}$ in the ground states are weak. However, photoluminescence and transient absorption spectroscopy studies revealed appreciable intramolecular interactions in the excited states of dyad **30**, leading to the formation of a long-lived radical ion pair ($La_2@C_{80}$)$^{\bullet-}$-(exTTF)$^{\bullet+}$.

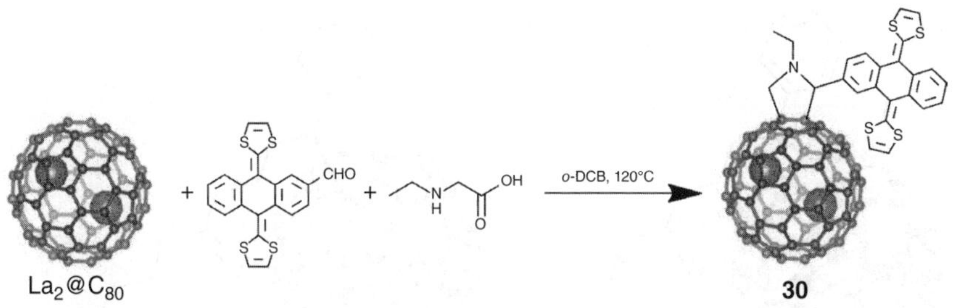

Figure 11.21 1,3-Dipolar cycloaddition reaction of azomethine ylides to La$_2$@C$_{80}$ forming La$_2$@C$_{80}$-exTTF dyad **30**

The first paramagnetic EMF carrying the exTTF unit was recently synthesized, based on the 1,3-dipolar cycloaddition reaction of azomethine ylides. To achieve this, exTTF-CHO and octylglycine were thermally reacted with La@C$_{82}$ giving rise to an inseparable mixture of two regioisomers in which the exTTF unit is present at the α-carbon of a pyrrolidine ring fused to the carbon cage. Photoinduced intramolecular electron transfer, for the first time to paramagnetic EMFs, was observed and verified by complementary spectroscopic experiments. Importantly, transient absorption spectroscopy measurements revealed the formation of oxidized exTTF species with the simultaneous generation of reduced La@C$_{80}$, while the radical ion pair (La@C$_{82}$)$^{•-}$-(exTTF)$^{•+}$ was formed and lived for more than 2 ns [105].

Long-range electron-transfer interactions of electron donor zinc-porphyrin chromophore and TNT fullerene Sc$_3$N@C$_{80}$ acceptor bridged via short and long oligophenylene-vinylene wires were investigated. The new dyads were fully characterized by NMR and MALDI-TOF mass spectrometry. Furthermore, based on complementary optical studies, including absorption and photoluminescence spectroscopy, together with electrochemistry, electron-transfer reactions between zinc porphyrin and Sc$_3$N@C$_{80}$ were revealed [106].

In general, TNT fullerenes are less reactive than empty fullerenes and classical EMFs due to the formal transfer of six electrons from the metal atoms to the fullerene cage, which leads to a closed shell electronic structure and an increase of the HOMO-LUMO gap with a corresponding reduction in reactivity. Since Sc$_3$N@C$_{80}$ is the most abundant species in the TNT fullerene family, it was functionalized by the well-known diazoalkane addition of methyl 4-benzoylbutyrate p-tosylhydrazone to furnish analog **31** (Figure 11.22), which resembles the most widely studied fullerene derivative, used as an acceptor in organic photovoltaic devices, namely the [6,6]-phenyl-C$_{61}$ butyric acid methyl ester (PC$_{60}$BM).

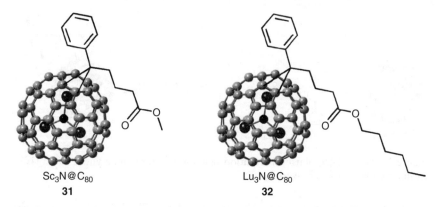

Sc$_3$N@C$_{80}$ Lu$_3$N@C$_{80}$
31   32

Figure 11.22   Analogs **31** and **32** as derived upon diazoalkane addition of alkyl 4-benzoylbutyrate p-tosylhydrazone to Sc$_3$N@C$_{80}$ and Lu$_3$N@C$_{80}$, respectively

Similarly, the Y$_3$N@C$_{80}$-PBM (phenyl butyric acid methyl ester) analog was also synthesized. However, due to insufficient material the photovoltaic properties of both materials are yet to be explored [107].

On the other hand, the availability of higher amounts of Lu$_3$N@C$_{80}$, not only allowed the formation of the corresponding derivative **32** (Figure 11.22) but also the exploration of its photovoltaic properties. The reduced energy offset of the molecular orbitals of Lu$_3$N@C$_{80}$ **32** to the donor regio-regular poly(3-hexylthiophene) (rrP3HT) reduces energy losses in the charge transfer process and increases the open circuit voltage as compared with that when PC$_{60}$BM was used as an acceptor. Thus, it is no surprise that photoconversion efficiency higher than 4% was registered, highlighting the potentiality of EMFs in general and TNT fullerenes in this particular case [108–110].

Because PC$_{60}$BM is considered as one of the most promising fullerene derivatives for photovoltaic applications and since lithium ion encapsulation drastically alters the electronic properties of the carbon cage in the novel [Li$^+$@C$_{60}$]PF$_6^-$ material, the corresponding functionalization is not surprising. Thus, chemical modification of the latter under typical diazoalkane conditions was afforded after HPLC purification [Li$^+$@PC$_{60}$BM] PF$_6^-$ as identified by mass spectrometry. In fact, two regioisomers were formed, attributed to the 5,6- and 6,6-adducts based on X-ray crystallography studies [111].

The [1+2] cycloaddition reaction of diazo precursors was applied to conjugate a porphyrin chromophore on Sc$_3$N@C$_{80}$. Moreover, zinc metallation of the TNT-based adduct resulted in the formation of Sc$_3$N@ C$_{80}$-ZnP dyad **33**, according to Figure 11.23 [112]. The structure of

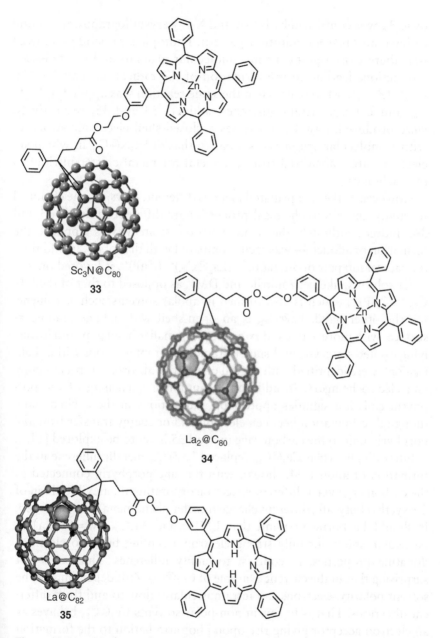

Figure 11.23 Porphyrin chromophores 33–35 conjugated on $Sc_3N@C_{80}$, $La_2@C_{80}$, and $La@C_{82}$, respectively

dyad 31 was confirmed by NMR and X-ray crystallography and found to have an open 6,6-addition pattern. Photophysical studies showed that there exist significant intramolecular ground- and excited-state interactions leading to the formation of the radical ion pair $(Sc_3N@C_{80})^{\bullet-}$-$(ZnP)^{\bullet+}$. Moreover, when the same reaction was applied to $La_2@C_{80}$ and $La@C_{82}$ metallofullerenes, adducts 34 and 35, respectively, were produced. $La_2@C_{80}$ possesses a closed-shell electronic structure with I_h molecular symmetry, similar to that of $Sc_3N@C_{80}$ in which six electrons are transferred from the metal (or metal cluster nitride) to the carbon cage.

However, as the encapsulated metal is different, La vs. Sc, the chemical reactivity and electrochemical properties are different in dyads 33 and 34. Indeed, although the same reaction strategy was applied, the formation of adduct 34 was facile compared with that of 33. In addition, the radical ion pair of 34, namely $(La_2@C_{80})^{\bullet-}$-$(ZnP)^{\bullet+}$, is favored only in polar solvents, like benzonitrile and DMF, as opposed to that of $(Sc_3N@C_{80})^{\bullet-}$-$(ZnP)^{\bullet+}$, which is also formed in apolar solvents such as toluene. On the other hand, $La@C_{82}$, is an open-shell with a huge anionic π-surface, with more active redox properties, smaller band gap, and lower-lying excited state. Ground-state intramolecular interactions within $La@C_{82}$-ZnP were identified with the aid of electron absorption spectroscopy and electrochemistry. In addition, quantitative quenching of the porphyrin emission signifies appreciable interactions in the excited state, though discrimination between electron and/or energy transfer from the porphyrin unit to the carbon cage in dyad 35 is yet to be explored [113].

Interestingly, when $Ce_2@C_{80}$ replaced $La_2@C_{80}$ metallofullerene in the formation of adduct 34, though with the zinc porphyrin connected to the carbon cage via a different spacer incorporating the flexible chain of 2-oxyethyl butyrate, different charge transfer phenomena were observed. It should be borne in mind that EMFs, and M_2C_{80} (M = Ce, La) in particular, show not only strong electron-accepting but also electron-donating properties, in contrast to empty fullerenes. Thus, it is not surprising that in the current example of $Ce_2@C_{80}$-ZnP, depending on the solvent polarity, electrons can intramolecularly flow to and from different directions. That is to say, in non-polar solvents $Ce_2@C_{80}$ behaves as an electron acceptor giving rise upon photoirradiation to the formation of the radical ion pair $(Ce_2@C_{80})^{\bullet-}$-$(ZnP)^{\bullet+}$. On the contrary, in polar solvents $Ce_2@C_{80}$ acts as an efficient electron donor leading to $(Ce_2@C_{80})^{\bullet+}$-$(ZnP)^{\bullet-}$ [114].

The one-electron oxidation potentials of TNT-based fullerenes are generally much lower than those of empty fullerenes. This property

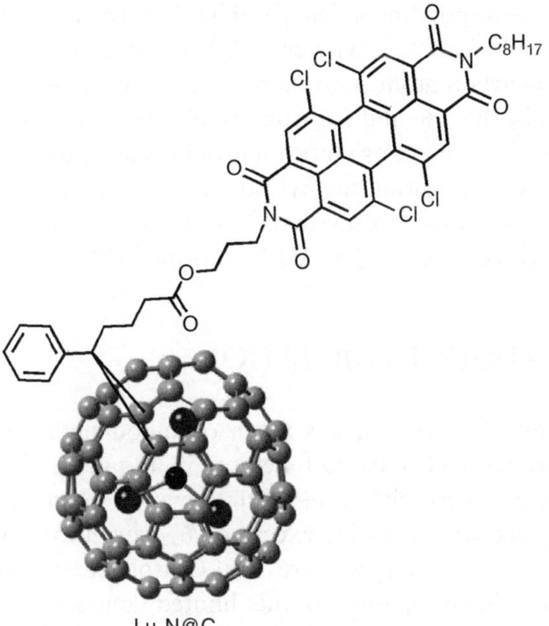

Lu₃N@C₈₀

36

Figure 11.24 Lu₃N@C₈₀-PDI dyad **36** as derived upon the [1+2] cycloaddition reaction of Lu₃N@C₈₀ with the light-harvesting and electron-acceptor 1,6,7,12-tetrachloro -3,4,9,20-perylenediimide (PDI) diazo compound

makes them better electron donor materials and simultaneously allows them to participate in photoinduced electron transfer schemes showing easier oxidations. Along these lines, a recent and representative example of utilizing TNT fullerenes as electron donors started with the [1+2] cycloaddition reaction of Lu₃N@C₈₀ with the light-harvesting and electron-acceptor 1,6,7,12-tetrachloro-3,4,9,20-perylenediimide (PDI) diazo compound, yielding Lu₃N@C₈₀-PDI dyad **36** (Figure 11.24). Based on MALDI-TOF and NMR measurements, only a single isomer of mono-adduct Lu₃N@C₈₀-PDI **36** was formed and found to have the 6,6-open addition pattern. The most intriguing feature of Lu₃N@C₈₀-PDI dyad **36** is that photoinduced electron transfer reactions evolve from the ground state of Lu₃N@C₈₀ to the excited state of PDI, thus disclosing new uses for TNT fullerenes as electron donors [115].

Another example of EMF acting as an electron donor is the recent report [116] in which La₂C₈₀ was modified and linked to the electron donor tetracyanoanthra-*p*-quinodimethane (TCAQ). For the transformation of La₂@C₈₀, the 1,3-dipolar cycloaddition reaction was chosen utilizing the corresponding aldehyde of TCAQ. Thus, a pyrrolidine ring was formed onto $La_2@C_{80}$, with the TCAQ unit as a substituent of the pyrrolidine α-carbon atom. Two mono-adducts were isolated via HPLC and found to be the 5,6- and 6,6-isomers. Photophysical studies revealed that although there are not significant ground state interactions between the two components within the hybrid material, in the excited state the reactivity differs. This is to say that upon $La_2@C_{80}$ photoexcitation the radical ion pair $(La_2@C_{80})^{\bullet+}$-$(TCAQ)^{\bullet-}$ is formed [116].

11.5   BIS-ADDUCT FORMATION

The formation of mono-adducts when o-quinodimethane diene intermediates react with TNT-based fullerenes in a [4+2] Diels-Alder cycloaddition reaction is established [49,50]. However, when $Gd_3N@C_{80}$ was allowed to react with a molar excess of 6,7-dimethoxy-3-isochromanone, bis-adduct derivatives were HPLC isolated as confirmed by MALDI-TOF. However, due to the limited amount of bis-adducts isolated, structural characterization, and determination of the location where the two isocromanone moieties exist has yet to be performed and clarified [117].

When the Bingel reaction of $La@C_{82}$ with diethyl bromomalonate in the presence of DBU was conducted at a slightly higher temperature, the isolation of a single isomer of bis-adduct was achieved. The ESR spectrum of bis-adduct showed a g-value and hyperfine coupling constant similar to those of the parent $La@C_{82}$. However, the electronic absorption spectrum of the bis-adduct was significantly differ from that of intact $La@C_{82}$, showing a characteristic band at 1135 nm and an onset in the NIR region at 1600 nm. The latter findings suggest that the bis-adduct of $La@C_{82}$ has a larger HOMO-LUMO band gap than the intact one, even though the open-shell structure has been retained [118].

Following the Bingel cyclopropanation reaction conditions, $Sc_3N@C_{78}$ with D_{3h} molecular symmetry was reacted with excess diethyl malonate in the presence of DBU. Although under those conditions $Sc_3N@C_{80}$ remains intact, mono- and bis-adducts of $Sc_3N@C_{78}$ were produced, thus clearly suggesting higher reactivity as compared with I_h-$Sc_3N@C_{80}$ [119]. Based on NMR measurements, the bis-malonate adduct had molecular symmetry. In addition, the characteristic absorption in the UV-Vis spectrum of $Sc_3N@C_{78}$ was missing in the bis-malonate derivative, indicating that its π-electron system is significantly altered. Moreover, electrochemistry studies suggest that the high regioselectivity for the formation of the C_{2v}-$Sc_3N@C_{78}$ is attributed to the strong influence of the internal trimetallic nitride cluster.

A dibenzyl adduct of $M_3N@C_{80}$ $(M = Sc, Lu)$ was synthesized with high regioselectivity via a photochemical reaction utilizing a large excess of benzyl bromide. X-ray crystallography together with ^{13}C NMR measurements unambiguously confirmed that the dibenzyl adduct of $M_3N@C_{80}$ is the 1,4-adduct. Moreover, UV-Vis absorption spectroscopy studies showed that the benzylic substituents affect the electronic structure of the C_{80} cage [120].

The perfluoroalkylation of $La@C_{82}$ was performed aiming to increase the solubility in highly fluorinated solvents. In this context, when a degassed solution of $La@C_{82}$ was irradiated by a UV lamp in the presence of excess perfluorooctyl iodide, a mixture of seven bis-perfluorooctyl adducts $La@C_{82}$-$(C_8F_{17})_2$ was produced. Each isomer was isolated in pure form with the aid of recycling HPLC and found to be paramagnetic, possessing characteristic hyperfine coupling constants and g-values [121].

11.6 SUPRAMOLECULAR FUNCTIONALIZATION

EMFs form host–guest complexes, though the field of EMF supramolecular chemistry is awaiting full exploration [122]. As interactions of heteroatoms such as N, O, and S with metallofullerenes is possible mainly via electron-transfer pathways, it is reasonable to expect the formation of complexes when $La@C_{82}$ is treated with azacrown ethers. Hence, upon mixing of $La@C_{82}$ with 1,4,7,10,13,16-hexaazacyclooctadecane 37 (Figure 11.25) in toluene, a precipitate is formed. The latter, which is

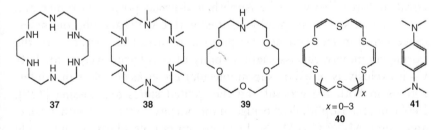

Figure 11.25 Heteroatom compounds with which $La@C_{82}$ forms supramolecular charge-transfer complexes: 1,4,7,10,13,16-hexaazacyclooctadecane **37**, 1,4,7,10,13, 16-hexamethyl-1,4,7,10,13,16-hexaazacyclooctadecane and mono-aza-18-crown-6 ether, **38** and **39**, respectively, thiacrown ether **40** and N,N,N',N'-tetramethyl-p-phenylenediamine **41**

soluble in polar solvents, shows a characteristic absorption band in the NIR region due to the anion $(La@C_{82})^-$ which evolves from an electron transfer process from the macrocycle to $La@C_{82}$. The same behavior, namely formation of precipitates and emergence of a new NIR band, is also observed upon treatment of $La@C_{82}$ with 1,4,7,10,13,16-hexameth yl-1,4,7,10,13,16-hexaazacyclooctadecane and mono-aza-18-crown-6 ether, **38** and **39**, respectively [123]. Moreover, facile electron transfer and complexation of $La@C_{82}$ with unsaturated thiacrown ether **40** (Figure 11.25) is also observed [124].

Along the same lines, complexation of $La@C_{82}$ with N,N,N',N'-tetramethyl-p-phenylenediamine (TMPD) **41** (Figure 11.25) was also achieved [125]. As the radical cation of TMPD$^{•+}$ is stable and has distinct absorptions in the visible region, titration experiments in which increasing amounts of TMPD **41** were added to $La@C_{82}$ revealed that electron transfer from TMPD to $La@C_{82}$ occurs. The latter was obvious by the progressive formation of the bands due to TMPD$^{•+}$ and $La@C_{82}^{•-}$. Furthermore, the intermolecular complexation and electron transfer behavior between dimetallofullerene $La_2@C_{80}$ and TMPD were studied and it was found that significant electronic interactions in the ground state exist [126].

On a different aspect of host–guest complexes with metallofullerenes, a cyclic porphyrin trimer was used to complex $La@C_{82}$. The presence of three porphyrin units preorganized to chelate to the same fullerene cage afforded very strong binding for $La@C_{82}$ [127]. Moreover, a cyclodimer porphyrin was utilized to host paramagnetic $La@C_{82}$ which was further transformed to a caged complex upon ring-closing olefin metathesis of its side chain alkene termini [128].

In addition, a dimeric porphyrin tweezer supramolecularly captured $La@C_{82}$ to generate superstructure **42** (Figure 11.26). Cooperative non-covalent forces between the isophthaloyl-based porphyrin dimer and $La@C_{82}$ were spectroscopically justified with UV-Vis and photolumines-cence titration experiments. Despite the photoactive character of the porphyrin moiety, the association ability of porphyrin tweezer for La@ C_{82} is exclusively based on supramolecular forces as any electronic inter-actions in both the ground and the excited states are missing [129]. Replacing the isophthaloyl bridge in the porphyrin tweezer with a calix-arene unit, $M_3N@C_{80}$ $(M = Sc, Lu)$ were supramolecularly bound, while electron transfer phenomena from the TNT fullerenes to the porphyrin tweezers were found to exist [130].

Considering that porphyrins play an important role in self-assembly, ZnP was used to encapsulate and self-assemble $Sc_3N@C_{80}$

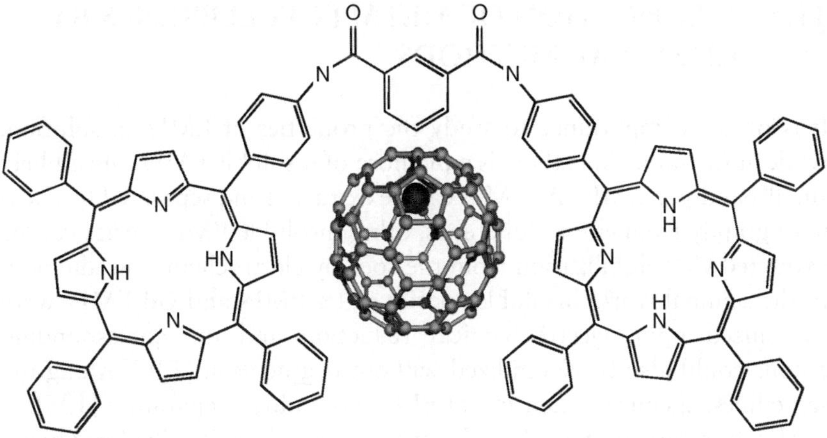

42

Figure 11.26 Dimeric porphyrin tweezer supramolecularly capturing La@C_{82}

nanorods. Thus, with the aid of cetyltrimethylammonium bromide surfactant in water the length and size distribution of the Sc$_3$N@C_{80}-ZnP nanorods was controlled. Furthermore, the fluorescence emission intensity of Sc$_3$N@C_{80}-ZnP was found to be strongly quenched compared with the emission of intact ZnP, thus suggesting that photoinduced electron transfer from the excited porphyrin to the TNT fullerene occurs [131].

Similar well-arrayed one-dimensional nanorods were also formed upon reaction of La@C_{82} with 2-adamantane-2,3-[3H]-diazirine. Nanorods of functionalized EMF La@C_{82}-Ad material **17** (cf. Figure 11.11) were prepared by a liquid–liquid interfacial precipitation method. Moreover, La@C_{82}-Ad material **17** showed p-type field effect transistor behavior and was found to orient with the nanorod axis perpendicular to the magnetic field, thus possessing a negative magnetic anisotropy [132].

Pyrrolidine functionalized M$_3$N@C_{80} (M = Er, Sc) TNT fullerenes carrying a dithiolane unit as an extension of the pyrrolidine α-carbon substituent, formed spontaneous molecular monolayers on gold surfaces. Particularly, the S atoms in the dithiolane group serve as an anchor, enabling efficient immobilization of the modified TNT fullerenes on Au(111) surfaces [133]. Notably, the optical activity of Er$_3$N@C_{80} is retained not only upon functionalization but also within a monolayer assembled on the surface, a property that could enable their incorporation in functional electronic devices.

11.7 PURIFICATION OF METALLOFULLERENES BY CHEMICAL METHODS

It is of great importance to study the properties of EMFs in solution, while at the same time, there is a plethora of insoluble EMFs, most likely small-band-gap EMFs. As EMFs can be extracted and separated by chromatography from empty fullerenes, other insoluble EMF species can be extracted via solubilization from the soot by electrochemical reduction. In the seminal work, insoluble species of La-EMFs and Gd-EMFs were solubilized upon electrochemical reduction and the corresponding anions could also be re-oxidized without degradation [134]. Along the same lines, chemical oxidation of EMFs also allows separation [135].

The first isolation of La@C_{74} was achieved as its dichlorophenyl adduct. In more detail, although La@C_{74} can be observed in the raw soot, its isolation was hampered, until the fullerene soot was extracted with 1,2,3-trichlorobenzene (TCB) [136]. Thus, it was spectroscopically proved and confirmed with MALDI-TOF that La@C_{74} reacts with the solvent used for extraction (i.e., TCB) and three isomers of La@C_{74}–$(C_6H_3Cl)_2$ are formed. This is the case where an unstable EMF with a radical character, such as La@C_{74}, can be trapped and stabilized in the form of a closed electronic shell, by the dichlorophenyl radical formed during reflux upon extraction of the soot. Extensive studies on the addition reaction of dichlorophenyl radical to La@C_{74} showed that actually two sets of adducts were produced, each of which contains three distinct isomers with different substitution patterns on the dichlorophenyl group [137].

Similarly, the non-IPR La@C_{72} was trapped and isolated from the raw soot as its dichlorophenyl adduct. Again, the La@C_{72}–$(C_6H_3Cl)_2$ has a closed shell electronic structure [138]. It is of no surprise that this unique derivatization strategy emerged as a powerful tool to extract and stabilize other missing EMFs. Thus, La@C_{80} was also isolated as La@C_{80}–$(C_6H_3Cl)_2$ possessing a closed shell [139]. Moreover, since this method is so powerful, not only missing EMFs could be isolated but also other new and higher EMFs could be obtained as dichlorophenyl adducts. Thus, a new La@C_{82} EMF was isolated with the cage having C_{3v} molecular symmetry. Interestingly, NMR and X-ray studies revealed that the dichlorophenyl group is single bonded to a triple-hexagon junction of the fullerene cage, so that the C_3 symmetry is preserved in the isolated adduct La@C_{82}–$(C_6H_3Cl)_2$ [140].

Regardless of the above-mentioned methodology with which EMFs can be isolated from the soot in the form of their dichlorophenyl adducts, the area is still in its infancy in that non-chromatographic techniques are needed to purify and isolate EMFs. Recently, a breakthrough was reported on the separation and purification of EMFs from empty fullerenes based on their reaction with $TiCl_4$. It was found that EMFs react rapidly with $TiCl_4$ Lewis acid forming insoluble complexes in toluene or CS_2, whereas most of the empty cages are almost inert under the applied conditions [141]. Importantly, very short reaction times are needed for the purification of EMFs from empty fullerenes, a fact that makes this methodology advantageous over the existing ones, thus ensuring large-scale preparation of purified EMFs.

Furthermore, in depth study of the complex EMFs form with $TiCl_4$ showed that the first oxidation potential of EMFs is crucial and governs the reactivity. Lower oxidation potential leads to higher reactivity with $TiCl_4$, with the threshold being around 0.62–0.72 V vs. Fc/Fc^+, thus EMFs – regardless of the encapsulated metal – with first oxidation potential lower than that threshold value can be efficiently separated and purified from empty fullerenes [142]. Importantly, electronic absorption studies showed the cation formation of EMFs upon treatment with $TiCl_4$, thus suggesting the oxidation of EMFs by $TiCl_4$, which in turn explains why EMFs with lower oxidation potential exhibit higher reactivity with $TiCl_4$. Also, metallic nitride and oxometallic fullerenes were found to efficiently form complexes with Lewis acids such as $AlCl_3$ and $FeCl_3$. Thus, they can be purified and isolated from the soot free of empty cages [143].

Lanthanide-based trimetallic nitride EMFs $M_3N@C_{80}$ (M = Er, Gd, Ho, Lu, Sc, Tb, Tm, Y) can be purified and isolated directly from the soot by applying a functionalization methodology in which the $M_3N@C_{80}$ are less reactive compared with the rest of the fullerenes in the soot. In this context, as $M_3N@C_{80}$ metallofullerenes show kinetic chemical stability compared with other EMFs and empty cages present in the as-produced soot, Diels-Alder functionalization with a cyclopentadiene-modified resin results in their purification. Importantly, as the Diels-Alder cycloaddition reaction is thermally reversible, the modified empty cages and conventional EMFs can be easily recovered upon treatment with maleic anhydride.

The advantage of this methodology is that $M_3N@C_{80}$ can be separated from classical EMFs and empty fullerenes, though isomers and other TNTs of different sizes have yet to be purified by extensive HPLC [144]. In addition, the so-called "stir and filter approach" took advantage of the higher reactivity of the D_{5h} isomer of $Sc_3N@C_{80}$ compared with that of the I_h isomer, not only to separate and purify the two isomers but also to remove empty cages and classical EMFs from the soot, when passed through a cyclopentadienyl and diamino functionalized silica [145,146]. Along the same lines, reaction of the crude fullerene soot with 9-methylanthracene, in the absence of any solvent, results in the exclusive functionalization of empty cages to their adducts, while leaving the TNT-based EMFs unreacted [147]. Finally, the two isomers of $Sc_3N@C_{80}$ with molecular symmetry D_{5h} and I_h can be also purified and isolated by electrochemical oxidation. This is because these two isomers have a significant difference in their first electrochemical oxidation potential [148].

REFERENCES

[1] Liu X, Akasaka T and Nagase S 2011 Chemistry of metallofullerenes: the role of metals *Chem. Commun.* **47** 5942

[2] Chaur M N, Melin F, Ortiz A L and Echegoyen L 2009 Chemical, electrochemical, and structural properties of endohedral metallofullerenes *Angew. Chem. Int. Ed.* **48** 7514

[3] Akasaka T, Kato T, Kobayashi K *et al.* 1995 Exohedral adducts of La@C_{82} *Nature* **374** 600

[4] Akasaka T, Nagase S, Kobayashi K *et al.* 1995 Exohedral derivatization of an endohedral metallofullerene Gd@C_{82} *J. Chem. Soc., Chem. Commun.* 1343

[5] Akasaka T, Nagase S, Kobayashi K *et al.* 1995 Synthesis of the first adducts of the dimetallofullerenes $La_2@C_{80}$ and $Sc_2@C_{84}$ by addition of a disilirane *Angew. Chem. Int. Ed.* **34** 2139

[6] Iiduka Y, Ikenaga O, Sakuraba A *et al.* 2005 Chemical reactivity of $Sc_3N@C_{80}$ and $La_2@C_{80}$ *J. Am. Chem. Soc.* **127** 9956

[7] Wakahara T, Iiduka Y, Ikenaga O *et al.* 2006 Characterization of the bis-silylated endofullerene $Sc_3N@C_{80}$ *J. Am. Chem. Soc.* **128** 9919

[8] Sato K, Kako M, Mizorogi N *et al.* 2012 Bis-silylation of $Lu_3N@I_h$-C_{80}: considerable variation in the electronic structures *Org. Lett.* **14** 5908

[9] Wakahara T, Yamada M, Takahashi S *et al.* 2007 Two-dimensional hopping of encapsulated La atoms in silylated $La_2@C_{80}$ *Chem. Commun.* 2680

[10] Yamada M, Nakahodo T, Wakahara T. *et al.* 2005 Positional control of encapsulated atoms inside a fullerene cage by exohedral addition *J. Am. Chem. Soc.* **127** 14570

[11] Yamada M, Wakahara T, Tsuchiya T *et al.* 2008 Location of the metal atoms in $Ce_2@C_{78}$ and its bis-silylated derivative *Chem. Commun.* 558

[12] Maeda Y, Miyashita J, Hasegawa T *et al.* 2005 Chemical reactivities of the cation and anion of M@C_{82} (M = Y, La, and Ce) *J. Am. Chem. Soc.* **127** 2143

[13] Maggini M, Scorrano G and Prato M 1993 Addition of azomethine ylides to C_{60}: synthesis, characterization, and functionalization of fullerene pyrrolidines *J. Am. Chem. Soc.* **115** 9798

[14] Tagmatarchis N, and Prato M 2003 The addition of azomethine ylides to [60]fullerene leading to fulleropyrrolidines *Synlett* **6** 768

[15] Cao B, Wakahara T, Maeda Y *et al*. 2004 Lanthanum endohedral metallofullero-pyrrolidines: synthesis, isolation and EPR characterization *Chem. Eur. J*. **10** 716

[16] Cardona C M, Litaygorodskiy A, Ortiz A *et al*. 2005 The first fulleropyrrolidine derivatives of $Sc_3N@C_{80}$: pronounced chemical shift differences of the germinal protons on the pyrrolidine ring *J. Org. Chem*. **70** 5092

[17] Cai T, Ge Z, Iezzi E B *et al*. 2005 Synthesis and characterization of the first trimetallic nitride template pyrrolidino endohedral metallofullerenes *Chem. Commun*. 3594

[18] Cai T, Slebodnick C, Xu L *et al*. 2006 A pirouette on a metallofullerene sphere: interconversion of isomers of N-tritylpyrrolidino I_h $Sc_3N@C_{80}$ *J. Am. Chem. Soc*. **128** 6486

[19] Cai T, Xu L, Anderson M R *et al*. 2006 Structure and enhanced reactivity rates of the D_{5h} $Sc_3N@C_{80}$ and $Lu_3N@C_{80}$ metallofullerene isomers: the importance of the pyracylene motif *J. Am. Chem. Soc*. **128** 8581

[20] Osuma S, Rodriguez-Fortea A, Poblet J M *et al*. 2012 Product formation in the Prato reaction on $Sc_3N@D_{5h}$-C_{80}: preference for [5,6]-bonds, and not pyracylenic bonds *Chem. Commun*. **48** 2486

[21] Cardona C M, Kitaygorodskiy A and Echegoyen L 2005 Trimetallic nitride endohedral metallofullerenes: reactivity dictated by the encapsulated metal cluster *J. Am. Chem. Soc*. **127** 10448

[22] Rodriguez-Fortea A, Campanera J M, Cardona C M *et al*. 2006 Dancing on a fullerene surface: isomerization of $Y_3N@(N$-ethylpyrrolidino-C_{80}) from the 6,6 to the 5,6 regioisomer *Angew. Chem. Int. Ed*. **45** 8176

[23] Cardona C M, Elliott B and Echegoyen L 2006 Unexpected chemical and electrochemical properties of $M_3N@C_{80}$ (M = Sc, Y, Er) *J. Am. Chem. Soc*. **128** 6480

[24] Campanera J M, Bo C and Poblet J M 2006 Exohedral reactivity of trimetallic nitride template (TNT) endohedral metallofullerenes *J. Org. Chem*. **71** 46

[25] Echegoyen L, Chancellor C J, Cardona C M *et al*. 2006 X-ray crystallographic and EPR spectroscopic characterization of a pyrrolidine adduct of $Y_3N@C_{80}$ *Chem. Commun*. 2653

[26] Chen N, Zhang E-Y, Tan K *et al*. 2007 Size effect of encaged clusters on the exohedral chemistry of endohedral fullerenes: a case study on the pyrrolidino reaction of $Sc_xGd_{3-x}NC_{80}$ (x = 0-3) *Org. Lett*. **9** 2011

[27] Aroua S and Yamakoshi Y 2012 Prato reaction of $M_3N@I_h$-C_{80} (M = Sc, Lu, Y, Gd) with reversible isomerization *J. Am. Chem. Soc*. **134** 20242

[28] Chen N, Fan L-Z, Tan K *et al*. 2007 Comparative spectroscopic and reactivity studies of $Sc_{3-x}Y_xNC_{80}$ (x = 0-3) *J. Phys. Chem. C* **111** 11823

[29] Olmstead M M, de Bettencourt-Dias A, Duchamp J C *et al*. 2001 Isolation and structural characterization of the endohedral fullerene $Sc_3N@C_{78}$ *Angew. Chem. Int. Ed*. **40** 1223

[30] Campanera J M, Bo C, Olmstead M M *et al*. 2002 Bonding within the endohedral fullerenes $Sc_3N@C_{78}$ and $Sc_3N@C_{80}$ as determined by density functional calculations and reexamination of the crystal structure of $\{Sc_3N@C_{78}\}\cdot Co(OEP)\}\cdot 1.5(C_6H_6)\cdot 0.3$ $(CHCl_3)$ *J. Phys. Chem. A* **106** 12356

[31] Krause M, Popov A and Dunsch L 2006 Vibrational structure of endohedral fullerene $Sc_3N@C_{78}$ (D_3h): evidence for a strong coupling between the Sc_3N cluster and C_{78} cage *Eur. J. Chem. Phys. Phys. Chem*. **7** 1734

[32] Park S S, Liu D and Hagelberg F 2005 Comparative investigation on non-IPR C_{68} and IPR C_{78} fullerenes encaging Sc_3N molecules *J. Phys. Chem. A* **109** 8865

[33] Cai T, Xu L, Gibson H W *et al.* 2007 $Sc_3N@C_{78}$: encapsulated cluster regiocontrol of adduct docking on an ellipsoidal metallofullerene sphere *J. Am. Chem. Soc.* **129** 10795

[34] Martin N, Altable M, Filippone S *et al.* 2006 Retro-cycloaddition reaction of pyrrolidinofullerenes *Angew. Chem. Int. Ed.* **45** 110

[35] Yamada M, Wakahara T, Nakahodo T *et al.* 2006 Synthesis and structural characterization of endohedral pyrrolidinodimetallofullerene: $La_2@C_{80}(CH_2)_2NTrt$, *J.Am. Chem. Soc.* **128** 1402

[36] Yamada M, Okamura M, Sato S *et al.* 2009 Two regioisomers of endohedral pyrrolidinodimetallofullerenes $M_2@I_h\text{-}C_{80}(CH_2)_2NTrt$ (M = La, Ce; Trt = trityl): control of metal atom positions by addition positions *Chem. Eur. J.* **15** 10533

[37] Iiduka Y, Wakahara T, Nakahodo T *et al.* 2005 Structural determination of metallofullerene Sc_3C_{82} revisited: a surprising finding *J. Am. Chem. Soc.* **127** 12500

[38] Wang T, Wu J, Xu W *et al.* 2010 Spin divergence induced by exohedral modification: ESR study of $Sc_3C_2@C_{80}$ fulleropyrrolidine *Angew. Chem. Int. Ed.* **49** 1786

[39] Inakuma M, Yamamoto E, Kai T *et al.* 2000 Structural and electronic properties of isomers of $Sc_2@C_{84}$ (I, II, III): ^{13}C NMR and IR/Raman spectroscopic studies *J. Phys. Chem. B* **104** 5072

[40] Lu X, Nakajima K, Iiduka Y *et al.* 2011 Structural elucidation and regioselective functionalization of an unexplored carbide cluster metallofullerene $Sc_2C_2@C_s(6)\text{-}C_{82}$ *J. Am. Chem. Soc.* **133** 19553

[41] Lukoyanova O, Cardona C M, Rivera J *et al.* 2007 "Open rather than closed" malonate methano-fullerene derivatives. The formation of methanofullereoid adducts of $Y_3N@C_{80}$ *J. Am. Chem. Soc.* **129** 10423

[42] Feng L, Nakahodo T, Wakahara T *et al.* 2005 A single bonded derivative of endohedral metallofullerene: $La@C_{82}CBr(COOC_2H_5)_2$ *J. Am. Chem. Soc.* **127** 17136

[43] Feng L, Wakahara T, Nakahodo T *et al.* 2006 The Bingel monoadducts of $La@C_{82}$: synthesis, characterization, and electrochemistry *Chem. Eur. J.* **12** 5578

[44] Chaur M N, Melin F, Athans A J *et al.* 2008 The influence of cage size on the reactivity of trimetallic nitride metallofullerenes: a mono- and bis-methanoadduct of $Gd_3N@C_{80}$ and a monoadduct of $Gd_3N@C_{84}$ *Chem. Commun.* 2665

[45] Alegret N, Chaur M N, Santos E *et al.* 2010 Bingel-Hirsch reactions on non-IPR $Gd_3N@C_{2n}$ (2n = 82, 84) *J. Org. Chem.* **75** 8299

[46] Stevenson S, Lee H M, Olmstead M M *et al.* 2002 Preparation and crystallographic characterization of a new endohedral, $Lu_3N@C_{80} \cdot 5$ (o-xylene), and comparison with $Sc_3N@C_{80} \cdot 5$ (o-xylene) *Chem. Eur. J.* **8** 4528

[47] Pinzon J R, Zuo T and Echegoyen L 2010 Synthesis and electrochemical studies of Bingel-Hirsch derivatives of $M_3N@I_h\text{-}C_{80}$ (M = Sc, Lu) *Chem. Eur. J.* **16** 4864

[48] Maeda Y, Miyashita J, Hasegawa T *et al.* 2005 Reversible and regioselective reaction to $La@C_{82}$ with cyclopentadiene *J. Am. Chem. Soc.* **127** 12190

[49] Iezzi E B, Duchamp J C, Harich K *et al.* 2002 A symmetric derivative of the trimetallic nitride endohedral metallofullerene, $Sc_3N@C_{80}$ *J. Am. Chem. Soc.* **124** 524

[50] Lee H M, Olmstead M M, Iezzi E *et al.* 2002 Crystallographic characterization and structural analysis of the first organic functionalization product of the endohedral fullerene $Sc_3N@C_{80}$ *J. Am. Chem. Soc.* **124** 3494

[51] Lu, X, Xu J, He X *et al.* 2004 Addition of benzyne to $Gd@C_{82}$ *Chem. Mater.* **16** 953

[52] Lu X, Nikawa H, Tsuchiya T *et al.* 2010 Nitrated benzyne derivatives of $La@C_{82}$: addition of NO_2 and its positional directing effect on the subsequent addition of benzynes *Angew. Chem. Int. Ed.* **47** 594

[53] Nakamura Y, Takano N, Nishimura T *et al.* 2001 First isolation and characterization of eight regioisomers for [60]fullerene-benzyne bisadducts *Org. Lett.* **3** 1193

[54] Li F F, Pinzon J R, Mercado B Q *et al.* 2011 [2+2] Cycloaddition reaction to $Sc_3N@I_h$-C_{80}. The formation of very stable [5,6]- and [6,6]-adducts *J. Am. Chem. Soc.* **133** 1563

[55] Wang G W, Liu T X, Jiao M *et al.* 2011 The cycloaddition reaction of I_h-$Sc_3N@C_{80}$ with 2-amino-4,5-diisopropoxybenzoic acid and isoamyl nitrite to produce an open-cage metallofullerene *Angew. Chem. Int. Ed.* **50** 4658

[56] Maeda Y, Matsunaga Y, Wakahara T *et al.* 2004 Isolation and characterization of a carbine derivative of La@C_{82} *J. Am. Chem. Soc.* **126** 6858

[57] E. Nishibori E, Iwata K, Sakata M *et al.* 2004 Anomalous endohedral structure of Gd@C_{82} metallofullerenes *Phys. Rev. B* **69** 113412/1-4

[58] Liu L, Gao B, Chu W *et al.* 2008 The structural determination of endohedral metallofullerene Gd@C_{82} by XANES *Chem. Commun.* 474

[59] Akasaka T, Kono T, Takematsu Y *et al.* 2008 Does Gd@C_{82} have an anomalous endohedral structure? Synthesis and single crystal X-ray structure of the carbene adduct *J. Am. Chem. Soc.* **130** 12840

[60] Liu X, Nikawa H, Feng L *et al.* 2009 Location of yttrium atom in Y@C_{82} and its influence on the reactivity of cage carbons *J. Am. Chem. Soc.* **131** 12066

[61] Takano Y, Aoyagi M, Yamada M *et al.* 2009 Anisotropic behavior of anionic Ce@C_{82} carbene adducts *J. Am. Chem. Soc.* **131** 9340

[62] Hachiya M, Nikawa H, Mizorogi N *et al.* 2012 Exceptional chemical properties of $Sc@C_{2v}(9)$-C_{82} probed with adamantylidene carbine *J. Am. Chem. Soc.* **134** 15550

[63] Cao B, Nikawa H, Nakahodo T *et al.* 2008 Addition of adamantylidene to $La_2@C_{78}$: isolation and single-crystal X-ray structural determination of the monoadducts *J. Am. Chem. Soc.* **130** 983

[64] Yamamoto E, Tansho M, Tomiyama T *et al.* 1996 ^{13}C-NMR study on the structure of isolated $Sc_2@C_{84}$ metallofullerene *J. Am. Chem. Soc.* **118** 2293

[65] Iiduka Y, Wakahara T, Nakajima K *et al.* 2007 Experimental and theoretical studies of the scandium carbide endohedral metallofullerene $Sc_2C_2@C_{82}$ and its carbine derivative *Angew. Chem. Int. Ed.* **46** 5562

[66] Kurihara H, Lu X, Iiduka Y *et al.* 2012 Chemical understanding of carbide cluster metallofullerenes: a case study on $Sc_2C_2@C_{2v}(5)$–C_{80} with complete X-ray crystallographic characterizations *J. Am. Chem. Soc.* **134** 3139

[67] Yamada M, Someya C, Wakahara T *et al.* 2008 Metal atoms collinear with the spiro carbon of 6,6-open adducts, $M_2@C_{80}(Ad)$ (M = La and Ce, Ad = Adamantylidene) *J. Am. Chem. Soc.* **130** 1171

[68] Liu X, Nikawa H, Tsuchiya T *et al.* 2008 Bis-carbene adducts of non-IPR $La_2@C_{72}$: localization of high reactivity around fused pentagons and electrochemical properties *Angew. Chem. Int. Ed.* **47** 8642

[69] Liu X, Nikawa H, Nakahodo T *et al.* 2008 Chemical understanding of a non-IPR metallofullerene: stabilization od encaged metals on fused-pentagon bonds in $La_2@C_{72}$ *J. Am. Chem. Soc.* **130** 9129

[70] Kobayashi K, Nagase S and Akasaka T 1996 Endohedral dimetallofullerenes $Sc_2@C_{84}$ and $La_2@C_{80}$. Are the metal atoms still inside the fullerene cages? *Chem. Phys. Lett.* **261** 502

[71] Akasaka T, Nagase S, Kobayashi K *et al.* 1997 ^{13}C and ^{139}La NMR studies of $La_2@C_{80}$: first evidence for circular motion of metal atoms in endohedral dimetallofullerenes *Angew. Chem. Int. Ed.* **36** 1643

[72] Nishibori E, Takata M, Sakata M *et al.* 2001 Pentagonal-dodecahedral La$_2$ charge density in [80-I_h]fullerene: La$_2$@C$_{80}$ *Angew. Chem. Int. Ed.* **40** 2998

[73] Shimotani H, Ito T, Iwasa Y *et al.* 2004 Quantum chemical study on the configurations of encapsulated metal ions and the molecular vibration modes in endohedral dimetallofullerene La$_2$@C$_{80}$ *J. Am. Chem. Soc.* **126** 364

[74] Ishitsuka M O, Sano S, Enoki H *et al.* 2011 Regioselective bis-functionalization of endohedral dimetallofullerene, La$_2$@C$_{80}$: extremal La–La distance *J. Am. Chem. Soc.* **133** 7128

[75] Suzuki T, Maruyama Y, Kato T *et al.* 1995 Chemical reactivity of a metallofullerene: EPR study of diphenylmethano-La@C$_{82}$ radicals *J. Am. Chem. Soc.* **117** 9606

[76] Shu C, Cai T, Xu L *et al.* 2007 Manganese(III)-catalyzed free radical reactions on trimetallic nitride endohedral metallofullerenes *J. Am. Chem. Soc.* **129** 15710

[77] Takano Y, Yomogida A, Nikawa H *et al.* 2008 Radical coupling of paramagnetic endohedral metallofullerene La@C$_{82}$ *J. Am. Chem. Soc.* **130** 16224

[78] Shistova N B, Popov A, Mackey M A *et al.* 2007 Radical trifluoromethylation of Sc$_3$N@C$_{80}$ *J. Am. Chem. Soc.* **129** 11676

[79] Shustova N B, Peryshkov D V, Kuvychko I V *et al.* 2011 Poly(perfluoroalkylation) of metallic nitride fullerenes reveal addition-pattern guidelines: synthesis and characterization of a family of Sc$_3$N@C$_{80}$(CF$_3$)n (n = 2–16) and their radical anions *J. Am. Chem. Soc.* **133** 2672

[80] Hao C, Liu Z, Guo X *et al.* 1997 Gas phase derivatization of endohedral metallofullerenes R@C$_{82}$ and R$_2$@C$_{80}$ (R = Nd, Ce) *Rapid Commun. Mass Spectrom.* **11** 1677

[81] Li F F, Rodriguez-Fortea A, Poblet J M and Echegoyen L 2011 Reactivity of metallic nitride endohedral metallofullerene anions: electrochemical synthesis of a Lu$_3$N @I_h-C$_{80}$ derivative *J. Am. Chem. Soc.* **133** 2760

[82] Feng L, Zhang X, Yu Z *et al.* 2002 Chemical, modification of Tb@C$_{82}$ by copper (I)-catalyzed cycloadditions *Chem. Mater.* **14** 4021

[83] Li X, Fan L, Liu D *et al.* 2007 Synthesis of a Dy@C$_{82}$ derivative bearing a single phosphorus substituent via a zwitterion approach *J. Am. Chem. Soc.* **129** 10636

[84] Yamada M, Minowa M, Sato S *et al.* 2011 Regioselective cycloaddition of La$_2$@I_h-C$_{80}$ with tetracyanoethylene oxide: formation of an endohedral dimetallofullerene adduct featuring enhanced electron-accepting character *J. Am. Chem. Soc.* **133** 3796

[85] Li J, Takeuchi A, Ozawa M *et al.* 1993 C$_{60}$ fullerol formation catalysed by quaternary ammonium hydroxides *J. Chem. Soc., Chem. Commun.* 1784

[86] Kato H, Suenaga K, Mikawa M *et al.* 2000 Synthesis and EELS characterization of water-soluble multi-hydroxyl Gd@C$_{82}$ fullerenols *Chem. Phys. Lett.* **324** 255

[87] Sun D, Huang H and Yang S 1999 Synthesis and characterization of a water-soluble endohedral metallofullerol, *Chem. Mater.* **11** 1003

[88] Mikawa M, Kato H, Okumura M *et al* 2001 Paramagnetic water-soluble metallofullerenes having the highest relaxivity for MRI contrast agents *Bioconjugate Chem.* **12** 510

[89] Zhang Y, Xiang J, Zhuang J *et al.* 2005 Synthesis of the first dihydroxyl adduct of Gd@C$_{82}$ *Chem. Lett.* **34** 1264

[90] Kato H, Kanazawa Y, Okumura M *et al.* 2003 Lanthanoid endohedral metallofullerenols for MRI contrast agents *J. Am. Chem. Soc.* **125** 4391

[91] Wilson L J, Cagle D W, Thrash T P *et al.* 1999 Metallofullerene drug design *Coord. Chem. Rev.* **190–192** 199

[92] Cagle D W, Kennel S J, Mirzadeh S et al. 1999 In vivo studies of fullerene-based materials using endohedral metallofullerene radiotracers Proc. Natl. Acad. Sci. USA 96 5182

[93] Iezzi E B, Cromer F, Stevenson P and Dorn H C 2002 Synthesis of the first water-soluble trimetallic nitride endohedral metallofullerols Synth. Met. 128 289

[94] Zhang E-Y, Shu C-Y, Feng L and Wang C-R 2007 Preparation and characterization of two new water-soluble endohedral metallofullerenes as magnetic resonance imaging contrast agents J. Phys. Chem. B 111 14223

[95] Fatouros P P, Corwin F D, Chen Z-Y et al. 2006 In vitro and in vivo imaging studies of a new endohedral metallofullerene nanoparticle Radiology 240 756

[96] Bolskar R D, Benedetto A F, Husebo L O et al. 2003 First soluble M@C_{60} derivatives provide enhanced access to metallofullerenes and permit in vivo evaluation of Gd@$C_{60}[C(COOH)_2]_{10}$ as a MRI contrast agent J. Am. Chem. Soc. 125 5471

[97] Lu X, Zhou X, Shi Z and Gu Z 2004 Nucleophilic addition of glycine esters to Gd@C_{82} Inorg. Chim. Acta 357 2397

[98] Shu C-Y, Gan L-H, Wang C-R et al. 2006 Synthesis and characterization of a new water-soluble metallofullerene for MRI contrast agents Carbon 44 496

[99] Shu C-Y, Ma X-Y, Zhang J-F et al. 2008 Conjugation of a water-soluble gadolinium endohedral fulleride with an antibody as a magnetic resonance imaging contrast agent Bioconjugate Chem. 19 651

[100] Rudolf M, Wolfraum S, Guldi D M et al. 2012 Emdohedral metallofullerenes – filled fullerene derivatives towards multifunctional reaction center mimics Chem. Eur. J. 18 5136

[101] Pinzon J R, Plonska-Brzezinska M E, Cardona C M et al. 2008 Sc_3N@C_{80}-Ferrocene electron-donor/acceptor conjugates as promising materials for photovoltaic applications Angew. Chem. Int. Ed. 47 4173

[102] Pinzon J R, Gasca D C, Sankaranarayanan S G et al. 2009 Photoinduced charge transfer and electrochemical properties of triphenylamine I_h-Sc_3N@C_{80} donor-acceptor conjugates J. Am. Chem. Soc. 131 7727

[103] Pinzon J R, Cardona C M, Herranz M A et al. 2009 Metal nitride cluster fullerene M_3N@C_{80} (M = Y, Sc) based dyads: synthesis and electrochemical, theoretical and photophysical studies Chem. Eur. J. 15 864

[104] Takano Y, Herranz M A, Martin N et al. 2010 Donor-acceptor conjugates of lanthanum endohedral metallofullerene and π-extended tetrathiafulvalene J. Am. Chem. Soc. 132 8048

[105] Takano Y, Obuchi S, Mizorogi N et al. 2012 Stabilizing ion and radical ion pair states in a paramagnetic endohedral metallofullerene/π-extended tetrathiafulvalene conjugate J. Am. Chem. Soc. 134 16103

[106] Wolfrum S, Pinzon J R, Molina-Ontoria A et al. 2011 Utilization of Sc_3N@C_{80} in long-range charge transfer reactions Chem. Commun. 47 2270

[107] Shu C, Xu W, Slebodnick C et al. 2009 Syntheses and structures of phenyl-C_{81}-butyric acid methyl esters (PCBM) from M_3N@C_{80} Org. Lett. 11 1753

[108] Ross R B, Cardona C M, Guldi D M et al. 2009 Endohedral fullerenes for organic photovoltaic devices Nat. Mater. 8 208

[109] Ross R B, Cardona C M, Swain F B et al. 2009 Tuning conversion efficiency in metallo endohedral fullerene-based organic photovoltaic devices Adv. Funct. Mater. 19 2332

[110] Liedtke M, Sperlich A, Kraus H et al. 2011 Triplet exciton generation in bulk-heterojunction solar cells based on endohedral fullerenes J. Am. Chem. Soc. 133 9088

[111] Matsuo Y, Okada H, Maruyama M *et al.* 2012 Covalently chemical modification of lithium ion-encapsulated fullerene: synthesis and characterization of [Li⁺@PC$_{60}$BM] PF$_6^-$ *Org. Lett.* **14** 3784

[112] Feng L, Gayathri Radhakrishnan S, Mizorogi N *et al.* 2011 Synthesis and charge-transfer chemistry of La$_2$@I$_h$-C$_{80}$/Sc$_3$N@I$_h$-C$_{80}$ – zinc porphyrin conjugates: impact of endohedral cluster *J. Am. Chem. Soc.* **133** 7608

[113] Feng L, Slanina Z, Sato S *et al.* 2011 Covalently linked porphyrin-La@C$_{82}$ hybrids: structural elucidation and investigation of intramolecular interactions *Angew. Chem. Int. Ed.* **50** 5909

[114] Guldi D M, Feng L, Radhakrishnan S G *et al.* 2010 A molecular Ce$_2$@I$_h$-C$_{80}$ switch – unprecedented oxidative pathway in photoinduced charge transfer reactivity *J. Am. Chem. Soc.* **132** 9078

[115] Feng L, Rudolf M, Wolfrum S *et al.* 2012 A paradigmatic change: linking fullerenes to electron acceptors *J. Am. Chem. Soc.* **134** 12190

[116] Takano Y, Obuchi S, Mizorogi N *et al.* 2012 An endohedral metallofullerene as a pure electron donor: intramolecular electron transfer in donor-acceptor conjugates of La$_2$@C$_{80}$ and 11,11,12,12-tetracyano-9,10-anthra-*p*-quinodimethane (TCAQ) *J. Am. Chem. Soc.* **134** 19401

[117] Stevenson S, Stephen R R, Amos T M *et al.* 2005 Synthesis and purification of a metallic nitride fullerene bisadduct: exploring the reactivity of Gd$_3$N@C$_{80}$ *J. Am. Chem. Soc.* **127** 12776

[118] Feng L, Tsuchiya T, Wakahara T *et al.* 2006 Synthesis and characterization of bisadduct of La@C$_{82}$ *J. Am. Chem. Soc.* **128** 5990

[119] Cai T, Xu L, Sun C *et al.* 2008 Selective formation of a symmetric Sc$_3$N@C$_{78}$ bisadduct: adduct docking controlled by an internal trimetallic nitride cluster *J. Am. Chem. Soc.* **130** 2136

[120] Sun C, Slebodnick C, Xu L *et al.* 2008 Highly regioselective derivatization of trimetallic nitride templated endohedral metallofullerenes via a facile photochemical reaction *J. Am. Chem. Soc.* **130** 17755

[121] Tagmatarchis N, Taninaka A and Shinohara H 2002 Production and EPR characterization of exohedrally perfluoroalkylated paramagnetic lanthanum metallofullerenes: (La@C$_{82}$)-(C$_8$F$_{17}$)$_2$ *Chem. Phys. Lett.* **355** 226

[122] Tsuchiya T, Akasaka T and Nagase S 2009 Construction of supramolecular systems based on endohedral metallofullerenes *Bull. Chem. Soc. Jpn.* **82** 171

[123] Tsuchiya T, Sato K, Kurihara H *et al.* 2006 Host-guest complexation of endohedral metallofullerene with azacrown ether and its application *J. Am. Chem. Soc.* **128** 6699

[124] Tsuchiya T, Kurihara H, Sato K *et al.* 2006 Supramolecular complexes of La@C$_{82}$ with unsaturated thiacrown ethers *Chem. Commun.* 3585

[125] Tsuchiya T, Sato K, Kurihara H *et al.* 2006 Spin-state exchange system constructed from endohedral metallofullerenes and organic donors *J. Am. Chem. Soc.* **128** 14418

[126] Tsuchiya T, Wielopolski M, Sakuma N *et al.* 2011 Stable radical anions inside fullerene cages: formation of reversible electron transfer systems *J. Am. Chem. Soc.* **133** 13280

[127] Gill-Ramirez G, Karlen S D, Shundo A *et al.* 2010 A cyclic porphyrin trimer as a receptor for fullerenes *Org. Lett.* **12** 3544

[128] Hajjaj F, Tashiro K, Nikawa H *et al.* 2011 Ferromagnetic spin coupling between endohedral metallofullerene La@C$_{82}$ and a cyclodimeric copper porphyrin upon inclusion *J. Am. Chem. Soc.* **133** 9290

[129] Pagona G, Economopoulos S P, Aono T *et al.* 2010 Molecular recognition of La@ C_{82} endohedral metallofullerene by an isophthaloyl-bridged porphyrin dimer *Tetrahedron Lett.* **51** 5896

[130] Grimm B, Schornbaum J, Cardona C M *et al.* 2011 Enhanced binding strengths of acyclic porphyrin hosts with endohedral metallofullerenes *Chem. Sci.* **2** 1530

[131] Xu Y, He C, Liu F *et al.* 2011 Hybrid hexagonal nanorods of metal nitride cluster-fullerene and porphyrin using a supramolecular approach *J. Mater. Chem.* **21** 13538

[132] Tsuchiya T, Kumashiro R, Tanigaki K *et al.* 2008 Nanorods of endohedral metallofullerene derivative *J. Am. Chem. Soc.* **130** 450

[133] Gimenez-Lopez M C, Gardener J A, Shaw A O *et al.* 2010 Endohedral metallofullerenes in self-assembled monolayers *Phys. Chem. Chem. Phys.* **12** 123

[134] Diener M D and Alford J M 2008 Isolation and properties of small-bandgap fullerenes *Nature* **393** 668

[135] Bolskar R D and Alford J M 2003 Chemical oxidation of endohedral metallofullerenes: identification and separation of distinct classes *Chem. Commun.* 1292

[136] Nikawa H, Kikuchi T, Wakahara T *et al.* 2005 Missing La@C_{74} *J. Am. Chem. Soc.* **127** 9684

[137] Lu X, Nikawa H, Kikuchi T *et al.* 2011 Radical derivatives of insoluble La@C_{74}: X-ray structures, metal positions, and isomerization *Angew. Chem. Int. Ed.* **50** 6356

[138] Wakahara T, Nikawa H, Kikuchi T *et al.* 2006 La@C_{72} having a non-IPR carbon cage *J. Am. Chem. Soc.* **128** 14228

[139] Nikawa H, Yamada T, Cao B *et al.* 2009 Missing metallofullerene with C_{80} cage *J. Am. Chem. Soc.* **131** 10950

[140] Akasaka T, Lu X, Kuga H *et al.* 2010 Dichlorophenyl derivatives of La@C_{3v}(7)-C_{82}: endohedral metal induced localization of pyramidalization and spin on a triple-hexagon junction *Angew. Chem. Int. Ed.* **49** 9715

[141] Akiyama K, Hamano T, Nakanishi Y *et al.* 2012 Non-HPLC rapid separation of metallofullerenes and empty cages with $TiCl_4$ Lewis acid *J. Am. Chem. Soc.* **134** 9762

[142] Wang Z, Nakanishi Y, Noda S *et al.* 2012 The origin and mechanism of non-HPLC purification of metallofullerenes with $TiCl_4$ *J. Phys. Chem. C* **116** 25563

[143] Stevenson S, Mackey M A, Pickens J E *et al.* 2009 Selective complexation and reactivity of metallic nitride and oxometallic fullerenes with Lewis acids and use as an effective purification method *Inorg. Chem.* **48** 11685

[144] Ge Z X, Duchamp J C, Cai T *et al.* 2005 Purification of endohedral trimetallic nitride fullerenes in a single, facile step *J. Am. Chem. Soc.* **127** 16292

[145] Stevenson S, Harich K, Yu H *et al.* 2006 Nonchromatographic "stir and filter approach" (SAFA) for isolating $Sc_3N@C_{80}$ metallofullerenes *J. Am. Chem. Soc.* **128** 8829

[146] Stevenson S, Mackey M A, Coumbe C E *et al.* 2007 Rapid removal of D_{5h} isomer using the "stir and filter approach" and isolation of large quantities of isomerically pure $Sc_3N@C_{80}$ metallic nitride fullerenes *J. Am. Chem. Soc.* **129** 6072

[147] Angeli C D, Cai T, Duchamp J C *et al.* 2008 Purification of trimetallic nitride templated endohedral metallofullerenes by a chemical reaction of congeners with eutectic 9-methylanthracene *Chem. Mater.* **20** 4993

[148] Elliott B, Yu L and Echegoyen L 2005 A simple isomeric separation of D_{5h} and I_h $Sc_3N@C_{80}$ by selective chemical oxidation *J. Am. Chem. Soc.* **127** 10885

12

Applications with Metallofullerenes

12.1 SOLAR CELLS

12.1.1 Metallofullerene-Based Solar Cells

A fundamental aspect of endohedral metallofullerenes (EMFs) is their applicability and performance in electron donor-acceptor systems [1]. Especially considering that (i) empty fullerenes, particularly C_{60} and its derivatives, exhibit low reorganization energy in charge-transfer phenomena, and high electron mobility and (ii) EMFs possess larger absorptive coefficients in the visible region and a low highest occupied molecular orbital-lowest unoccupied molecular orbital (HOMO-LUMO) energy gap, it is of no surprise that as soon as the chemical reactivity and functionalization methodologies of EMFs were established and developed, novel donor–acceptor dyads would be synthesized. The first covalent linkage of an electron donor to a trimetallic nitride template (TNT) endohedral fullerene was achieved by performing the 1,3-dipolar cycloaddition reaction of azomethine ylide, as derived upon thermal condensation of ferrocene carboxylaldehyde and sarcosine, to $Sc_3N@C_{80}$ (Figure 12.1).

After column and high-performance liquid chromatography (HPLC) purification, mono-adduct $Sc_3N@C_{80}$-Fc pyrrolidine derivative **1** was isolated as proved by MALDI-TOF (time-of-flight) mass spectrometry [2]. Considering that (i) in the I_h symmetrical $Sc_3N@C_{80}$ two different C=C double bonds, namely at 6,6- and 5,6-ring junctions, are present, (ii) from the [1]H

Endohedral Metallofullerenes: Fullerenes with Metal Inside, First Edition.
Hisanori Shinohara and Nikos Tagmatarchis.
© 2015 John Wiley & Sons, Ltd. Published 2015 by John Wiley & Sons, Ltd.

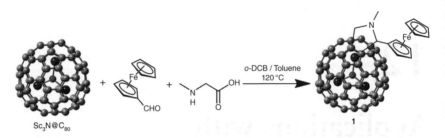

Figure 12.1 1,3-Dipolar cycloaddition reaction of azomethine ylides to $Sc_3N@C_{80}$ yielding $Sc_3N@C_{80}$-Fc dyad

and ^{13}C NMR measurements two different pyrrolidine carbons exist, and (iii) no isomerization is observed after prolonged heating, adduct **1** should possess a fused pyrrolidine ring at a 5,6-ring junction. Structure **1** was further proved by electrochemistry measurements, as its cyclic voltammetry showed a reversible reduction which is a fingerprint for the mono-adduct 5,6-structures, whereas the 6,6- ones exhibit irreversible behavior. As far as electron-transfer interactions are concerned, photoexcitation of **1** resulted in a rapid decay of the so-formed singlet excited state. Moreover, significant changes were observed in the differential absorption spectrum of **1**, with the formation of new bands in the visible and near infrared (NIR) regions.

The new band appearing in the visible region is due to the one-electron oxidized ferrocenium ion. Importantly, the new NIR band evolved resembles the one observed for a radiolytically and spectroelectrochemically derived one-electron reduced $Sc_3N@C_{80}$ species. The radical ion pair $(Sc_3N@C_{80})^{\bullet-}$-$Fc^{\bullet+}$ **1** is formed, thus opening new avenues for applications of TNT-based fullerenes in photovoltaics.

Along the same lines, two isomeric 5,6-pyrrolidine donor–acceptor dyads of $Sc_3N@C_{80}$ with triphenylamine as the electron donor were prepared. Materials **2** and **3** were obtained upon reaction of $Sc_3N@C_{80}$ with (i) 4-(N-diphenylamino)-benzaldehyde and sarcosine and (ii) formaldehyde and N-(benzyl-4-diphenylaminophenyl)glycine, respectively (Figure 12.2). The electrochemical properties of **2** and **3** were evaluated by cyclic voltammetry and it was found that they both exhibit three reversible reductions – characteristic for the 5,6-adduct – and two irreversible oxidation processes for **2** and three for **3**. Insight into the charge-transfer interactions within dyads **2** and **3** arises from transient absorption measurements. Briefly, it was found that (i) when triphenylamine is the substituent of the N atom in the pyrrolidine ring, in dyad **2**, a significantly longer lived radical ion pair is yielded, as compared with dyad **3** in which the triphenylamine is the substituent of the α-carbon in the pyrrolidine ring, and (ii) both dyads **2** and **3** have longer lived charged separated states

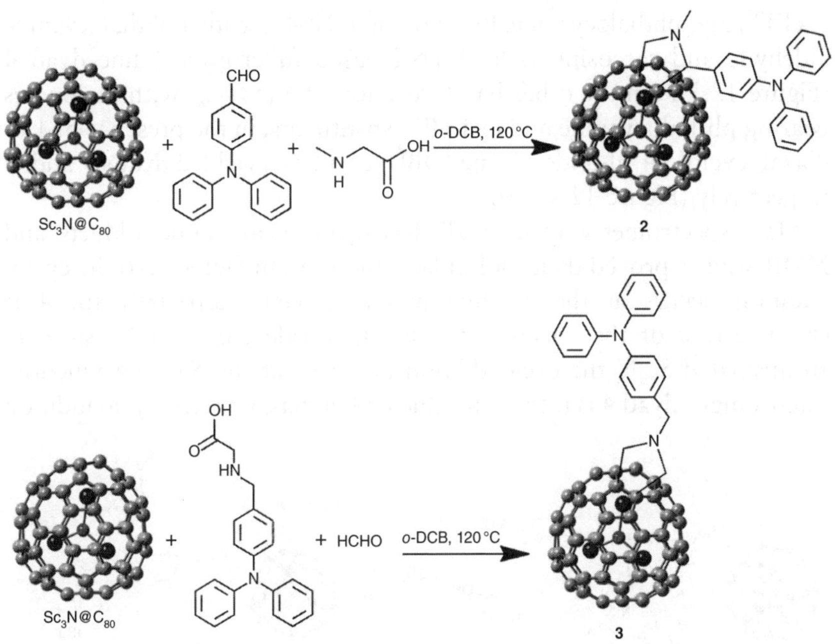

Figure 12.2 1,3-Dipolar cycloaddition reaction of azomethine ylides to $Sc_3N@C_{80}$ yielding $Sc_3N@C_{80}$-triphenylamine donor–acceptor dyads in which the triphenylamine unit is a substituent at the α-carbon in dyad **2** and the nitrogen of the pyrrolidine ring in dyad **3**

Table 12.1 Redox potentials in o-dichlorobenzene and charge-separation (k_{CS}) in THF and charge-recombination (k_{CR}) in benzonitrile rate constants of $Sc_3N@C_{80}$-based triphenylamine dyads **2** and **3** compared with those of intact $Sc_3N@C_{80}$

Materials	$E^3_{red.}$ (V)	$E^2_{red.}$ (V)	$E^1_{red.}$ (V)	$E^1_{ox.}$ (V)	$E^2_{ox.}$ (V)	$E^3_{ox.}$ (V)	k_{CS} (s^{-1})	k_{CR} (s^{-1})
2	−2.23	−1.50	−1.10	0.39	1.06	—	3.4×10^{10}	4.5×10^8
3	−2.29	−1.59	−1.21	0.32	0.63	0.99	1.9×10^{10}	$< 3.3 \times 10^8$
$Sc_3N@C_{80}$	−2.37	−1.62	−1.26	0.59	1.09	—	—	—

All potentials are corrected relative to Fc/Fc⁺.

than their corresponding C_{60} analogs [3]. The reduction/oxidation (redox) potentials and photophysical charge-separation and charge-recombination rate constants for dyads **2** and **3** are collected and presented in Table 12.1.

En route to investigating charge-transfer interactions of TNT-based fullerenes with organic electron donors, a series of $Y_3N@C_{80}$ donor–acceptor dyads were synthesized via the 1,3-dipolar cycloaddition reaction of azomethine ylides or the Bingel cyclopropanation reaction, with two different electron donors, namely extended tetrathiafulvalene (exTTF) and phthalocyanine. Reaction of $Y_3N@C_{80}$ with a phthalocyanine aldehyde and sarcosine yielded TNT-based fulleropyrrolidine dyad **4** (Figure 12.3). On the other hand, reaction of $Y_3N@C_{80}$ with malonates bearing phthalocyanine and exTTF susbstituents, in the presence of 1,8-diazabicyclo[5,4,0]-undec-7-ene (DBU) as a base, yielded dyads **5** and **6**, respectively (Figure 12.4) [4].

Mass spectrometry verified all **4–6** dyads were mono-adducts and NMR studies proved their molecular structures. In fact, in **4–6** the cyclo-addition occurs at the 6,6-ring junction, which especially for **4** is characteristic of the encapsulated metal nitride cluster Y_3N, since in encapsulated Sc_3N the cycloaddition proceeds at the 5,6-ring junction. Interestingly, dyad **4** is rather unstable and undergoes retro-cycloaddition

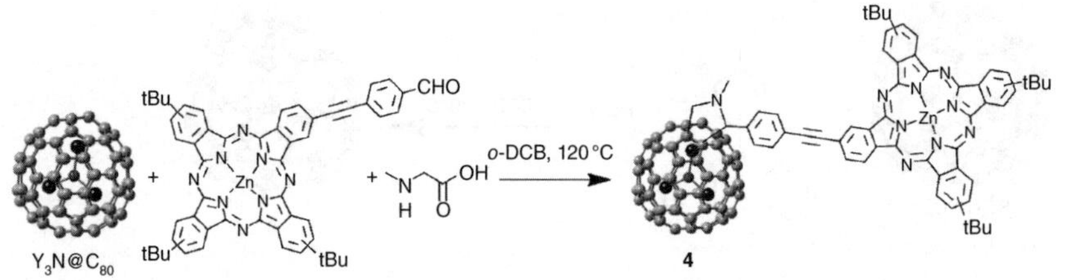

Figure 12.3 1,3-Dipolar cycloaddition reaction of in-situ generated azomethine ylides by thermal condensation of methyl glycine and a phthalocyanine-based alde-hyde to $Y_3N@C_{80}$ yielding $Y_3N@C_{80}$-Pc dyad **4**

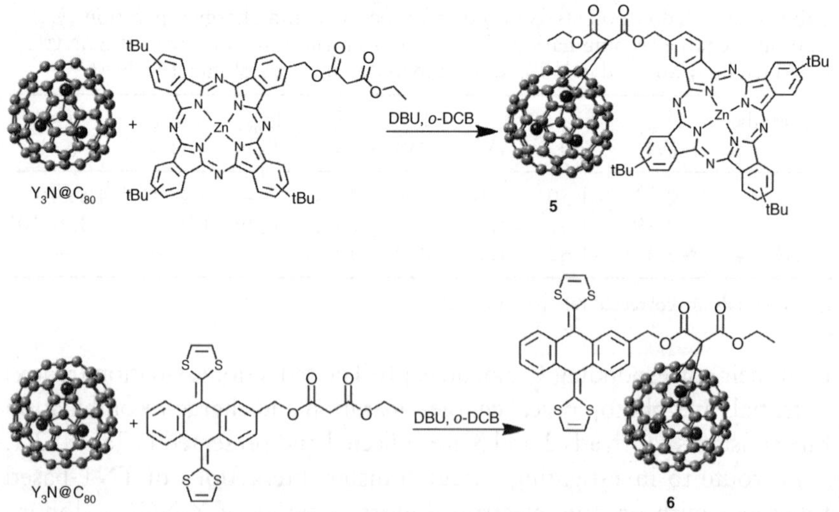

Figure 12.4 Bingel cyclopropanation reaction on $Y_3N@C_{80}$ yielding TNT-based donor–acceptor $Y_3N@C_{80}$-Pc dyad **5** and $Y_3N@C_{80}$-exTTF dyad **6**

reaction at room temperature. Photoinduced electron transfer interactions from the organic addends to the TNT fullerene were studied by means of photoluminescence and transient absorption spectroscopy and confirmed with the aid of theoretical calculations.

As $Y_3N@C_{80}$-exTTF dyad **6** proved to be quite unstable and the exTTF moiety easily oxidized to the corresponding anthraquinone, a new exTTF adduct based on $La_2@C_{80}$ metallofullerene was synthesized. Following the established strategy of 1,3-dipolar cycloaddition for the functionalization of metallofullerenes via the in-situ formation of azomethine ylides, exTTF was incorporated as a substituent of the α-carbon in the fused pyrrolidine ring on $La_2@C_{80}$ [5]. The fused pyrrolidine moiety in $La_2@C_{80}$-exTTF adduct **7** (Figure 12.5) was cycloadded on the 5,6-double bond, in accordance with other similarly modified metallofullerenes. Combination of absorption spectroscopy and electrochemistry measurements showed that electronic interactions between exTTF and $La_2@C_{80}$ in the ground states are weak. The redox potentials of $La_2@C_{80}$-exTTF dyad **7** compared with those of exTTF-CHO and $La_2@C_{80}$ together with the charge-separation in tetrahydrofuran (THF) and charge-recombination in benzonitrile rate constants of **7** are shown in Table 12.2. However, photoluminescence and transient

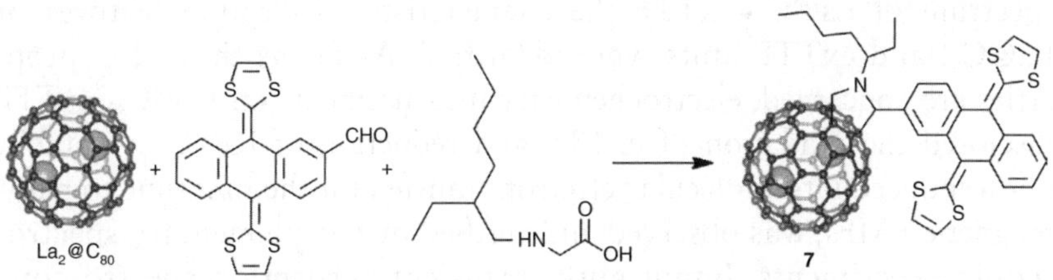

Figure 12.5 1,3-Dipolar cycloaddition reaction of azomethine ylides to $La_2@C_{80}$ yielding donor–acceptor $La_2@C_{80}$-exTTF dyad **7**

Table 12.2 Redox potentials of $La_2@C_{80}$-exTTF dyad **7** compared with those of exTTF-CHO and intact $La_2@C_{80}$ and charge-separation (k_{CS}) in THF and charge-recombination (k_{CR}) in toluene rate constants of **7**

Materials	$E^3_{red.}$ (V)	$E^2_{red.}$ (V)	$E^1_{red.}$ (V)	$E^1_{ox.}$ (V)	$E^2_{ox.}$ (V)	$E^3_{ox.}$ (V)	$E^4_{ox.}$ (V)	k_{CS} (s⁻¹)	k_{CR} (s⁻¹)
$La_2@C_{80}$-exTTF (7)	−2.23	−1.73	−0.45	0.00	0.19	0.58	0.95	3×10^{10}	3×10^{8}
$La_2@C_{80}$	−2.13	−1.71	−0.31	0.56	0.95	—	—	—	—
exTTF-CHO	—	—	—	0.06	—	—	—	—	—

All potentials are corrected relative to Fc/Fc⁺.

absorption spectroscopy studies revealed appreciable intramolecular interactions in the excited states of dyad 7, leading to the formation of a long-lived radical ion pair $(La_2@C_{80})^{\bullet-}$-$(exTTF)^{\bullet+}$.

The first paramagnetic EMF carrying the exTTF unit was recently synthesized, based on the 1,3-dipolar cycloaddition reaction of azomethine ylides. To achieve this, exTTF-CHO and octylglycine were thermally reacted with $La@C_{82}$ giving rise to an inseparable mixture of two regioisomers in which the exTTF unit is present at the α-carbon of a pyrrolidine ring fused to the carbon cage. The presence of a broad and asymmetric octet signal on the electron paramagnetic resonance (EPR) spectrum of $La@C_{82}$-exTTF suggested an open-shell electronic structure, similar to that of intact $La@C_{82}$. Moreover, the aforementioned EPR signal was nicely fitted by combined spectrum consisting of two simulated octets in 1:1 ratio. In addition, NMR spectroscopy of the electrolytically reduced $La@C_{82}$-exTTF material revealed the presence of two signals with half intensity, corresponding to the methylene protons of the fused pyrrolidine ring in $La@C_{82}$-exTTF. Both the EPR and NMR results suggest the presence of an equimolar isomeric mixture in $La@C_{82}$-exTTF [6]. Furthermore, the ground-state electronic properties of $La@C_{82}$-exTTF were probed by absorption spectroscopy. To this end, in the ultraviolet-visible (UV-Vis) spectrum of $La@C_{82}$-exTTF the characteristic absorption features of $La@C_{82}$ and exTTF units were identified. As far as the redox properties are concerned, electrochemistry measurements on $La@C_{82}$-exTTF disclosed the oxidation of exTTF and reduction of $La@C_{82}$. Notably, photoinduced intramolecular electron transfer, for the first time to paramagnetic EMFs, was observed and verified by complementary spectroscopic experiments. Importantly, transient absorption spectroscopy measurements revealed the formation of oxidized exTTF species with the simultaneous generation of reduced $La_2@C_{80}$, while the radical ion pair $(La@C_{82})^{\bullet-}$-$(exTTF)^{\bullet+}$ was formed and lived for more than 2 ns [6].

Long-range electron-transfer interactions of electron donor zinc-porphyrin chromophore and TNT fullerene $Sc_3N@C_{80}$ acceptor bridged via short and long oligophenylene-vinylene wires were investigated. The new dyads were fully characterized by NMR and MALDI-TOF mass spectrometry. Furthermore, based on complementary optical studies, including absorption and photoluminescence spectroscopy, together with electrochemistry, electron-transfer reactions between zinc porphyrin and $Sc_3N@C_{80}$ were revealed [7].

12.1.2   Metallofullerenes Incorporating Phenyl Butyric Acid Methyl Ester Solar Cells

In general, TNT fullerenes are less reactive than empty fullerenes and classical EMFs due to the formal transfer of six electrons from the metal atoms to the fullerene cage, which leads to a closed shell electronic structure and an increase of the HOMO-LUMO gap with a corresponding reduction in reactivity. Since $Sc_3N@C_{80}$ is the most abundant species in the TNT fullerene family, it was functionalized by the well-known diazoalkane addition of methyl 4-benzoylbutyrate p-tosylhydrazone to furnish analog 8 (Figure 12.6), which resembles the most widely studied fullerene derivative, used as an acceptor in organic photovoltaic devices, namely the [6,6]-phenyl-C_{61} butyric acid methyl ester ($PC_{60}BM$). Similarly, the $Y_3N@C_{80}$-PBM (phenyl butyric acid methyl ester) analog was also synthesized. However, due to insufficient material the photovoltaic properties of both materials are yet to be explored [8].

On the other hand, the availability of higher amounts of $Lu_3N@C_{80}$, not only allowed the formation of the corresponding derivative 9 but also the exploration of its photovoltaic properties. The reduced energy offset of the molecular orbitals of $Lu_3N@C_{80}$ 9 to the donor regioregular poly(3-hexylthiophene) (rrP3HT) reduces energy losses in the charge transfer process and increases the open circuit voltage compared with that when $PC_{60}BM$ was used as an acceptor. Thus, it is no surprise that photoconversion efficiency higher than 4% was registered,

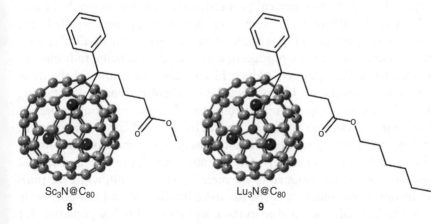

Sc₃N@C₈₀ — 8

Lu₃N@C₈₀ — 9

Figure 12.6 Structures of analogs 8 and 9 as formed upon diazoalkane addition of alkyl 4-benzoylbutyrate p-tosylhydrazone to $Sc_3N@C_{80}$ and $Lu_3N@C_{80}$, respectively

highlighting the potentiality of EMFs in general and TNT fullerenes in this particular case [9–11].

Because $PC_{60}BM$ is considered as one of the most promising fullerene derivatives for photovoltaic applications and since lithium ion encapsulation drastically alters the electronic properties of the carbon cage in the novel $[Li^+@C_{60}]PF_6^-$ material, the corresponding functionalization is not surprising. Thus, chemical modification of the latter under typical diazoalkane conditions was afforded after HPLC purification $[Li^+@ PC_{60}BM]PF_6^-$ as identified by mass spectrometry. In fact, two regioisomers were formed, attributed to the 5,6- and 6,6-adducts based on X-ray crystallography studies [12].

The [1+2] cycloaddition reaction of diazo precursors was applied to conjugate a porphyrin chromophore on $Sc_3N@C_{80}$. Moreover, zinc metallation of the TNT-based adduct resulted in the formation of $Sc_3N@ C_{80}$-ZnP dyad 10, according to Figure 12.7 [13]. The structure of dyad 10 was confirmed by NMR and X-ray crystallography and found to have an open 6,6-addition pattern. Photophysical studies showed that there exist significant intramolecular ground- and excited-state interactions leading to the formation of the radical ion pair $(Sc_3N@C_{80})^{\bullet-}$-$(ZnP)^{\bullet+}$. Moreover, when the same reaction was applied to $La_2@C_{80}$ and $La@C_{82}$ metallofullerenes, adducts 11 and 12, respectively, were produced. $La_2@ C_{80}$ possesses a closed-shell electronic structure with I_h molecular symmetry, similar to that of $Sc_3N@C_{80}$, in which six electrons are transferred from the metal (or metal cluster nitride) to the carbon cage.

However, as the encapsulated metal is different, La vs. Sc, the chemical reactivity and electrochemical properties are different in dyads 11 and 12. Indeed, although the same reaction strategy was applied, the formation of adduct 11 was facile compared with that of 12. Actually, three isomers of 12 were produced, with one of them being fully characterized to be the 6,6-open adduct. Electrochemical studies of the three structural isomers of 12 conducted in o-dichlorobenzene containing Bu_4NPF_6 as electrolyte revealed four one-electron oxidation processes and four one-electron reduction steps. The first and last oxidation steps involve the $La@C_{82}$ species, while the second and third oxidations are due to the porphyrin moiety. On the other hand, the first and second reduction steps are assigned to the metallofullerene unit, while the third reduction corresponds to both the metallofullerene and the porphyrin and the last reduction is due to the porphyrin. All redox potentials for dyads 10–12 are collected and presented in Table 12.3.

In addition, the formation of the radical ion pair for $(La_2@C_{80})^{\bullet-}$-$(ZnP)^{\bullet+}$ 10 and $(Sc_3N@C_{80})^{\bullet-}$-$(ZnP)^{\bullet+}$ 11, are formed in both apolar and

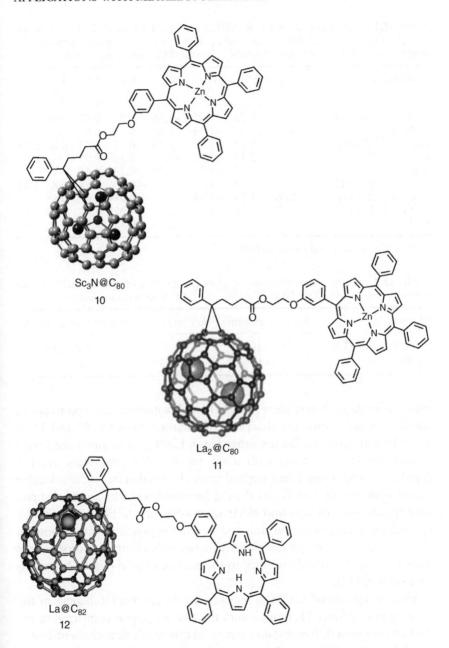

Figure 12.7 Structures of donor–acceptor $Sc_3N@C_{80}$-ZnP dyad **10**, $La_2@C_{80}$-ZnP dyad **11**, and $La@C_{82}$-H_2P dyad **12**

Table 12.3 Redox potentials for $Sc_3N@C_{80}$-ZnP dyad **10**, $La_2@C_{80}$-ZnP dyad **11**, and the three structural isomers of $La@C_{82}$-H_2P dyad **12** compared with the free porphyrins ZnP, H_2P, and intact $Sc_3N@C_{80}$ and $La_2@C_{80}$ metallofullerenes

Material	$E^4_{red.}$ (V)	$E^3_{red.}$ (V)	$E^2_{red.}$ (V)	$E^1_{red.}$ (V)	$E^1_{ox.}$ (V)	$E^2_{ox.}$ (V)	$E^3_{ox.}$ (V)	$E^4_{ox.}$ (V)
$Sc_3N@C_{80}$-ZnP (**10**)	—	−2.26	−1.92	−1.34	0.37	0.55	0.64	1.10
$Sc_3N@C_{80}$	—	—	−1.91	−1.33	0.59	1.09	—	—
$La_2@C_{80}$-ZnP (**11**)	—	−2.15	−1.79	−0.39	0.33	0.52	0.65	0.88
$La_2@C_{80}$	—	−2.13	−1.71	−0.31	0.56	0.95	—	—
ZnP	—	—	−2.27	−1.94	0.39	0.65	—	—
$La@C_{82}$-H_2P (I) (**12**)	−2.07	−1.74	−1.41	−0.48	−0.03	0.58	0.89	1.13
$La@C_{82}$-H_2P (II) (**12**)	−2.06	−1.75	−1.41	−0.48	−0.04	0.57	0.89	1.06
$La@C_{82}$-H_2P (III) (**12**)	−2.07	−1.71	−1.36	−0.45	−0.02	0.61	0.89	1.08
H_2P	—	—	−2.07	−1.75	0.52	0.94	1.16	—

All potentials are corrected relative to Fc/Fc^+.

Table 12.4 Charge-recombination (k_{CR}) rate constants of the radical ion pair formed for $Sc_3N@C_{80}$-ZnP **10** and $La_2@C_{80}$-ZnP **11** in different solvents

Material	k_{CR} (s^{-1}) (toluene)	k_{CR} (s^{-1}) (THF)	k_{CR} (s^{-1}) (benzonitrile)
$Sc_3N@C_{80}$-ZnP (**10**)	2.4×10^{10}	—	2.0×10^{10}
$La_2@C_{80}$-ZnP (**11**)	4.3×10^9	5.8×10^9	—

polar solvents such as toluene, and THF and benzonitrile, respectively. In Table 12.4 are shown the charge-recombination rates for **10** and **11** in the different solvents. On the other hand, $La@C_{82}$ is an open-shell with a huge anionic π-surface, with more active redox properties, smaller band gap, and lower-lying excited state. Ground-state intramolecular interactions within $La@C_{82}$-H_2P were identified with the aid of electron absorption spectroscopy and electrochemistry. In addition, quantitative quenching of the porphyrin emission signifies appreciable interactions in the excited state, though discrimination between electron and/or energy transfer from the porphyrin unit to the carbon cage in dyad **12** is yet to be explored [14].

Interestingly, when $Ce_2@C_{80}$ replaced $La_2@C_{80}$ metallofullerene in the formation of adduct **11**, though with the zinc porphyrin connected to the carbon cage via a different spacer incorporating the flexible chain of 2-oxy-ethyl butyrate, different charge transfer phenomena were observed. It should be borne in mind that EMFs and M_2C_{80} (M = Ce, La) in particular, show not only strong electron-accepting but also electron-donating properties, in contrast to empty fullerenes. Thus, it is not surprising that in the current example of $Ce_2@C_{80}$-ZnP, depending on the solvent polarity, electrons can intramolecularly flow to and from different directions. That is to say, in non-polar solvents $Ce_2@C_{80}$ behaves as an electron acceptor giving rise upon photoirradiation to the formation of the radical ion pair $(Ce_2@C_{80})^{\bullet-}-(ZnP)^{\bullet+}$. On the contrary, in polar solvents $Ce_2@C_{80}$ acts as an efficient electron donor leading to $(Ce_2@C_{80})^{\bullet+}-(ZnP)^{\bullet-}$ [15].

The one-electron oxidation potentials of TNT-based fullerenes are generally much lower than those of empty fullerenes. This property makes them better electron donor materials and simultaneously allow them to participate in photoinduced electron transfer schemes showing easier oxidations. Along these lines, a recent and representative example of utilizing TNT fullerenes as electron donors started with the [1+2] cycloaddition reaction of $Lu_3N@C_{80}$ with the light-harvesting and electron-acceptor 1,6,7,12-tetrachloro-3,4,9,20-perylenediimide (PDI) diazo compound, yielding $Lu_3N@C_{80}$-PDI dyad 13 (Figure 12.8).

Based on MALDI-TOF and NMR measurements, only a single isomer of mono-adduct $Lu_3N@C_{80}$-PDI 13 was formed and found to have the 6,6-open addition pattern. The redox potentials for $Lu_3N@C_{80}$-PDI 13 compared with those for the reference material C_{60}-PDI 13 and PDI itself

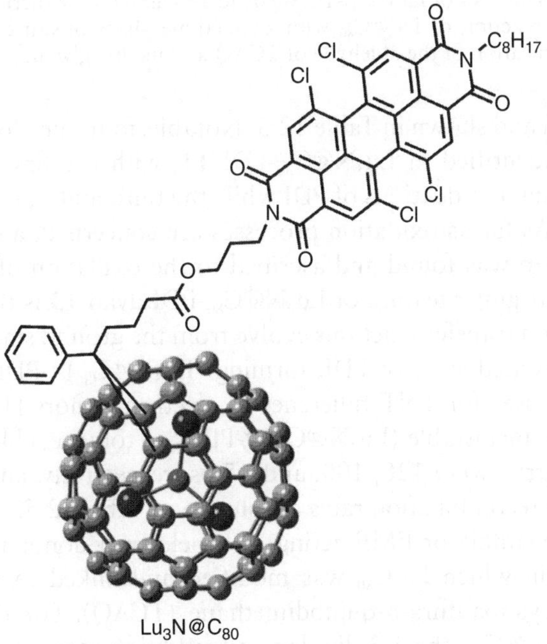

$Lu_3N@C_{80}$

13

Figure 12.8 Structure of donor–acceptor $Lu_3N@C_{80}$-PDI dyad 13

Table 12.5 Redox potentials of $Lu_3N@C_{80}$-PDI dyad **13** compared with those of reference C_{60}-PDI and PDI and charge-recombination (k_{CR}) rate constants of **13** in different solvents

Material	$E^4_{red.}$ (V)	$E^3_{red.}$ (V)	$E^2_{red.}$ (V)	$E^1_{red.}$ (V)	$E^1_{ox.}$ (V)	k_{CR} (s^{-1}) (PhMe)	k_{CR} (s^{-1}) (C_6H_5Cl)	k_{CR} (s^{-1}) (BnCN)
$Lu_3N@C_{80}$-PDI (**13**)	−1.87	−1.41	−1.08	−0.86	0.57	8×10^9	1×10^{10}	2×10^{12}
C_{60}-PDI	—	−1.52	−1.13	−0.89	1.21	—	—	—
PDI	—	—	−1.12	−0.89	—	—	—	—

All potentials are corrected relative to Fc/Fc$^+$.

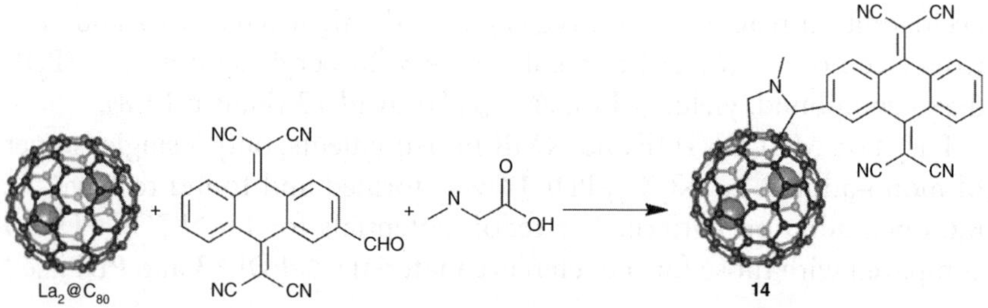

Figure 12.9 Synthesis of $La_2@C_{80}$-TCAQ dyad **14** based on the thermal 1,3-dipolar cycloaddition reaction of $La_2@C_{80}$ with azomethine ylides in-situ generated upon thermal condensation of the aldehyde of TCAQ and methyl glycine

are collected and shown in Table 12.5. Notably, four one-electron reductions were identified in $Lu_3N@C_{80}$-PDI **13**, with the first and second corresponding to reductions of PDI while the third and fourth are due to $Lu_3N@C_{80}$. As far as oxidation processes are concerned, a one-electron oxidation step was found and ascribed to the oxidation of $Lu_3N@C_{80}$. The most intriguing feature of $Lu_3N@C_{80}$-PDI dyad **13** is that photoinduced electron transfer reactions evolve from the ground state of $Lu_3N@C_{80}$ to the excited state of PDI, forming $(Lu_3N@C_{80})^{\bullet+}$-PDI$^{\bullet-}$, thus disclosing new uses for TNT fullerenes as electron donors [16]. The lifetimes for the metastable $(Lu_3N@C_{80})^{\bullet+}$-PDI$^{\bullet-}$ in toluene, chlorobenzene, and benzonitrile were 120, 100, and 45 ps, respectively, and the calculated charge-recombination rates are shown in Table 12.5.

Another example of EMF acting as an electron donor is the recent report [17] in which La_2C_{80} was modified and linked to the electron donor tetracyanoanthra-p-quinodimethane (TCAQ). For the transformation of $La_2@C_{80}$, the 1,3-dipolar cycloaddition reaction was chosen utilizing the corresponding aldehyde of TCAQ and methyl glycine (Figure 12.9). Thus, a pyrrolidine ring was formed onto $La_2@C_{80}$, with the TCAQ unit as a substituent of the α-pyrrolidine carbon atom, forming $La_2@C_{80}$-TCAQ dyad **14**. Two mono-adducts were isolated via HPLC and based on careful NMR and electrochemistry studies found to be the 5,6- and 6,6-isomers. Cyclic voltammetry and differential pulse voltammetry measurements allowed the redox properties of $La_2@C_{80}$-TCAQ dyad **14** to be identified. Thus, the first one-electron oxidation of $La_2@C_{80}$ and first one-electron reduction of TCAQ were identified at 0.24 and −0.91 V, respectively. Photophysical studies revealed that although there were not significant ground-state interactions between the two components within the hybrid material, in the excited state the reactivity was different. This is to say that upon $La_2@C_{80}$ photoexcitation the radical ion pair $(La_2@C_{80})^{\bullet+}$-$(TCAQ)^{\bullet-}$ is formed [17].

12.2 BIOMEDICAL ASPECTS OF WATER-SOLUBLE METALLOFULLERENES

12.2.1 Metallofullerene-Based MRI Contrast Agents

A series of water-soluble EMF based on Gd and Pr encapsulated metals – metallofullerenols $Pr@C_{82}O_m(OH)_n$ – were prepared and characterized. The synthesis involves treatment either with aqueous sodium hydroxide and tetrabutylammonium hydroxide as a phase transfer catalyst [18,19] or nitric acid and subsequent hydrolysis, thus, initially involving a NO_2 radical addition process [20]. The $Gd@C_{82}(OH)_x$, gadolinium fullerenols, were investigated as magnetic resonance imaging (MRI) contrast agents. The *in-vitro* water proton relaxivity of $Gd@C_{82}(OH)_x$ found to be significantly higher than that of the commercially available MRI agent. This was also confirmed *in vivo* in the lung, liver, spleen, and kidney of mice [21] (Figure 12.10).

Measurements on a purified (>99.9%) $Gd@C_{82}(OH)_n$ by Mikawa *et al.* [21] gave high R_1 relaxivity (81 l mmol^{-1} s^{-1} at pH = 7.5) which is more than 20 times greater than for commercial Gd-MRI contrast agents such as Gd-diethylenetriaminepentaacetic acid (Gd-DTPA) (4.3 l mmol^{-1} s^{-1}). This is encouraging for the prospects of the future use of water-soluble gadolinium metallofullerene.

At this point it should be emphasized that it is also possible to control the addition of hydroxyl groups onto the metallofullerene and, in particular, the preparation of a dihydroxyl adduct of $Gd@C_{82}$ was reported. According to that report, when a 1,2-dichlorobenzene solution of the metallofullerene was treated with an aqueous hydrogen peroxide

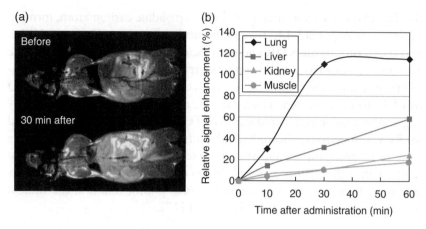

Figure 12.10 (a) T_1-weighted MRI of CD F1 mice before and 30 minutes after i.v. administration of Gd@C$_{82}$(OH)$_{40}$ via the tail vein at a dose of 5 μmol Gd kg^{-1} [which was 1/20th of a typical clinical dose of Gd-DTPA (100 μmol Gd kg^{-1})] and (b) its time-dependent signal intensity change in various organs. MRI conditions: 4.7 T Unity INOVA (Varian), at TR/TE = 300/11 ms. At 30 minutes after i.v. administration of Gd@C$_{82}$(OH)$_{40}$, enormous signal enhancement was observed in the lung, liver, spleen, and kidney. (a) Reproduced with permission from Ref. [21]. Copyright 2001 American Chemical Society

solution, in the absence of air, the Gd@C$_{82}$(OH)$_2$ adduct was obtained as confirmed by MALDI-TOF mass spectrometry [22].

A further systematic study was performed involving the investigation of lanthanide metallofullerenols, M@C$_{82}$(OH)$_x$ (M = La, Ce, Gd, Dy, Er) to obtain insight into the proton relaxation mechanism of these novel forms as MRI contrast agents. Overall, it was found that these metallofullerenols possess longitudinal and transverse relaxivities for water protons significantly higher than those of the corresponding commercially available ones. The strong relaxivities of those lanthanoid-based metallofullerenols are rationalized in terms of dipole–dipole relaxation together with a substantial decrease in the overall molecular rotational motion [23].

In addition, when water-soluble holmium metallofullerenols were prepared and neutron irradiated, different biodistribution than that of other fullerenes was observed, thus suggesting that certain derivatized metallofullerenes can be selectively tissue-targeted [24]. Moreover, ^{166}Ho@ C$_{82}$(OH)$_x$ as a radiotracer analog was intravenously administered to mice and found to be selectively localized in the liver but with slow clearance, as well as taken up by bones without clearance. Thus, the feasibility of using water-soluble metallofullerenols as radiotracers to elucidate their behavior in animals was shown [25].

The first water-soluble trimetallic nitride endohedral metallofullerols $Sc_3N@C_{80}(OH)_{10}O_{10}$ were synthesized from polyanionic radical intermediates. Briefly the procedure followed to obtain water-soluble adducts of $Sc_3N@C_{80}$ involves treatment of toluene solution of the metallofullerene with Na metal under an inert atmosphere and subsequent exposure of the reaction mixture to air and water. Initially, a precipitate from the toluene solution is formed containing polyanionic radical species, which is then oxidized by air in water to yield a golden-colored aqueous solution. These TNT-based metallofullerols were characterized by a variety of traditional spectroscopic methods including IR, XPS, and MALDI-TOF mass spectrometry [26]. Along the same lines, following similar experimental protocols, $Sc_xGd_{3-x}@C_{80}O_{12}(OH)_{26}$ $(x = 1, 2)$ were also synthesized. It is interesting to note that the production yield of $Sc_xGd_{3-x}@C_{80}$ $(x = 1, 2)$ is 10 and 50 times higher compared with that of $Gd@C_{82}$ and $Gd_3N@C_{80}$, thus highlighting their potential as novel MRI contrast agents [27].

In addition, water-soluble $Gd_3N@C_{80}$ and $Lu_3N@C_{80}$ functionalized with poly(ethyleneglycol) and multihydroxyl groups were prepared. The new $M_3N@C_{80}(OH)_x\{C[COO(CH_2CH_2O)_{114}CH_3]_2\}$ $(M = Gd, Lu)$ were found to have T_1 and T_2 relaxivity values superior to those of the conventional gadolinium-containing MRI agent. The effectiveness was demonstrated by means of *in vitro* relaxivity and MRI studies, infusion experiments with agarose gel and *in vivo* rat brain studies. Notably, an inherent advantage of the water-soluble TNT-based adducts is that the water molecules are shielded by the fullerene cage from direct interaction with the metal atoms [28].

In an attempt to utilize the highly insoluble and thus intractable from the fullerene soot $Gd@C_{60}$ fraction, its chemical derivatization was performed. In such a way, new water-soluble and air-stable $Gd@C_{60}[C(COOH)_2]_{10}$ was prepared. The functionalization reaction involves multiple cycloadditions of bromomalonates, according to the well-established methodology of Bingel cyclopropanation. Thus, $Gd@C_{60}[C(COOCH_2CH_3)_2]_{10}$ was formed which was then hydrolyzed to the corresponding acid $Gd@C_{60}[C(COOH)_2]_{10}$. The aqueous solubility of the latter allowed its evaluation as MRI contrast agent and relaxometry measurements revealed relaxivity values superior to those derived from the commercially available gadolinium chelate-based one [29].

In a different reaction strategy for the functionalization of EMFs, nucleophilic addition of glycine to esters to $Gd@C_{82}$ was investigated. Based on MALDI-TOF mass spectrometry studies, the formation of multiple adducts possessing glycine esters or both glycine esters and hydroxyl groups was found [30]. In addition, nucleophilic addition of an alkaline solution of β-alanine to $Gd@C_{82}$ was also performed and resulted in the production of water-soluble Gd@C$_{82}$(O)$_x$(OH)$_y$(NHCH$_2$COOH)$_z$ **15** (Figure 12.11).

The structure of the new adduct was confirmed by standard analytical techniques, including IR, XPS, and mass spectrometry [31]. Importantly, the free carboxylic acid moiety in **15** can be further derivatized, following classical peptide synthesis conditions. Thus, conjugation of the antibody of the green fluorescence protein (GFP) to **15** was achieved yielding a new gadolinium-based modified metallofullerene as a model for a tumor targeted imaging agent. In that model system, the activity of the antiGFP conjugate was conveniently recognized by GFP thus leading to the performance of more direct and facile *in vitro* experiments than those involving tumor antibodies. In addition, the **15**-antiGFP conjugate was found to exhibit higher water proton relaxivity than that of **15**, thus suggesting that **15**-antiGFP shows not only targeting potential but also exhibits high efficiency as a MRI contrast agent [32].

The advantages of EMFs as MRI contrast agents have been documented for a decade. Actually, EMFs show great ability to reduce the spin relaxation time of water protons compared with commercially available gadolinium-based chelates. Furthermore, the carbon cage can prevent any leakage of the toxic gadolinium in tissues and organs. In Table 12.6, the relaxivity values (T_1) of some polyhydroxylated EMFs are shown and compared with the T_1 of Gd-DTPA, the gadolinium chelate complex currently used in MRI.

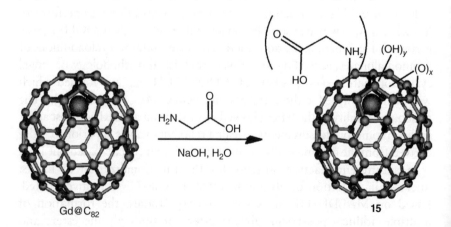

Figure 12.11 Nucleophilic addition of an alkaline solution of β-alanine to Gd@C$_{82}$ furnishing water-soluble Gd@C$_{82}$(O)$_x$(OH)$_y$(NHCH$_2$COOH)$_z$ **15**

Table 12.6 Relaxivity values for some polyhydroxylated EMFs compared with the currently used Gd-DTPA complex

	Relaxivity ($mM^{-1} s^{-1}$)	References
Functionalized EMFs		
$Gd@C_{60}(OH)_x$	83.2–97.7 (1.4 T)	[33,34]
$Gd@C_{60}[C(COOH)_2]_{10}$	15–24 (1.4 T)	[33]
$Gd@C_{82}(O)_6(OH)_{16}(NHCH_2CH_2CO\text{-}antiGFP)_5$	12 (0.35 T)	[32]
$Gd@C_{82}(O)_6(OH)_{16}(NHCH_2CH_2COOH)_8$	9.1 (1.5 T)	[31]
	8.1 (0.35 T)	
$Gd_3N@C_{80}[DIPEG5000(OH)_x]$	102 (0.35 T)	[18]
	143 (2.4 T)	
	32 (9.4 T)	
$ScGd_2N@C_{80}(O)_{12}(OH)_{16}$	20.7 (14.1 T)	[27]
$Sc_2GdN@C_{80}(O)_{12}(OH)_{16}$	17.6 (14.1 T)	[27]
$Gd@C_{82}(OH)_{40}$	81 (1 T)	[21]
Gadolinium chelate complex currently used		
Gd-DTPA	3.9 (1 T)	[35]

12.2.2 Metallofullerenes for Radiotracers

The advantages and uniqueness of radiochemical techniques in metallofullerene studies have been pointed out by the Tokyo Metropolitan group [36–40] and Cagle *et al.* [41]. The HPLC behavior of metallofullerenes of 14 lanthanide elements was studied with the use of radiotracers [40]. From the radiochromatographic HPLC elution behavior, the 14 lanthanide elements forming metallofullerenes were found to be grouped into two, that is, Sm, Eu, Tm, and Yb as one group and the rest of the elements as another. The former elements have 2+ divalent charge state in the C_{82} cage, whereas the latter elements have 3+ trivalent charge state. Radioactive endohedral $^7Be@C_{60}$ can also be detected using radiochemical and radiochromatographic techniques in solvents [42].

The distribution of metallofullerenes among organs when they are administered to an animal can be best studied by radioisotopically labeled fullerenes. The HPLC fractions of ^{140}La-labeled $La@C_{82}$ and $La_2@C_{80}$ were made into an emulsion and injected into the hearts of rats. The rats were dissected after 24 hours and the radioactivity remaining in each organ was determined by γ-ray spectrometry. Large portions of lanthanum fullerenes were found in the liver and blood [37,38]. Although the physiological meaning of the results is not clear at present, it certainly suggests possible biological applications of metallofullerene chemistry.

Kikuchi *et al.* [36] showed that the endohedral form of metallofullerenes was not affected by the recoil energy of the metal atom resulting from the emission of electrons in the β-decay in which nuclear reaction and decay processes are related to ^{159}Ga@C$_{82}$, ^{161}Tb@C$_{82}$, and ^{153}Gd@C$_{82}$. Successful encapsulation of radioactive atoms inside the fullerene cage will widen the potential use of metallofullerenes not only in materials science and technology but also in biological and even medical science.

Akiyama *et al.* [43] synthesized and isolated some light actinide metallofullerenes such as U@C$_{82}$, Np@C$_{82}$, and Am@C$_{82}$ and found the U atom in the C$_{82}$ cage is in the 3+ state, showing almost exactly the same UV–Vis–NIR (ultraviolet-visible-near infrared) absorption spectrum as that of the trivalent lanthanide C$_{82}$ metallofullerenes such as La@C$_{82}$ and Gd@C$_{82}$.

REFERENCES

[1] Rudolf M, Wolfraum S, Guldi D M *et al.* 2012 Endohedral metallofullerenes – filled fullerene derivatives towards multifunctional reaction center mimics *Chem. Eur. J.* **18** 5136

[2] Pinzon J R, Plonska-Brzezinska M E, Cardona C M *et al.* 2008 Sc$_3$N@C$_{80}$-ferrocene electron-donor/acceptor conjugates as promising materials for photovoltaic applications *Angew. Chem. Int. Ed.* **47** 4173

[3] Pinzon J R, Gasca D C, Sankaranarayanan S G *et al.* 2009 Photoinduced charge transfer and electrochemical properties of triphenylamine I_h-Sc$_3$N@C$_{80}$ donor-acceptor conjugates *J. Am. Chem. Soc.* **131** 7727

[4] Pinzon J R, Cardona C M, Herranz M A *et al.* 2009 Metal nitride cluster fullerene M$_3$N@C$_{80}$ (M = Y, Sc) based dyads: synthesis and electrochemical, theoretical and photophysical studies *Chem. Eur. J.* **15** 864

[5] Takano Y, Herranz M A, Martin N *et al.* 2010 Donor-acceptor conjugates of lanthanum endohedral metallofullerene and π-extended tetrathiafulvalene *J. Am. Chem. Soc.* **132** 8048

[6] Takano Y, Obuchi S, Mizorogi N *et al.* 2012 Stabilizing ion and radical ion pair states in a paramagnetic endohedral metallofullerene/π-extended tetrathiafulvalene conjugate *J. Am. Chem. Soc.* **134** 16103

[7] Wolfrum S, Pinzon J R, Molina-Ontoria A *et al.* 2011 Utilization of Sc$_3$N@C$_{80}$ in long-range charge transfer reactions, *Chem. Commun.*, **47** 2270

[8] Shu C, Xu W, Slebodnick C *et al.* 2009 Syntheses and structures of phenyl-C$_{81}$-butyric acid methyl esters (PCBM) from M$_3$N@C$_{80}$ *Org. Lett.* **11** 1753

[9] Ross R B, Cardona C M, Guldi D M *et al.* 2009 Endohedral fullerenes for organic photovoltaic devices *Nat. Mater.* **8** 208

[10] Ross R B, Cardona C M, Swain F B *et al.* 2009 Tuning conversion efficiency in metallo endohedral fullerene-based organic photovoltaic devices, *Adv. Funct. Mater.*, **19** 2332

[11] Liedtke M, Sperlich A, Kraus H *et al.* 2011 Triplet exciton generation in bulk-heterojunction solar cells based on endohedral fullerenes *J. Am. Chem. Soc.* **133** 9088

[12] Matsuo Y, Okada H, Maruyama M *et al.* 2012 Covalently chemical modification of lithium ion-encapsulated fullerene: synthesis and characterization of [Li⁺@PC₆₀BM] PF₆⁻ *Org. Lett.* **14** 3784

[13] Feng L, Gayathri Radhakrishnan S, Mizorogi N *et al.* 2011 Synthesis and charge-transfer chemistry of La₂@I$_h$-C₈₀/Sc₃N@I$_h$-C₈₀ – zinc porphyrin conjugates: impact of endohedral cluster *J. Am. Chem. Soc.* **133** 7608

[14] Feng L, Slanina Z, Sato S *et al.* 2011 Covalently linked porphyrin-La@C₈₂ hybrids: structural elucidation and investigation of intramolecular interactions *Angew. Chem. Int. Ed.* **50** 5909

[15] Guldi D M, Feng L, Radhakrishnan S G *et al.* 2010 A molecular Ce₂@I$_h$-C₈₀ switch – unprecedented oxidative pathway in photoinduced charge transfer reactivity *J. Am. Chem. Soc.* **132** 9078

[16] Feng L, Rudolf M, Wolfrum S *et al.* 2012 A paradigmatic change: linking fullerenes to electron acceptors *J. Am. Chem. Soc.* **134** 12190

[17] Takano Y, Obuchi S, Mizorogi N *et al.* 2012 An endohedral metallofullerene as a pure electron donor: intramolecular electron transfer in donor-acceptor conjugates of La₂@C₈₀ and 11,11,12,12-tetracyano-9,10-antra-*p*-quinodimethane (TCAQ) *J. Am. Chem. Soc.* **134** 19401

[18] Li J, Takeuchi A, Ozawa M *et al.* 1993 C₆₀ fullerol formation catalysed by quaternary ammonium hydroxides *J. Chem. Soc., Chem. Commun.* 1784

[19] Kato H, Suenaga K, Mikawa M *et al.* 2000 Synthesis and EELS characterization of water-soluble multi-hydroxyl Gd@C₈₂ fullerenols *Chem. Phys. Lett.* **324**, 255

[20] Sun D, Huang H and Yang S 1999 Synthesis and characterization of a water-soluble endohedral metallofullerol *Chem. Mater.* **11** 1003

[21] Mikawa M, Kato H, Okumura M *et al.* 2001 Paramagnetic water-soluble metallofullerenes having the highest relaxivity for MRI contrast agents *Bioconjug. Chem.* **12** 510

[22] Zhang Y, Xiang J Zhuang J *et al.* 2005 Synthesis of the first dihydroxyl adduct of Gd@C₈₂ *Chem. Lett.* **34** 1264

[23] Kato H, Kanazawa Y, Okumura M *et al.* 2003 Lanthanoid endohedral metallofullerenols for MRI contrast agents *J. Am. Chem. Soc.* **125** 4391

[24] Wilson L J, Cagle D W, Thrash T P *et al.* 1999 Metallofullerene drug design *Coord. Chem. Rev.* **190–192** 199

[25] Cagle D W, Kennel S J, Mirzadeh S *et al.* 1999 In vivo studies of fullerene-based materials using endohedral metallofullerene radiotracers *Proc. Natl. Acad. Sci. USA* **96** 5182

[26] Iezzi E B, Cromer F, Stevenson P and Dorn H C 2002 Synthesis of the first water-soluble trimetallic nitride endohedral metallofullerols *Synth. Met.* **128** 289

[27] Zhang E-Y, Shu C-Y, Feng L and Wang C R 2007 Preparation and characterization of two new water-soluble endohedral metallofullerenes as magnetic resonance imaging contrast agents *J. Phys. Chem. B* **111** 14223

[28] Fatouros P P, Corwin F D, Chen Z-Y *et al.* 2006 In vitro and in vivo imaging studies of a new endohedral metallofullerene nanoparticle *Radiology* **240** 756

[29] Bolskar R D, Benedetto A F, Husebo L O *et al.* 2003 First soluble M@C₆₀ derivatives provide enhanced access to metallofullerenes and permit in vivo evaluation of Gd@C₆₀[C(COOH)₂]₁₀ as a MRI contrast agent *J. Am. Chem. Soc.* **125** 5471

[30] Lu X, Zhou X, Shi Z and Gu Z 2004 Nucleophilic addition of glycine esters to Gd@C₈₂ *Inorg. Chim. Acta* **357** 2397

[31] Shu C-Y, Gan L-H, Wang C-R *et al*. 2006 Synthesis and characterization of a new water-soluble metallofullerene for MRI contrast agents *Carbon* **44** 496

[32] Shu C-Y, Ma X-Y, Zhang J F *et al*. 2008 Conjugation of a water-soluble gadolinium endohedral fulleride with an antibody as a magnetic resonance imaging contrast agent *Bioconjug. Chem.* **19** 651

[33] Laus S, Sitharaman B, Toth E *et al*. 2005 Destroying gadofullerene aggregates by salt addition in aqueous solution of $Gd@C_{60}(OH)_x$ and $Gd@C_{60}[C(COOH_2)]_{10}$ *J. Am. Chem. Soc.* **127** 9368

[34] Laus S, Sitharaman B, Toth E *et al*. 2007 Understanding paramagnetic relaxation phenomena for water-soluble gadofullerenes *J. Phys. Chem. C* **111** 5633

[35] Caravan P, Ellison J J, McMurry T J and Lauffer R B 1999 Gadolinium(III) chelates as MRI contrast agents: structure, dynamics, and applications *Chem. Rev.* **99** 2293

[36] Kikuchi K, Kobayashi K, Sueki K *et al*. 1994 Encapsulation of radioactive ^{159}Gd and ^{161}Tb atoms in fullerene cages *J. Am. Chem. Soc.* **116** 9775

[37] Kobayashi K, Nagase S and Akasaka T 1995 A theoretical study of C_{80} and $La_2@C_{80}$ *Chem. Phys. Lett.* **245** 230

[38] Kobayashi K, Kuwano M, Sueki K *et al*. 1995 Activation and tracer techniques for study of metallofullerenes *J. Radioanal. Nucl. Chem.* **192** 81

[39] Ohtsuki T, Masumoto K, Kikuchi K and Sueki K 1995 Production of radioactive fullerene families using accelerators *Mater. Sci. Eng. A* **217–218** 38

[40] Sueki K, Akiyama K, Yamauchi T *et al*. 1997 New lanthanoid metallofullerenes and their HPLC elution behavior *Fullerene Sci. Technol.* **5** 1435

[41] Cagle D W, Thrash T P, Wilson L J *et al*. 1996 *Fullerenes: Recent Advances in the Chemistry and Physics of Fullerenes and Related Materials*, Vol. 3, eds K Kadish and R Ruoff (Pennington, NJ: Electrochemical Society), pp. 854–868

[42] Ohtsuki T, Masumoto K, Ohno K *et al*. 1996 Insertion of Be atoms in C_{60} fullerene cages: $Be@C_{60}$ *Phys. Rev. Lett.* **77** 3522

[43] Akiyama K, Zhao Y, Sueki K *et al*. 2001 Isolation and characterization of light actinide metallofullerenes *J. Am. Chem. Soc.* **123** 181

13

Growth Mechanism

13.1 CARBON CLUSTERS: A ROAD TO FULLERENE GROWTH

It is extremely important to know how fullerenes/metallofullerenes are formed in high temperature plasma of arc discharge or laser vaporization (cf. Chapter 2), because elucidation of the growth mechanism may lead to high yield syntheses/fabrications of the metallofullerenes. Unfortunately, so far experimental techniques to probe growth processes of fullerenes have been very limited due to the very high temperature (4000–10 000 °C) reactions inherently involved. Moreover, it is still not known whether or not the growth of the metallofullerenes is similar to that of empty fullerenes because of the presence of metal atoms (which definitely play some "catalytic" roles in the formation). Even so, some important progress on the growth mechanism of the metallofullerenes has been reported by several research groups during the past couple of decades, which will be discussed in this chapter.

One of the earliest ideas for fullerene formation was the curling up of graphite sheets [1,2]. Smalley proposed the so-called "pentagon road" formation mechanism as a variant of the curling up of graphite sheets, in which pentagons appear at a certain stage of the growth so as to automatically close the carbon cage [3]. As one of the earliest growth mechanisms, the pentagon road formation had been thought reasonable since graphite sheets already possess hexagons needed for the fullerene growth. However, a series of experimental studies done by Bowers and co-workers seem to have ruled out such a graphitic-like formation mechanism [4–6]

Endohedral Metallofullerenes: Fullerenes with Metal Inside, First Edition.
Hisanori Shinohara and Nikos Tagmatarchis.
© 2015 John Wiley & Sons, Ltd. Published 2015 by John Wiley & Sons, Ltd.

as described in this section. Also, an arc-discharge experiment with ^{13}C-enriched graphite rods provided a totally statistical $^{13}C/^{12}C$ distribution in the C_{60} produced, indicating that the evaporation of carbon atoms (not graphic sheets) is occurring in the arc carbon vapor.

To obtain information on the growth mechanism of fullerenes, we then first have to know how carbon clusters are formed and how these cluster "ingredients" can transform into fullerenes (and the metallofullerenes when metal atoms coexist in high temperature carbon vapor). In this context, one of the particular interests is how and at what size the fullerene can start to form and become a stable structure.

Carbon clusters are known to possess specific stable structures depending on their cluster size. In fact, the Exxon's first cluster beam experiment [7] and the Sussex–Rice University experiment [1] clearly show a bimodal cluster size distribution, that is, an even–odd size distribution for C_n ($2<n<28$) and an even size distribution for C_n ($30<n<120$) as shown in Figure 13.1. It is known that the most stable structures of the carbon

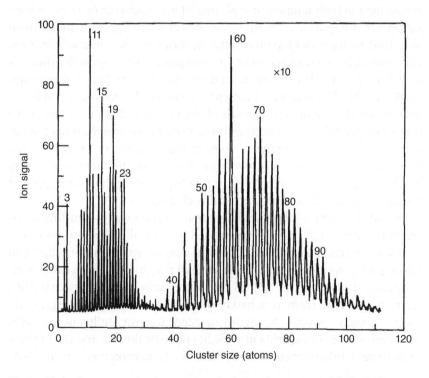

Figure 13.1 Laser-vaporization supersonic cluster-beam time-of-flight (TOF) mass spectrum of various carbon clusters. Together with a magic number feature of C_{60}, a bimodal cluster size distribution is clearly seen. Reproduced with permission from Ref. [7] AIP Publishing LLC

clusters C_n (10<n<30) are either monocyclic rings or polycyclic rings whereas C_n clusters of size smaller than n = 10 have linear-chain structures according to photodetachment experiments (i.e., electron affinity measurements) as shown in Figure 13.2 [8]. The electron affinity of the carbon clusters varies alternately depending on even–odd cluster size.

Then, it is necessary to determine from what size the closed-shell fullerene structures having five- and six-membered rings begin to form? One of the most powerful experimental techniques to probe this is the so-called gas-phase ion-chromatography developed by Bowers and co-workers [4–6]. The fundamental principle of this technique is similar to the conventional liquid chromatography where molecules are separated by "retention time" through the column. In gas-phase chromatography, molecules/clusters (even with structural isomers) are separated via the column (drift-cell) filled with He gas of low pressures (typically 2–5 Torr). The separation can be achieved since the collision cross-section of cluster ions and He atoms differs sensitively depending on the cluster size and shape (structure). Various sizes of carbon clusters are prepared by laser vaporization of graphite and are mass-selected, which are then introduced into the drift cell. This provides very detailed cluster size and the isomer dependence on the final retention time observed.

Bowers and co-workers observed that, for positively charged carbon clusters, there are six different kinds of cluster series of structures: (i) linear (2<n<10); (ii) ring I (6<n<40); (iii) ring II (20<n<40); (iv) ring III

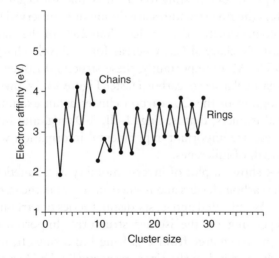

Figure 13.2 Vertical electron affinity of neutral carbon clusters as measured by the observed photodetachment thresholds of the negative cluster ions. Reproduced with permission from Ref. [8], Elsevier

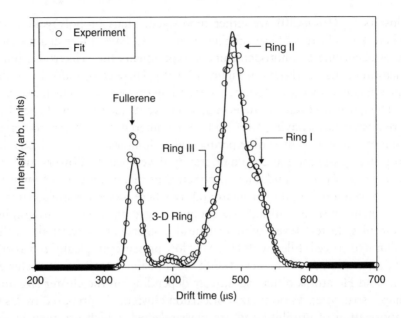

Figure 13.3 An ion chromatogram of C_{36}^+ showing the relative abundance of five structural isomers of a C_{36}^+ carbon cluster. From short to long drift times: fullerene, 3-D ring, ring III, ring II, and ring I. Reproduced with permission from Ref. [5]. Copyright 1993 American Chemical Society

($30<n<40$); (v) three-dimensional (3-D) ring ($30<n<40$); and (vi) fullerenes ($50 < n$) [5]. It is interesting to note here that for negatively charged carbon clusters the ring structures are dominantly observed up to C_{60}. In fact, the fullerene-structured C_{60} is less than 20% of the entire C_{60} clusters for negatively charged C_{60}, whereas for positively charged C_{60} it is more than 95%. More importantly, these structural isomers oftentimes coexist for a particular size of carbon cluster. Figure 13.3 shows a gas-phase ion chromatogram of the C_{36}^+ carbon cluster as an example, in which five structural isomers, that is, rings I–III, 3-D ring, and fullerenes, are observed. These are important experimental information when considering the growth of fullerenes.

Figure 13.4 shows a plot of inverse mobility as a function of cluster size for all the carbon clusters and isomers observed in the laser-vaporized cluster beam. As indicated above, six distinct series of carbon clusters are observed depending on the isomer structures. Importantly, fullerene strictures begin to emerge from C_{30}. Ring I and rings II, III structures are monocyclic and polycyclic rings, respectively. The fact that the 3-D ring family has a slope similar to the fullerene family is evidence that it is composed of three-dimensional objects. Very interestingly, Bowers and

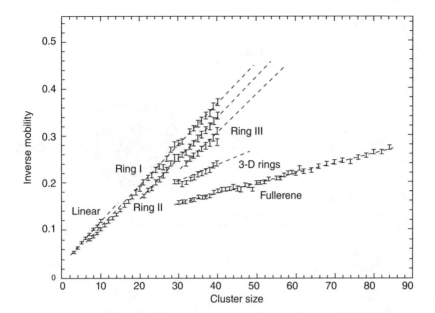

Figure 13.4   Inverse mobility vs. cluster size plot of laser-vaporized carbon clusters of graphite. The centroid of the ring peak moved to shorter times as cluster size increased, indicating relative growth of the more compact polycyclic ring systems. Ring I and rings II, III represent monocyclic and polycyclic rings, respectively. Reproduced with permission from Ref. [5]. Copyright 1993 American Chemical Society

co-workers reported that ring I, II, III structures can easily transform into fullerene structures upon heating induced by collisions with He buffer gas [6]. Figure 13.5 shows pictorically some of these transitions. When C_{40}^{+} rings are collisionally heated above the threshold for isomerization to a fullerene, rapid bond breaking and rearrangement occurs and the cluster begins to relax to its most stable form of fullerenes in this size domain. This experimental evidence strongly suggests that the ring to fullerene isomerization mechanism helps explain the somewhat surprising propensity for C_{60} formation in carbon arc-discharge plasma.

13.2   ROLES PLAYED BY METAL ATOMS IN THE FULLERENE GROWTH

In the preceding section, we saw how fullerenes can be formed from carbon clusters having various types of isomer structures and how the fullerene (structure) formation process starts gradually around the size

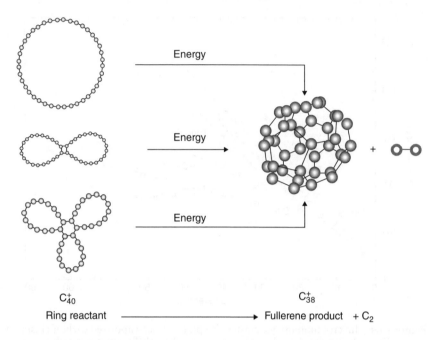

C_{40}^+ C_{38}^+

Ring reactant ────────────────────→ Fullerene product $+ C_2$

Figure 13.5 Transformation of cyclic carbon clusters into fullerenes upon thermal heating. Three monocyclic, bicyclic, and tricyclic C_{40}^+ ions form C_{38}^+ fullerene plus C_2 when collisionally heated above the isomerization barrier for fullerene formation. The bicyclic and tricyclic rings have lower isomerization barriers to fullerene formation than the monocyclic ring. Reprinted by permission from Macmillan Publishers Ltd: [6]. Copyright 1993 Rights Managed by Nature Publishing Group

region of C_{30}. A similar fullerene formation threshold has been found for endohedral metallofullerenes (EMFs). Dunk *et al.* [9] found that the C_{28} fullerene can be stabilized by encapsulation with an appropriate metal atom in carbon vapor as the smallest fullerene under laser-vaporization conditions of metal-impregnated graphite rods. Figure 13.6 shows the relative abundance of Ti@C_n with respect to the number of carbon atoms (n). The Ti@C_{28} cluster has a particularly high abundance compared with that of the smallest empty cage of C_{32}. The distribution shows that only Ti@C_{44} exhibits higher abundance, indicating that Ti@ C_{28} is particularly stable. The observed special stability of Ti@C_{28} with respect to the empty C_{28} clearly shows that only a single Ti metal atom may play a significant and/or catalytic role in the fullerene formation. Dunk *et al.* [9] also studied the structural stability of Ti@C_{28} via collision-induced dissociation (CID) with buffer gases, where Ti@C_{28} undergoes many collisions with He or Ar to achieve high internal energies above the threshold energy for dissociation, and all resulting

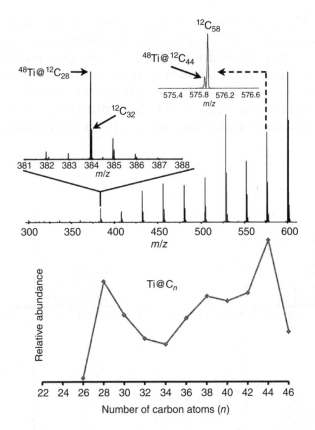

Figure 13.6 High-resolution Ti@C_n cluster (positive ions) mass spectrum and abundance generated from vaporization of a Ti-doped graphite composite rod under conditions that most efficiently generate fullerenes. Ti@C_{28} is particularly enhanced in the Ti@C_n family. Note that no empty-cage C_{28} is observed. Reproduced with permission from Ref. [9]. Copyright 2012 American Chemical Society

products are detected. They found that Ti@C_{28} remains totally intact without any loss of the Ti atom after many collisions with He/Ar, and Ti@C_{28} does not fragment by C or C_3 (but does by the so-called fullerene C_2 loss process), suggesting that Ti@C_{28} possesses a closed-cage fullerene structure. Interestingly, it was found that, in a similar CID experiment, C_{27} fragments predominantly by C_3 loss, demonstrating that it is a ring structure rather than a fullerene as described in the previous section.

A similar stability of C_{28} encapsulating a metal atom is reported for U@C_{28} [10,11]. The observed very strong formation of U@C_{28} provides a rare opportunity to probe the growth and initial formation of small metallofullerenes.

Figure 13.7 shows a series of mass spectra, showing that $U@C_{28}$ is particularly stable in the early growth of uranofullerenes [9]. Initially, $U@C_{28}$ is weakly observed without the presence of larger $U@C_{2n}$. The smallest fullerene, $U@C_{28}$, forms in greater abundance at higher pressure (Figure 13.7b,c), however, and larger EMFs are not present or begin only to very weakly form. Figure 13.7d shows that $U@C_{28}$ becomes much more dominant when vaporization of the target is performed at higher

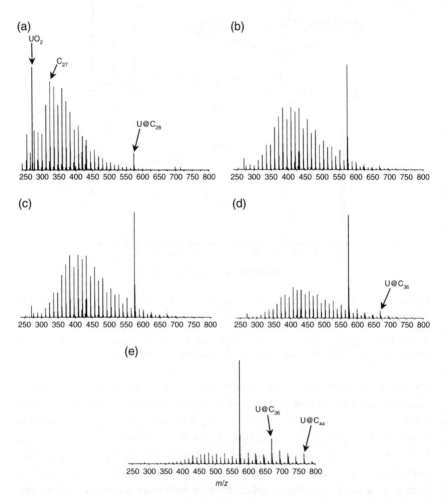

Figure 13.7 $U@C_n$ cluster (positive ions) mass spectrum and abundance generated from vaporization of a U-doped graphite composite rod under increasing He buffer gas pressure. $U@C_{28}$ is particularly enhanced in the $U@C_n$ family and is demonstrated to form before larger $U@C_n$ EMFs. The laser is fired after the initial opening of the pulsed valve at several intervals of time to capture various stages of growth (70 psi He gas pressure, 800μs pulse width): (a)400, (b)600, (c)700, (d)800, and (e) 1000μs. Reproduced with permission from Ref. [9]. Copyright 2012 American Chemical Society

He gas pressure. The larger endohedral uranofullerenes are still formed weakly as evidenced by the low abundance of $U@C_{36}$. Evidently, the presence of U atoms(s) leads to the preferential formation of closed-cage fullerene structures even in smaller fullerene size than C_{60}. Also, as described in the later sections, the fact that the enhanced formation of $U@C_{28}$ starts before larger $U@C_n$ fullerenes strongly suggests that $U@C_{28}$ is likely the precursor or "gateway" fullerene to the larger $U@C_n$.

13.3 TOP-DOWN OR BOTTOM-UP GROWTH?

Because of the inherent difficulty to directly probe the growth process of macroscopically obtainable metallofullerenes such as $M@C_{82}$ (M = metal atom), detailed experimental investigations on the formation mechanism have so far been very limited. One of the main reasons for this is that the formation of metallofullerenes is basically occurring under plasma conditions at very high temperatures such as 4000–10 000 K either by laser-ablation or arc-discharge synthesis. Consequently, few experimental studies and theoretical investigations on the growth mechanism have been reported to date.

In earlier studies, several fullerene growth mechanisms have been proposed to explain how the high symmetry fullerene molecules can form from high-temperature plasma of carbon atoms, ions, and molecules. These are the "pentagon road" [3,12–14], the "fullerene road" [15], the "ring-stacking" [16,17], and the "ring coalescence" [5,18] mechanisms.

There are two primary important routes for growth of EMFs: (i) top-down formation, that is, a graphite fragment or giant carbon clusters originating from the target is directly involved and (ii) bottom-up growth, that is, formed initially from small carbon clusters and atomic carbon. In the following, both mechanisms are discussed based on the recent experimental and theoretical studies.

13.3.1 Top-Down Growth

Top-down formation processes have been proposed to explain fullerene synthesis over the past decade. In one of the earliest studies, the so-called C_2-loss process within the framework of "shrink-wrapping" of fullerenes, where C_2 is ejected from fullerenes followed by annealing of the resulting smaller fullerene, has experimentally been shown in a laser-photofragmentation study of fullerenes [19]. In this study, the ability of such a shrinking system to find the special stability of C_{60}^+ by photofragmentation was also demonstrated.

Theoretical calculations support the presence of such C_2-loss shrink-wrapping processes. For example, by means of quantum chemical dynamics simulation, Irle *et al.* [20, 21] proposed that spontaneous organization of very large carbon cages occurs first, after which C_2 elimination subsequently produces smaller fullerenes such as C_{60}. The study indicates that carbon clusters of size greater than 60 atoms are readily formed, which anneal to giant fullerenes, and then these fullerenes shrink to smaller fullerenes. Unfortunately, the simulation could not be carried to long enough times for the actual shrinking to reach C_{60}, but they propose that this shrink-wrapping process ultimately forms C_{60} [22].

Figure 13.8 shows snapshots of one of the trajectories which includes cluster nucleation, ring condensation growth, and cage closure. They suggested the necessity of locally high concentrations of carbon to initiate cluster formation, but high density is not a sufficient condition as their success rate is only 4/25 (roughly 16%) [20]. Based on the simulations, they suggested that either C_{60} molecules are gradually formed from these giant fullerenes by loss of small carbon fragments, which

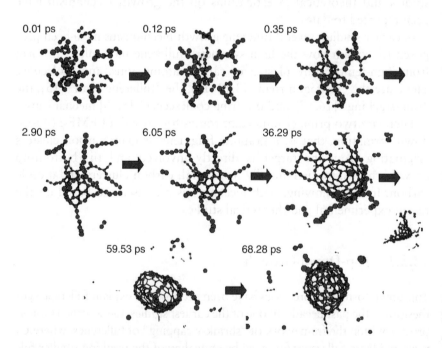

Figure 13.8 Snapshots of a trajectory. The frame at 36.29 ps shows a view into the bowl and a side view (smaller). Reproduced with permission from Ref. [20]. Copyright 2003 American Chemical Society

should be a very slow process due to the sturdiness of large fullerene cages, or C_{60} is initially formed under different conditions, such as lower carbon densities, or lower temperature. One of the main difficulties that these computer calculation/simulation studies inherently possess, particularly for calculating plasma phenomena such as laser-ablation and arc-discharge carbon plasma, is that unrealistically high carbon densities are required to obtain fullerene formation compared with the average experimental densities of carbon. This is due to the limited simulation time domain (normally up to picoseconds at the longest) available for calculations, whereas the typical time required for experimental fullerene formation is micro- to milliseconds.

Chuvilin et al. [23] demonstrated that a top-down graphene-to-fullerene conversion is possible under particular experimental conditions. They found that aberration-corrected transmission electron microscopy (TEM) directly visualizes, in real time, a structural transformation process of fullerene formation from a graphene sheet (cf. Figure 13.9). They stated that this becomes possible, because the energy of the van der Waals interaction of a fullerene with the substrate is significantly reduced compared with a flat graphene flake due to the reduced surface area in contact with the underlying graphene sheet. They also used quantum chemical modeling to explain four critical steps in this top-down sort of mechanism of fullerene formation: (i) loss of carbon atoms at the edge of graphene, leading to (ii) the formation of pentagons, which (iii) triggers the curving of graphene into a bowl-shaped structure, and which (iv) subsequently zips up its open edges to form a closed fullerene structure. Loss of carbon atoms at the edge of graphene is a key initial step in the graphene-to-fullerene transformation. Carbon atoms at the edge of a graphene flake are labile, because only two bonds connect them to the rest of the structure [23].

This transformation from a graphene sheet to fullerenes can be realized by electron-beam irradiation, which, in fact, is very much different from the conventional fullerene syntheses of high-temperature laser ablation or arc discharge. The energy density of focused electron beams from TEM is usually orders of magnitude higher than those of laser ablation and arc discharge. This means the structural transformation from graphene to fullerenes induced by the TEM electron beam does not necessarily ensure what is happening in the actual fullerene/metallofullerene syntheses. Also, they stated that the initial size of the graphene flake is important, because it determines the size of the fullerene cage that can be formed. If the flake is too large, in the region of several hundreds of carbon atoms, there will be a significant energetic penalty during the curving step (Figure 13.9d) associated with the van der Waals interactions

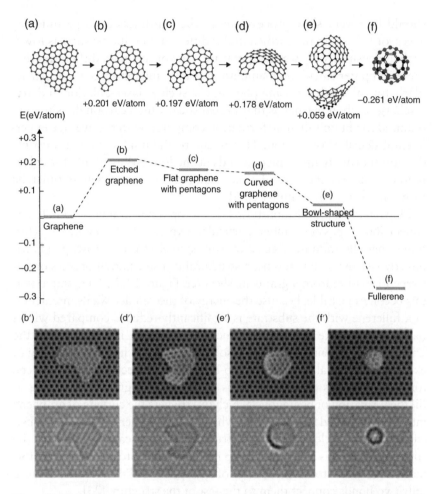

Figure 13.9 Quantum chemical modeling of the four critical stages of fullerene formation from a small graphene flake. (a–d) Loss of carbon atoms at the edge (a → b); formation of pentagons (b → c); curving of the flake (c → d); formation of new bonds, leading to zipping of the flake edges (d → e). (e) Top and side views of a bowl-shaped intermediate structure. Stabilization energies (in electron volts per carbon atom) of the intermediate structures and resultant fullerene C_{60} (f), relative to the flat defect-free flake of graphene shown in a, are presented pictorially and graphically. (b′,d′–f′) Top views of the graphene flake (b′), curved graphene intermediates (d′,e′), and the fullerene C_{60} molecule (f′) adsorbed on the underlying graphene substrate and simulated TEM images corresponding to each structure showing how they would appear in TEM experiments. Reprinted by permission from Macmillan Publishers Ltd: [23]. Copyright 2010 Rights Managed by Nature Publishing Group

between the underlying graphene sheet and the flake. Etching of edge carbon by the electron beam is the key mechanism for the graphene to fullerene structural transformation. From the standpoint of the actual fullerene/metallofullerene syntheses, this is a very special case.

Dorn and co-workers isolated and characterized a metallofullerene that is proposed to be a key intermediate in the top-down formation of EMFs from graphite [24]. They proposed molecular structural evidence for a kind of top-down mechanism based on metal carbide metallofullerenes $M_2C_2@C_1(51383)$-C_{84} (M = Y, Gd). They thought that the asymmetric $C_1(51383)$-C_{84} cage with destabilizing fused pentagons is a preserved "missing link" in the top-down mechanism (cf. Figure 13.10) and that well-established rearrangement steps can form many well-known, high-symmetry, IPR (isolated pentagon rule)-allowed fullerene structures.

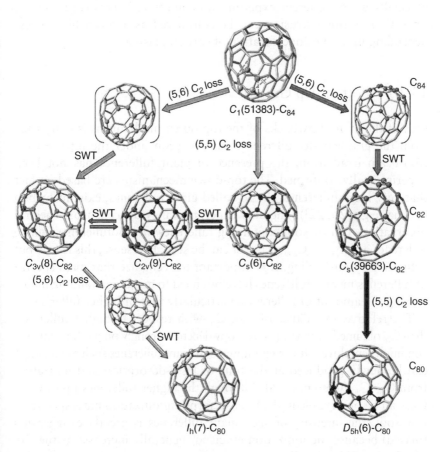

Figure 13.10 The fullerene structural rearrangement map starting from the missing link $C_1(51383)$-C_{84} cage. Many well-known metallofullerene cages are involved in this process. Depending on the size and charge of the encapsulated atom(s) or clusters, the cage may prefer a certain sequence of the transformation map. Reprinted by permission from Macmillan Publishers Ltd: [25]. Copyright 2013 Rights Managed by Nature Publishing Group

They argued that this growth process accounts for the majority of solvent-extractable metallofullerenes.

This growth mechanism is much different from and superior to those so far proposed in that it is based on detailed molecular structures experimentally determined for solvent-extractable metallofullerenes. However, a simple loss of C_2 from $M_2C_2@C_{2n}$ to form $M_2C_2@C_{2n-2}$ is not a "true" top-down mechanism as Irle et al. [21] proposed theoretically, in which hot-giant fullerenes are mother fullerenes in producing virtually all of the fullerenes (cf. Figure 13.8). Also, Zhang et al. [25] stated that "although our current study suggests a top-down mechanistic pathway, we acknowledge that the bottom-up mechanism cannot be excluded, especially under different experimental conditions". This is reasonable since C_2 insertion/extrusion can be considered as a reversible process depending upon the concentration of carbon plasma.

13.3.2   Bottom-Up Growth

One of the main drawbacks of the top-down mechanisms is that, aside from the graphene-to-fullerene transformation under the intense electron-beam irradiation, the presence of giant fullerenes has not been experimentally confirmed. The top-down mechanisms are based on fact that very large fullerenes, the so-called giant fullerenes, exist in carbon plasma (e.g., [20, 22]). The only experimental evidence is mass spectral observations of raw soot containing fullerenes/metallofullerenes, where fullerenes as large as C_{200} or larger can be seen. In a sense, this is true but actually very misleading. It is important to recognize that mass spectra of fullerenes/metallofullerenes have often led to misunderstanding in discussing the amount of fullerenes, particularly that of higher fullerenes.

The relative abundance of, say, C_{60} with respect to higher fullerenes than C_{76} obtained by mass spectroscopy depends strongly on the "ionization" conditions employed. It is well-known that giant fullerenes such as C_{100}–C_{200} can be newly produced at the time of laser-desorption and ionization from C_{60}–C_{90} fullerenes [26]. Although the higher fullerenes up to C_{100}–C_{200} can be observable in the laser-desorption/ionization mass spectra of raw soot, the intensity of such giant fullerenes is greatly exaggerated (biased) because the ionization efficiency generally increases as the size of fullerenes/metallofullerenes increases.

Figure 13.11 shows two types of TOF mass spectra of fullerenes produced by the (imperfect) combustion method [27]. The mass spectra were used to determine approximately how much of the fullerene

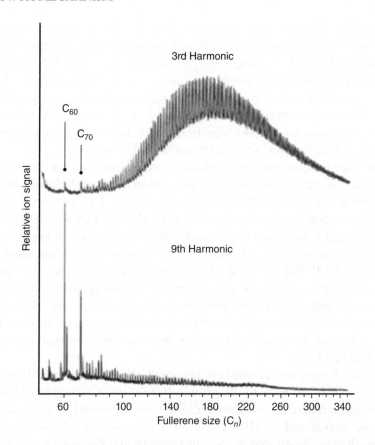

Figure 13.11 TOF positive ion mass spectra of fullerenes produced by the combustion method. The YAG 3rd harmonic (355 nm) ionization after another YAG 3rd desorption, and the YAG 9th harmonic (118 nm) ionization after YAG 3rd desorption are shown. Note that the intensities of higher/giant fullerenes are extremely exaggerated in the top spectrum. Reproduced with permission from Ref. [26], Elsevier

product was giant fullerenes and how much was C_{60}. The two mass spectra are presented are referred to as "YAG 3rd" or "YAG 9th." In both cases, sample desorption was performed by a YAG 3rd harmonic (355 nm) pulse. For the YAG 3rd spectrum, the desorption pulse was also the ionization pulse. For the YAG 9th spectrum, ions created by the desorption pulse were removed by a pulsed electric field, whereas the desorbed neutral fullerenes were ionized by a second pulse of YAG 9th harmonic photons (118 nm). The so-called small gap fullerenes can be ionized with one YAG 3rd photon, whereas C_{60} and other fullerenes can require up to three YAG 3rd photons to be ionized. This dramatically biases the relative signal intensity of YAG 3rd mass spectra toward extreme exaggeration of higher/giant fullerenes relative abundance with respect to C_{60} and C_{70}, where absorption of a single YAG 9th photon is sufficient to ionize any fullerenes. Raebiger *et al.* [26] stated that YAG 9th mass spectra therefore provide a much more accurate depiction of the fullerene distribution within the sample. Considering this important study, the presence of (any significant abundance) of giant fullerenes, say, C_{100}–C_{200}, has not yet been experimentally observed.

In contrast, any bottom-up growth mechanism does not require the presence of such giant fullerenes since fullerenes/metallofullerenes can transform into larger fullerenes by the inclusion of C_2 and other source of carbon clusters. Even in the very early days of fullerene research, it was already known that C_{60} is produced by atomic carbon vapor in the arc-discharge synthesis through $^{12}C/^{13}C$ isotope scrambling measurements [28,29]. Ebbesen *et al.* [29] unambiguously identified that in the arc-discharge synthesis the distribution of ^{13}C among the C_{60} molecules follows exactly Poisson statistics for every ratio of unmixed ^{12}C and ^{13}C in the feed of these arc-plasma experiments. This strongly suggests that C_{60} is formed by bottom-up mechanisms starting from atomic carbon (cf. Figure 13.12).

Dunk, Kroto, and co-workers proposed a bottom-up growth model, the so-called "closed network growth (CNG) mechanism," of fullerenes and metallofullerenes, where larger fullerenes are formed by incorporating atomic carbon and C_2 [10]. The growth processes have been elucidated through experiments that probe direct growth of fullerenes upon exposure to laser-ablated carbon vapor, which is analyzed by the state-of-the-art Fourier transform ion cyclotron resonance (FT-ICR) mass spectrometry.

They examined the growth of C_{60} and other higher fullerenes by direct exposure to carbon vapor produced from graphite, in which a fullerene-containing carbon target is subjected to a single laser shot in a pulsed supersonic cluster source. Upon laser irradiation, the graphite (or amorphous carbon target) is vaporized into atomic carbon and carbon clusters, whereas the C_{60} is softly desorbed into this carbon vapor under a flow of helium. The C_{60} is then reacted with carbon vapor, and the species produced then transferred to a high-resolution FT-ICR mass spectrometer for analysis. The resulting ions are further probed by CID (collision-induced dissociation). They think that the positive ions are representative of the distribution of neutral fullerenes.

Soft laser ablation of pure C_{60} in the absence of carbon vapor (Figure 13.13a) produces mostly C_{60}. In contrast, enhanced formation of larger clusters than C_{60} corresponding to C_{60+2n} including C_{70} are

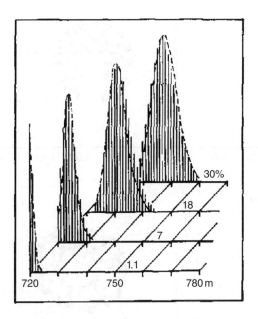

Figure 13.12 Observed mass spectra for C_{60} formed for various percentages of amorphous ^{13}C in the feed. The corresponding Poisson distribution curves are superimposed. For clarity, the distributions are normalized to peak height. Reproduced with permission from Ref. [29], Elsevier

seen after exposure of C_{60} to carbon vapor as shown in Figure 13.13b. Fullerenes smaller than C_{60} do not form (Figure 13.13b), suggesting that the growth process is bottom-up. Figure 13.13c shows the typical distribution of fullerenes produced by ablation of a pure graphite rod at higher buffer gas pressure than in Figure 13.3b. Fullerenes smaller than C_{60} are formed in high abundance from the pure graphite case, whereas they are absent in vapor. The distribution of larger fullerenes than C_{60} formed in Figure 13.13b,c is essentially reproduced. They proposed that this similarity supports an identical growth mechanism in both experimental conditions [10].

The formation mechanism of EMFs can provide strong experimental confirmation that the fullerene cages must remain closed during incorporation of C and C_2 species, since the endohedrally situated atom(s) will be lost if the carbon cage stays open. This means that the encapsulated metal plays as an *in situ* probe to characterize the growth mechanism of fullerenes/metallofullerenes.

For example, $La_2@C_{80}$ with an I_h-C_{80} cage encapsulating two rotating La atoms [30–33] is exposed to carbon vapor generated from laser ablation of graphite. Figure 13.14 shows that the growth to much larger $La_2@C_{80+2n}$

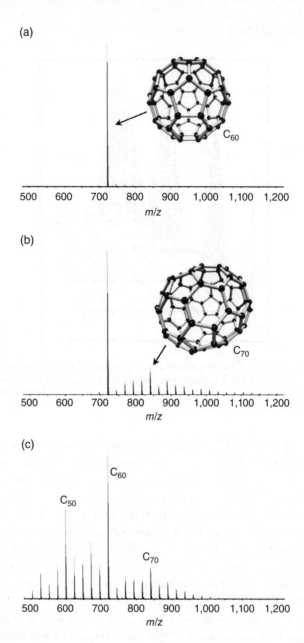

Figure 13.13 FT-ICR mass spectra of laser-ablated C_{60} with and without exposure to carbon vapor. (a) Positive ions produced by laser irradiation of a pure C_{60} target by a single laser pulse, Nd:YAG, 532 nm, in the presence of helium. (b) Larger fullerenes formed by ablation of a C_{60}-graphite target under the same conditions as (a). Note, the helium pressure is lower than that required for fullerene generation from pure graphite. All larger fullerenes formed in (b) are generated exclusively from gas-phase reactions of desorbed C_{60} with carbon vapor produced from graphite. (c) The fullerene distribution produced by ablation of a pure graphite target at a higher helium pressure than in (b). Reprinted by permission from Macmillan Publishers Ltd: [10]. Copyright 2012 Rights Managed by Nature Publishing Group

metallofullerenes (up to around $La_2@C_{120}$) is clearly observed after exposure of $La_2@C_{80}$ to carbon vapor. $La_2@C_{100}$ has been structurally characterized and reported to possess a tubular carbon fullerene cage [34]. The large metallofullerene $Sm_2@C_{104}$ has also been shown to possess

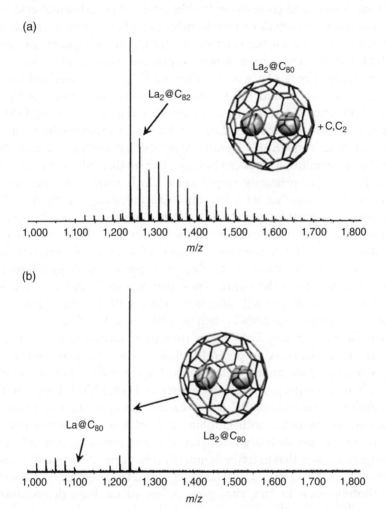

Figure 13.14 Direct evidence of the bottom-up growth of EMFs. (a) Larger $La_2@C_{2n}$ endohedral fullerenes are formed after exposure of $La_2@C_{80}$ to carbon vapor by laser ablation (50 mJ cm^{-2}) of a $La_2@C_{80}$-coated graphite rod. (b) Laser irradiation (50 mJ cm^{-2}) of $La_2@C_{80}$ without exposure to carbon vapor. An encapsulated La atom can be ejected to form $La@C_{80}$ and C_2 loss-fragment ions. When exposed to carbon vapor, this small amount of $La@C_{80}$ formed will grow by the CNG mechanism (rather than fragment) to produce the very weak $La@C_{80+2n}$ signals in (a). Reprinted by permission from Macmillan Publishers Ltd: [10]. Copyright 2012 Rights Managed by Nature Publishing Group

a tubular structure [35]. The $La_2@C_{80+2n}$ fullerenes formed by incorporating C and C_2 may exist as basically tubular structures, and prolonged exposure to carbon vapor should further extend their size by the bottom-up growth [10].

Dunk, Kroto, and co-workers further reported experimental evidence for a bottom-up growth of metallofullerenes [36]. Figure 13.15 shows fullerene cage behavior and reactivity of $Pr@C_{82}$ (isomer I) in the presence of high-temperature carbon vapor evaporated from graphite and in a low-pressure of buffer He gas as in their $La_2@C_{80}$ study described above. $Pr@C_{82}$ is shown to undergo C_2 insertion reactions to form $Pr@C_{84}$ in high abundance. Interestingly, an inverse reaction by C_2 loss to $Pr@C_{80}$ essentially does not occur. As shown in Figure 13.15, extensive C_2 insertion reactions form EMFs significantly larger than $Pr@C_{84}$, including very large metallofullerenes ($>M@C_{100}$). The smaller fullerenes observed than $Pr@C_{82}$ are primarily empty ones which result from bottom-up growth of C_{80} together with low abundance $Pr@C_{80}$ and $Pr@C_{78}$. The fullerenes C_{80}, $Pr@C_{78}$, and $Pr@C_{80}$ can be attributed to result from an unavoidable laser fragmentation of the vaporization laser and the $Pr@C_{82}$ starting material. Importantly, smaller $M@C_{2n}$ ($2n<76$) metallofullerenes, although abundantly generated by vaporization of graphite, are entirely absent from the figure. They also reported similar bottom-up formation of medium-sized fullerenes to larger EMFs with other metal atoms encapsulated in $M@C_{82}$ such as $Y@C_{82}$ and $Yb@C_{82}$.

The bottom-up metallofullerene formation mechanism that Dunk, Kroto, and co-workers proposed [9,10,36] seems to explain well many experimental observations, including the long-standing issue of "isotope scrambling experiments" previously described [28,29,37]. Those studies establish that amorphous ^{13}C is statistically incorporated into fullerene cages when combined with graphite and subsequently evaporated by means of the arc-discharge or laser-ablation synthesis method. This strongly indicates that graphite is initially transformed into small carbon species such as C and C_2, which then react to ultimately generate metallofullerenes. In fact, that process has so far been demonstrated repeatedly during the past couple of decades considering the large number of reports which utilize formation of ^{13}C-enriched metallofullerenes for ^{13}C NMR spectroscopic study [38].

However, it is important to note that a C_2-loss event can occur even in a bottom-up reaction process and thus it is necessary for a comprehensive understanding of fullerene/metallofullerene formation [36]. In fact, the formation of the so-called carbide metallofullerenes (cf. Chapter 5) seems to occur via C_2 loss from metallofullerenes as well as C_2 encapsulation

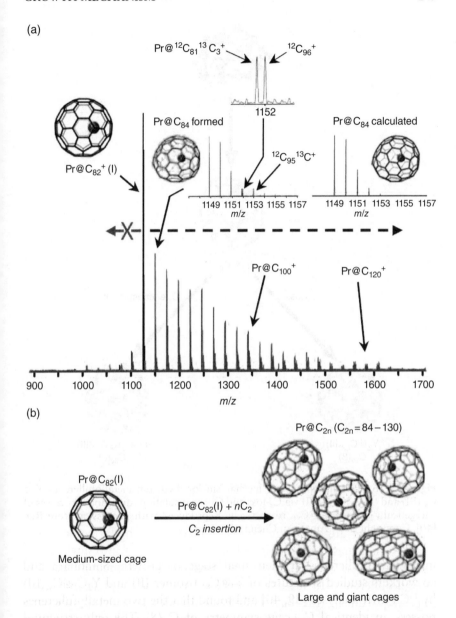

Figure 13.15 Bottom-up growth of large mono-EMFs from Pr@C_{2v}-C_{82}. (a) FT-ICR mass spectrum of cluster cations after reaction of Pr@C_{82} with carbon vapor evaporated from graphite in a low-pressure He atmosphere. (b) C_2 incorporation reaction scheme with possible structures of the larger EMFs. The starting material is shown in left side, whereas bottom-up formation products are shown in right side. Reprinted by permission from Macmillan Publishers Ltd: [36]. Copyright 2014 Rights Managed by Nature Publishing Group

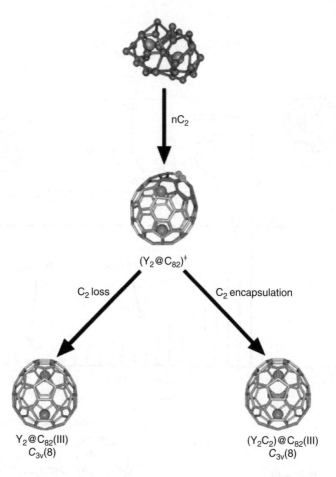

Figure 13.16 Bottom-up growth mechanism for diyttrium metallofullerenes $Y_2@$ $C_{82}(III)$ and $(Y_2C_2)@C_{82}(III)$ via C_2-loss and C_2-encapsulation (digestion) processes of energetically excited $Y_2@C_{84}$, respectively. Reproduced with permission from Ref. [40]. Copyright 2004 American Chemical Society

into metallofullerenes at their final stage of growth. Shinohara and co-workers studied structures of $Y_2@C_{82}$ (isomer III) and $Y_2C_2@C_{82}(III)$ by ^{13}C NMR analyses [39, 40] and found that the two metallofullerenes possess an identical C_{82} cage symmetry of $C_{3v}(8)$. The only structural difference between $(Y_2C_2)@C_{82}(III)$ and $Y_2@C_{82}(III)$ is the presence and absence of C_2 species in the cage, respectively.

The existence of $(Y_2C_2)@C_{82}(III)$ and $Y_2@C_{82}(III)$ with exactly the same cage structure can strongly support the idea that $Y_2@C_{84}$ is much less stable than $(Y_2C_2)@C_{82}(III)$ and $Y_2@C_{82}(III)$. In fact, they found that the abundance of the pure diyttrium C_{84} metallofullerene $Y_2@C_{84}$ is substantially lower than that of C_{82}-based diyttrium fullerenes such as $Y_2@C_{82}$ and $(Y_2C_2)@C_{82}$. They proposed that at the final stage of the growth $Y_2@C_{84}$ preferentially evaporates C_2 (C_2 loss) either outward or inward of the cage to stabilize the fullerenes, which leads to pure $Y_2@C_{82}$ and carbide $(Y_2C_2)@C_{82}$ metallofullerenes, respectively (cf. Figure 13.16). Such evaporating/cooling processes of metallofullerenes via encapsulation of C_2 species inside the cages have actually been observed in various scandium metallofullerenes such as $Sc_2@C_{84}$ and $Sc_3@C_{82}$ by gas-phase ion chromatographic experiments [41].

Although the fundamental issue on the growth process of fullerenes/metallofullerenes, namely top-down or bottom-up, has not been completely clarified yet, as discussed in this chapter, the bottom-up growth mechanism seems much more likely because of the direct experimental observation of C_2-incorporation (digestion) into fullerenes/metallofullerenes to form larger fullerenes. Further studies are needed, however, as to whether or not the bottom-up growth is equally valid in arc-discharge and combustion synthetic conditions as in the laser ablation described here.

REFERENCES

[1] Kroto H W, Heath J R, O'Brien S C et al. 1985 C_{60}: Buckminsterfullerene Nature 318 162
[2] Heath J R, O'Brien S C, Zhang Q et al. 1985 Lanthanum complexes of spheroidal carbon shells J. Am. Chem. Soc. 107 7779
[3] Smalley R E 1992 Self-assembly of the fullerenes Acc. Chem. Res. 25 97
[4] Helden G, Hsu M T, Kemper P R and Bowers M T 1991 Structures of carbon cluster ions from 3 to 60 atoms: linears to rings to fullerenes J. Chem. Phys. 95 3835
[5] Helden G, Hsu M T and Bowers M T 1993 Carbon cluster cations with up to 84 atoms: structures, formation mechanism, and reactivity J. Phys. Chem. 97 8182
[6] Helden G, Gotts N G and Bowers M T 1993 Experimental evidence for the formation of fullerenes by collisional heating of carbon rings in the gas phase Nature 363 60
[7] Rohlfing E A, Cox D M and Kaldor A 1984 Production and characterization of supersonic carbon cluster beams J. Chem. Phys. 81 3322
[8] Yang S, Taylor K J, Craysraft M J, et al. 1988 UPS of 2–30-atom carbon clusters: Chains and rings Chem. Phys. Lett. 144 431
[9] Dunk P W, Kaiser N K, Mulet-Gas M et al. 2012 The smallest stable fullerene, M@ C_{28} (M = Ti, Zr, U): stabilization and growth from carbon vapor J. Am. Chem. Soc. 134 9380
[10] Dunk P W, Kaiser N K, Hendrickson C L et al. 2012 Closed network growth of fullerenes Nat. Commun. 3 855
[11] Guo T, Diener M D, Chai Y et al. 1992 Uranium stabilization of C_{28}: a tetravalent fullerene Science 257 1661

[12] Kroto H W and Mckay K 1988 The formation of quasi-icosahedral spiral shell carbon particles *Nature* **331** 328

[13] Haufler R E, Chai Y, Chibante L P F *et al*. 1991 *Cluster-Assembled Materials*, Vol. 206, eds R S Averback, J Bernhoc and D L Nelson (Pittsburgh, PA: Materials Research Society), pp. 627–637

[14] Goroff N S 1996 Mechanism of fullerene formation *Acc. Chem. Res.* **29** 77

[15] Heath J R 1991 Synthesis of C_{60} from small carbon clusters: a model based on experiment and theory. *ACS Symp Ser.* **481** 1

[16] Wakabayashi T and Achiba Y 1992 A model for the C_{60} and C_{70} growth mechanism *Chem. Phys. Lett.* **190** 465

[17] Wakabayashi T, Shiromaru H, Kikuchi K and Achiba Y 1993 A selective isomer growth of fullerenes *Chem. Phys. Lett.* **201**, 470

[18] Hunter J, Fye J and Jarrold M F 1993 Annealing C_{60}^{+}: synthesis of fullerenes and large carbon rings *Science* **260** 784

[19] O'Briens S C, Heath J R, Curl R F and Smalley R E 1988 Photophysics of buckminsterfullerene and other carbon cluster ions *J. Chem. Phys.* **88** 220

[20] Irle S, Zheng G S, Elstner M and Morokuma K 2003 From C_2 molecules to self-assembled fullerenes in quantum chemical molecular dynamics *Nano Lett.* **3**, 1657

[21] Irle S, Zheng G S, Wang Z and Morokuma K 2006 The C_{60} formation puzzle "solved": QM/MD simulations reveal the shrinking hot giant road of the dynamic fullerene self-assembly mechanism *J. Phys. Chem. B* **110** 14531

[22] Curl R F, Lee M K and Scuseria G E 2008 C_{60} Buckminsterfullerene high yields unraveled *J. Phys. Chem. A* **112** 11951

[23] Chuvilin A, Kaiser U, Bichoutskaia E *et al*. 2010 Direct transformation of graphene to fullerene *Nat. Chem.* **2** 450

[24] Zhang J, Stevenson S and D. Harry 2013 Trimetallic nitride template endohedral metallofullerenes: discovery, structural characterization, reactivity, and applications *Acc. Chem. Res.* **46** 1548

[25] Zhang J, Bowler F, Bearden D W *et al*. 2013 A missing link in the transformation from asymmetric to symmetric metallofullerene cages implies a top-down fullerene formation mechanism *Nat. Chem.* **5** 880

[26] Raebiger J W, Alford J M, Bolskar R D and Diener M D 2011 Chemical redox recovery of giant, small-gap and other fullerenes *Carbon* **49** 37

[27] Takehara H, Fujiwara H, Arikawa M *et al*. 2005 Scale fullerene production by combustion synthesis *Carbon* **43** 311

[28] Hawkins J M, Meyer A, Loren S and Nunlist R 1991 Statistical incorporation of $^{13}C_2$ units into C_{60} *J. Am. Chem. Soc.* **113** 9394

[29] Ebbesen T W, Tabuchi J and Tanigaki K 1992 The mechanistics of fullerene formation *Chem. Phys. Lett.* **191** 336

[30] Alvarez M M, Anz S J and Whetten R L 1991 The unusual electron spin resonance of fullerene C_{60} anion radical *J. Am. Chem. Soc.* **113** 2780

[31] Akasaka T, Nagase S, Kobayashi K *et al*. 1997 ^{13}C and ^{139}La NMR studies of $La_2@$$C_{80}$: first evidence for circular motion of metal atoms in endohedral dimetallofullerenes *Angew. Chem., Int. Ed. Engl.* **36** 1643

[32] Nishibori E, Takata M, Sakata M *et al*. 2001 Pentagonal-dodecahedral La_2 charge density in [80-I_h]fullerene: $La_2@C_{80}$ *Angew. Chem. Int. Ed.* **40** 2998

[33] Shimotani H, Ito T, Iwasa Y *et al.* 2004 Quantum chemical study on the configurations of encapsulated metal ions and the molecular vibration modes in endohedral dimetallofullerene La$_2$@C$_{80}$ *J. Am. Chem. Soc.* **126** 364

[34] Beavers C M, Jin H X, Yang H *et al.* 2011 Very large, soluble endohedral fullerenes in the series La$_2$C$_{90}$ to La$_2$C$_{138}$: isolation and crystallographic characterization of La$_2$@D$_5$(450)-C$_{100}$ *J. Am. Chem. Soc.* **133** 15338

[35] Mercado B Q, Jiang A, Yang H *et al.* 2009 Isolation and structural characterization of the molecular nanocapsule Sm$_2$@D$_{3d}$(822)-C$_{104}$ *Angew. Chem. Int. Ed.* **48** 9114

[36] Dunk P W, Mulet-Gas M, Nakanishi Y *et al.* 2014 Bottom-up formation of endohedral mono-metallofullerenes is directed by charge transfer *Nat. Commun.* **5** 5844

[37] Tan Y Z, Li J, Zhu F *et al.* 2010 Chlorofullerenes featuring triple sequentially fused pentagons *Nat. Chem.* **2** 269

[38] Popov A A, Yang S and Dunsch L 2013 Endohedral fullerenes *Chem. Rev.* **113** 5989

[39] Inoue T, Tomiyama T, Sugai T and Shinohara H 2003 Spectroscopic and structural study of Y$_2$C$_2$ carbide encapsulating endohedral metallofullerene: (Y$_2$C$_2$)@C$_{82}$ *Chem. Phys. Lett.* **382** 226

[40] Inoue T, Tomiyama T, Sugai T *et al.* 2004 Trapping a C$_2$ radical in endohedral metallofullerenes: synthesis and structures of (Y$_2$C$_2$)@C$_{82}$ (isomers I, II, and III) *J. Phys. Chem. B* **108** 7573

[41] Sugai T, Inakuma M, Hudgins R *et al.* 2001 Structural studies of Sc metallofullerenes by high-resolution ion mobility measurements *J. Am. Chem. Soc.* **123** 6427

14

M@C$_{60}$: a Big Mystery and a Big Challenge

14.1 WHAT HAPPENS TO M@C$_{60}$?

It had been expected, before the first macroscopic production and extraction of La@C$_{82}$ [1], that metallofullerenes based on the C$_{60}$ cage would be the most abundant metallofullerenes that were prepared in macroscopic amounts, as was the case in empty fullerenes. This is simply because C$_{60}$ is the most abundant fullerene which can be easily produced by either the arc-discharge or the laser-furnace method (cf. Section 2.1). In fact, an earlier gas-phase experiment on the production of carbon clusters containing lanthanum via the laser-vaporization cluster-beam technique [2] indicated that La@C$_{60}$ is a prominent "magic number" species among various La@C$_n$ ($44 < n < 80$) clusters as shown in Figure 1.1.

A series of gas-phase ion chromatographic studies on metallofullerenes in the gas phase done by Jarrold and co-workers [3–8] has presented important clues on the stability and growth mechanism of endohedral metallofullerenes (EMFs). They showed that laser vaporization of a La$_2$O$_3$/graphite rod produces a number of LaC$_{60}^+$ and a variety of different isomers including the endohedral La@C$^+$ in which the La atom seems to be bound to polycyclic polyyne rings [6] and also to be networked into the fullerene cage [8]. Interestingly, when heated, nearly all of the different ring isomers convert spontaneously into EMFs, trapping the metal atom inside the cage with high efficiency (>98%) [4].

Endohedral Metallofullerenes: Fullerenes with Metal Inside, First Edition.
Hisanori Shinohara and Nikos Tagmatarchis.
© 2015 John Wiley & Sons, Ltd. Published 2015 by John Wiley & Sons, Ltd.

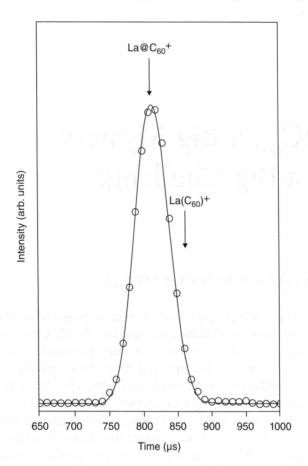

Figure 14.1 Drift-time distribution recorded for LaC_{60}^+ with injection energy of 400 eV. The curve shows the drift-time distribution calculated from the transport equation for ions in the drift tube. The arrows show the expected drift times for endohedral $La@C^+_{60}$ and exohedral $La(C_{60})$ complexes. The mobility was calculated using a simple hard-sphere collision model. Reprinted by permission from Macmillan Publishers Ltd: [4]. Copyright 1994 Rights Managed by Nature Publishing Group

Figure 14.1 shows a drift-time distribution recorded for LaC_{60}^+ with injection energy of 400 eV. The main peak corresponds to the endohedral $La@C_{60}^+$. Based on these results it was proposed that in the first step of the thermal annealing process the La atom acts as a nucleation center and the carbon rings arrange themselves around the La atom before converting into a fullerene cage.

The extraction of the $M@C_{60}$-type (M = metal) metallofullerene has been difficult, however, because all of the $M@C_{60}$ metallofullerenes so far produced in soot have not been soluble in normal fullerene solvents such as toluene and CS_2. A similar thing is true for $M@C_{70}$-type of metallofullerenes which also have not been solvent-extracted, purified, and characterized.

Several M@C$_{60}$-type metallofullerenes have, however, been extracted by solvents such as pyridine and aniline. Ca@C$_{60}$ has been extracted by pyridine [9,10] and aniline [11]. Other M@C$_{60}$ (M = La, Y, Ba, Ce, Pr, Nd, Gd) metallofullerenes have been similarly extracted from soot by aniline [12,13]. Since pyridine and aniline are not suited as HPLC (high-performance liquid chromatography) solvent (eluent) for purification, purification and isolation of the M@C$_{60}$-type metallofullerenes has been extremely difficult. The inability to purify such metallofullerenes has prevented any detailed structural and electronic studies on these fullerenes. A similar unconventional stability and solubility property has been observed for Li@C$_{60}$ which was prepared by Li$^+$ ion implantation onto C$_{60}$ thin films [14–16]. Bolskar et al. [17] have produced a water-soluble Gd@ C$_{60}$ derivative, Gd@C$_{60}$[C(COOH)$_2$]$_{10}$. Diener and Alford [18] reported separation of Gd@C$_{60}$ although the purification was not possible.

Despite these difficulties, Ogawa et al. [19] performed the first purification of an M@C$_{60}$-type metallofullerene, Er@C$_{60}$. The isolation though not complete was done by using a combined technique of vacuum sublimation of soot containing Er@C$_{60}$ together with higher erbium metallofullerenes Er@C$_{2n}$ (70<2n <120), which was followed by a Buckyclutcher HPLC purification (Section 2.3) with a 100% aniline eluent. Er@C$_{60}$ may be greatly stabilized in aniline solution by forming charge-transfer complexes of the type (Er@ C$_{60}$)–(aniline)$_n$. A similar result was reported for Eu@C$_{60}$ by Inoue et al. [20].

At present, it is still not known why the M@C$_{60}$-type metallofullerenes behave quite differently from the conventional M@C$_{82}$-type fullerenes in terms, for example, of the solubility property. This may correlate to high reactivity of M@C$_{60}$ toward moisture and/or air: M@C$_{60}$ can possibly form weak complexes to stabilize themselves only with pyridine or aniline through their nonbonding electrons. Another possible rationale is that the carbon cage structures of the M@C$_{60}$ so far produced somehow do not satisfy the IPR (isolated pentagon rule), which again leads to high chemical reactivity of the species. To fully understand "the M@C$_{60}$ mystery," the structural characterization/determination of purified M@C$_{60}$ metallofullerenes using X-ray diffraction is definitely required.

Recently, Wang et al. [21] reported the synthesis, solvent extraction, and HPLC isolation of trifluoromethylated Y@C$_{70}$, that is, Y@C$_{70}$(CF$_3$)$_3$, another "missing metallofullerene". They used a modified DC arc-discharge method for producing the functionalized Y@C$_{70}$ metallofullerene, which drastically alter the electronic properties of the metallofullerene.

It is convenient to introduce additional gaseous or solid reagents into the arc-discharge chamber to produce new types of fullerenes or metallofullerenes. For example, trifluoromethyl derivatives of C_{60} were produced using polytetrafluoroethene (PTFE) as a source for functional groups during arc discharge [22].

As shown schematically in Figure 14.2, Wang *et al.* placed PTFE near the area where arc discharge occurs [21]. Because of the high temperature of the

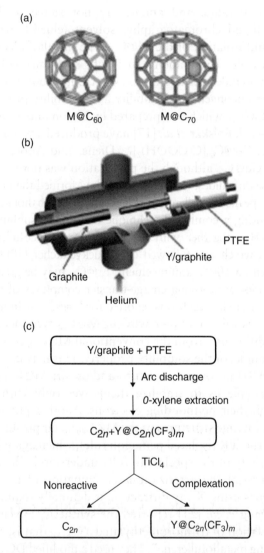

Figure 14.2 (a) Molecular structural models for missing metallofullerenes M@C_{60} and M@C_{70}. (b) The arc-discharge apparatus. (c) Production, extraction, and purification process for yttrium metallofullerenes functionalized with trifluoromethyl groups. Reproduced with permission from Ref. [21]. Copyright Wiley-VCH Verlag GmbH & Cpo. KGaA

arc zone, PTFE is evaporated together with the metal/graphite rod during arc discharge. They found that as a consequence, a series of trifluoromethyl derivatives of insoluble metallofullerenes M@C$_{2n}$(CF$_3$)$_m$ (e.g., $2n$ = 60, 70, 72, or 74) are formed effectively. Surprisingly, these derivatives, including those of M@C$_{60}$ and M@C$_{70}$, are totally soluble and stable in organic solvents such as toluene and CS$_2$, which is important for further purification and characterization. On the basis of theoretical results, the stability of the trifluoromethyl derivatives of Y@C$_{2n}$, Y@C$_{70}$(CF$_3$)$_3$ in particular can be attributed to their closed-shell configurations and the enlarged highest occupied molecular orbital-lowest unoccupied molecular orbital (HOMO-LUMO) gaps in reference to that of the pristine Y@C$_{70}$.

As shown in Figure 14.3, Y@C$_{70}$ has an open-shell configuration and thus possesses a very small band gap as for Y@C$_{60}$, indicating a high chemical reactivity. Density functional theory (DFT) calculations also reveal that the HOMO-LUMO gaps of five probable structural isomers are in the range of 0.26–0.48 eV. Y@C$_{70}$(CF$_3$) having a HOMO-LUMO gap of 0.48 eV is depicted in Figure 14.3b. Wang et al. [21] examined some of the probable isomers and found that one isomer has a HOMO-LUMO gap of 0.98 eV (Figure 14.3c), which is close to the estimated gap for the experimentally isolated Y@C$_{70}$(CF$_3$)$_3$. The LUMO level of Y@C$_{70}$(CF$_3$)$_3$ (–3.90 eV) is very close to that of Y@C$_{70}$(CF$_3$) (–3.94 eV). The HOMO level of Y@C$_{70}$(CF$_3$)$_3$ (–4.88 eV) is much lower than that of Y@C$_{70}$(CF$_3$) (–4.42 eV), indicating a higher stability for Y@C$_{70}$(CF$_3$).

The differences in the HOMO level together with the chemical stability may explain the easy accessibility of Y@C$_{70}$(CF$_3$)$_3$ compared with Y@C$_{70}$(CF$_3$) as observed in the experiments. The observed enhanced stability of the trifluoromethyl derivatives of Y@C$_{2n}$ can well be explained by the closed-shell configurations and the enlarged HOMO-LUMO gaps.

The method developed by Wang et al. [21] can also be applied (basically) to the solvent extraction and the final isolation of the missing M@C$_{60}$ metallofullerenes together with other small band-gap metallofullerenes, which have not yet been realized.

14.2 A BIG CHALLENGE: SUPERCONDUCTIVE METALLOFULLERENES

One of the most exciting and enticing topics on metallofullerenes is whether or not a superconductive metallofullerene could exist as a high-temperature fullerene superconductor such as alkaline-doped C$_{60}$ (K$_3$C$_{60}$, Rb$_3$C$_{60}$, RbCs$_2$C$_{60}$, Cs$_3$@C$_{60}$, etc.) [23,24]. A recent study indicates the

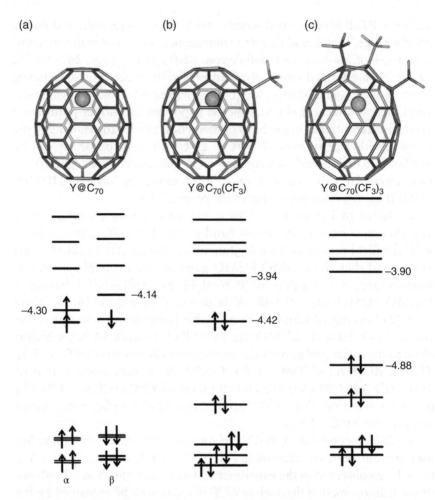

Figure 14.3 DFT-optimized structures and the calculated molecular orbital energy levels (in electronvolts) for (a) pristine Y@C$_{70}$, (b) Y@C$_{70}$(CF$_3$), and (c) Y@C$_{70}$(CF$_3$)$_2$. For the Y@C$_{70}$(CF$_3$)$_3$ molecule, one of the addition sites is the same as that in Y@C$_{70}$(CF$_3$); the other two addition sites are at the top of the cage. Reproduced with permission from Ref. [21].Copyright Wiley-VCH Verlag GmbH & Cpo. KGaA

superconductivity of alkali-doped Ar@C$_{60}$ [25] although the critical temperatures were lower than previously expected.

Trivalent M^{3+}@C$_{60}$$^{3-}$ (M = typically trivalent lanthanide metal atoms such as La, Y, Ce, Er, Gd) metallofullerenes might exhibit superconductive behavior in low temperatures, because, as described in Section 14.1, the electronic structure of M^{3+}@C$_{60}$$^{3-}$ might resemble that of the superconductive fulleride K^{3+}C$_{60}$$^{3-}$ (where the triply degenerate half-filled LUMO should play a crucial role in the emergence of superconductivity), provided that the point symmetry of M^{3+}@C$_{60}$$^{3-}$ retains its original symmetry of C$_{60}$-I_h [26].

In order to realize this, all we have to do first is to try to obtain purified M@C$_{60}$ materials which have not been experimentally available so far due to their high reactivity toward other fullerenes and amorphous carbons present in the original arc-processed soot (cf. Section 14.1). However, it is well known that the amount of M@C$_{60}$ is generally much higher than that of the conventional M@C$_{82}$ in original raw soot, so that the future purification of M@C$_{60}$ is, we think, highly probable (cf. Section 14.1). This is definitely a worthwhile challenge.

14.3 FUTURE PROSPECTS

Lanthanide metallofullerenes are novel forms of fullerene-related materials which were first obtained in their purified form in January 1993 [27]. Since then, as detailed in the earlier chapters, numerous investigations into various aspects of metallofullerenes have been intensively carried out by various research groups. These studies have revealed unique structural and novel electronic properties of the metallofullerenes.

Even so, there still remain many potentially important and intriguing problems which should be clarified in the not too distant future. For instance, the solid state properties such as electric conductivity and magnetic behavior are not well known apart for a few metallofullerenes. Only a limited number of such studies have been reported. Organic functionalizations of the metallofullerenes (Chapter 11) will be an important direction for the synthesis of further novel materials based on EMFs because, as Akasaka et al. [28–30] have found, metallofullerenes are generally more reactive, either thermally or photochemically, than the corresponding empty fullerenes, due to the small HOMO-LUMO gaps.

Metallofullerenes will also become an important nanostructured material for future nanoscale electronic devices, because the band gaps of EMFs, for example, can be varied between 1.0 eV and 0.2 eV depending on the fullerene size, the kind of metal atom(s) involved as well as the number of metal atoms encapsulated.

Since the amount of metallofullerenes so far produced and isolated has been small, studies on the solid state properties of metallofullerenes have been limited. However, scaling-up of production and isolation of metallofullerenes is possible by using a large-scale arc-discharge system as well as by using a preparative HPLC system with a large column

(30 mm diameter × 250 mm length) together with the $TiCl_4$ separation technique of metallofullerenes from empty fullerenes (Section 2.4). With this system, up to 100 mg of purified metallofullerenes is obtained already in a week or so.

In addition, biomedical applications of EMFs will become extremely important in relation to tracer chemistry in biological systems, and await further studies. In any case, EMFs will continue to tantalize physicists, chemists, and materials scientists for years to come.

REFERENCES

[1] Chai Y, Guo T, Jin C *et al*. 1991 Fullerenes with metals inside *J. Phys. Chem.* **95** 7564
[2] Heath J R, O'Brien S C, Zhang Q *et al*. 1985 Lanthanum complexes of spheroidal carbon shells *J. Am. Chem. Soc.* **107** 7779
[3] Hunter J, Fye J and Jarrold M F 1993 Annealing C_{60}^+: synthesis of fullerenes and large carbon rings *Science* **260** 784
[4] Clemmer D E, Shelimov K B and Jarrold M F 1994 Gas-phase self-assembly of endohedral metallofullerenes *Nature* **367** 718
[5] Clemmer D E, Hunter J M, Shelimov K B and Jarrold M F 1994 Physical and chemical evidence for metallofullerenes with metal atoms as part of the cage *Nature* **372** 248
[6] Clemmer D E, Hunter J M, Shelimov K B and Jarrold M F 1994 Bonding of metals to carbon rings: $LaCn^+$ isomers with La^+ inserted and attached to the ring *J. Am. Chem. Soc.* **116** 5971
[7] Clemmer D E and Jarrold M F 1994 Metal-containing carbon clusters: structures, isomerization, and formation of $NbCn^+$ clusters *J. Am. Chem. Soc.* **117** 8841
[8] Shelimov K B, Clemmer D E and Jarrold M F 1994 Structures and formation of small $LaCn^+$ metallofullerenes *J. Phys. Chem.* **98** 12819
[9] Wang L S, Alford J M, Chai Y *et al*. 1993 The electronic structure of Ca@C_{60} *Chem. Phys. Lett.* **207** 354
[10] Wang L S, Alford J M, Chai Y *et al*. 1993 Photoelectronspectroscopy and electronic structure of Ca@C_{60} *Z. Phys. D.* **26** S297
[11] Kubozono Y, Ohta T, Hayashibara T *et al*. 1995 Preparation and extraction of Ca@C_{60} *Chem. Lett.* 457
[12] Kubozono Y, Noto T, Ohta T *et al*. 1996 Extractions of Ca@C_{60} and Sr@C_{60} with aniline *Chem. Lett.* 1061
[13] Kubozono Y, Maeda H, Takabayashi Y *et al*. 1996 Isolation and characterization of C_{60}. *J. Am. Chem. Soc.* **118** 6998
[14] Tellgmann R, Krawez N, Hertel I V and Campbell E E B 1996 *Fullerenes and Fullerene Nanostructures*, eds H Kuzmany, J Fink, M Mehring and S Roth (London: World Scientific), pp. 168–172
[15] Campbell E E B, Tellgmann R, Krawez N and Hertel I V 1997 Production and LDMS characterisation of endohedral alkali–fullerene films *J. Phys. Chem. Solids* **58** 1763

[16] Gromov A, Ostrovskii D, Lassesson A *et al.* 2003 Fourier transform infrared and Raman spectroscopic study of chromatographically isolated Li@C$_{60}$ and Li@C$_{70}$ *J. Phys. Chem. B* **107** 11290

[17] Bolskar R D, Benedetto A F, Husebo L O *et al.* 2003 First soluble M@C$_{60}$ derivatives provide enhanced access to metallofullerenes and permit in vivo evaluation of Gd@C$_6$[C(COOH)$_2$]$_{10}$ as a MRI contrast agent *J. Am. Chem. Soc.* **125** 5471

[18] Diener M D and Alford J M 1998 Isolation and properties of small-bandgap fullerenes *Nature* **393** 668

[19] Ogawa T, Sugai T and Shinohara H 2000 Isolation and characterization of Er@C$_{60}$ *J. Am. Chem. Soc.* **122** 3538

[20] Inoue T, Kubozono Y, Kashino S *et al.* 2000 Electronic structure of Eu@C$_{60}$ studied by XANES and UV-VIS absorption spectra *Chem. Phys. Lett.* **316** 381

[21] Wang Z, Nakanishi Y, Noda S *et al.* 2013 Missing small-bandgap metallofullerenes: their isolation and electronic properties *Angew. Chem. Int. Ed.* **52** 11770

[22] Fritz H P and Hiemeyer R 1995 Formation in-situ of perfluoroalkylated fullerenes *Carbon* **33** 1601

[23] Rosseinsky M J and Prassides M 2010 Materials science: hydrocarbon superconductors *Nature* **464** 39

[24] Prassides K 2010 Superconductivity: interfaces heat up *Nat. Mater.* **9** 96

[25] Takeda A, Yokoyama Y, Ito S *et al.* 2006 Superconductivity of doped Ar@C$_{60}$ *Chem. Commun.* 912

[26] Bethune D S, Johnson R D, Salem J R *et al.* 1993 Atoms in carbon cages: the structure and properties of endohedral fullerenes *Nature* **366** 123

[27] Shinohara H, Yamaguchi H, Hayashi N *et al.* 1993 Isolation and spectroscopic properties of scandium fullerenes (Sc$_2$@C$_{74}$, Sc$_2$@C$_{82}$, and Sc$_2$@C$_{84}$) *J. Phys. Chem.* **97** 4259

[28] Akasaka T, Nagase S, Kobayashi K *et al.* 1995 Synthesis of the first adducts of the dimetallofullerenes La$_2$@C$_{80}$ and Sc$_2$@C$_{84}$ by addition of a disilirane *Angew. Chem. Int. Ed. Engl.* **34** 2139

[29] Akasaka T, Kato T, Kobayashi K *et al.* 1995 Exohedral adducts of La@C$_{82}$ *Nature* **374** 600

[30] Akasaka T, Nagase S, Kobayashi K *et al.* 1995 Exohedral derivatization of an endohedral metallofullerene Gd@C$_{82}$ *J. Chem. Soc. Chem. Commun.* 1343

Index

Endohedral Metallofullerenes: Fullerenes with Metal Inside, First Edition.
Hisanori Shinohara and Nikos Tagmatarchis.
© 2015 John Wiley & Sons, Ltd. Published 2015 by John Wiley & Sons, Ltd.